CW01301588

Fundamentals of Aerospace Engineering

An introductory course to aeronautical engineering

Fundamentals of Aerospace Engineering

An introductory course to aeronautical engineering

Manuel Soler.
Assistant Professor,
Universidad Carlos III.

Manuel Soler [Ed.].
Printed by Create Space.
Madrid, January 2014.

All contents of this books are subject to the following license
except when explicitly specified the opposite.

CREATIVE COMMONS (CC BY NC SA)

This work is licensed under a Creative Commons
Attribution-NonCommercial-ShareAlike 3.0 Unported License.

You are free to copy, distribute, and publicly communicate this creation as long as[*]:

- (BY) You give credit to the author and editor in the terms herein specified.
- (NC) You do not use it for commercial purposes.
- (SA) You license your derivative creations under the identical terms.

[*] Some of these conditions might not apply if you obtain explicit authorization by the owner of the rights. Notice also that this license applies to all contents except when explicitly licensed under other licenses as it is the case for those passages based on Wikipedia sources (licensed under CC-BY-SA) and some figures.

Interior design and layout by Manuel Fernando Soler Arnedo using LaTeX and font iwona.
Cover design by Francisco Javier García Caro and Manuel Fernando Soler Arnedo,

First edition January, 2014
© Manuel Fernando Soler Arnedo, 2014
© Manuel Soler [Ed.] 2014

ISBN-13: 978-14-937277-5-9
ISBN-10: 1493727753
Printed by CREATE SPACE

*Dedicado a Reme,
con amor virtuoso.*

Education is not the piling on of learning, information, data, facts, skills, or abilities–that's training or instruction–but is rather a making visible what is hidden as a seed [...] To be educated, a person doesn't have to know much or be informed, but he or she does have to have been exposed vulnerably to the transformative events of an engaged human life [...] One of the greatest problems of our time is that many are schooled but few are educated.

Thomas More,
The Education of the Heart.

Lo que más necesitan, aun los mejores de nuestros buenos estudiantes, es mayor intensidad de vida, mayor actividad para todo, en espíritu y cuerpo: trabajar más, sentir más, pensar más, querer más, jugar más, dormir más, comer más, lavarse más, divertirse más. **Francisco Giner de los Ríos.**

El espíritu de la casa, como llamaban los residentes al esfuerzo para transmitir la mejor tradición española de educación liberal, quedaba reflejado en una cierta forma de vida construida en torno a la responsabilidad personal, el trabajo, la búsqueda de la excelencia, el culto a la amistad y el ocio creativo, con el fin de que el esfuerzo particular se viera proyectado en la sociedad [...] El espíritu de la casa simboliza su modelo de educación integral, basado en la tolerancia, el pluralismo y el diálogo entre distintas disciplinas de las artes y las ciencias, entre diferentes generaciones, y entre tradición y modernidad [...] **Nota sobe *El Espíritu de la casa*.**

Exposición 100 años de la residencia de estudiantes.
Residencia de estudiantes, Madrid.

Una niña apunto de cumplir 10 años declara: cuando sea mayor, me inscribiré en el partido más cruel. Si mi partido está en el poder, no tendré nada que temer; y si es el otro sufriré menos puesto que es el partido menos cruel el que me perseguirá [...] Conozco ese razonamiento [...] **Albert Camus, *Carnets*, Vol. 3.**

Este es el verdadero problema: suceda lo que suceda, yo siempre te defenderé contra el pelotón de fusilamiento, pero tú te verás obligado a aprobar que me fusilen. **Episodio vital de Albert Camus: palabras a un antiguo resistente, recién afiliado al partido comunista.**

Valía más perecer con la justicia que triunfar con la injusticia. **Albert Camus, en algún momento de la postguerra.**

La libertad no es un estado natural del hombre, es una conquista perpetua frente a la naturaleza, la sociedad y uno mismo. **Albert Camus en el semanario *L'Express*, 1955.**

Albert Camus,
Notas sacadas de *Revista Turia*, num. 107.

About the Author

Manuel Soler received a Bachelor's Degree in Aeronautical and Aerospace Engineering (5-Year B.Sc, 07), a Master's degree in Aerospace Science and Technology (M.Sc, 11), both from the Universidad Politécnica de Madrid, and a Doctorate Degree in Aerospace Engineering (Ph.D, 13) from the Universidad Rey Juan Carlos, Madrid. He developed his early professional career in companies of the aeronautical sector. In 2008, he joined the Universidad Rey Juan Carlos, where he was a lecturer in the area of aerospace engineering. Since January 2014, Manuel Soler is Assistant Professor at the Universidad Carlos III de Madrid, where he teaches undergraduate and graduate courses in the field of aerospace engineering and conducts research activities. He has been a visiting scholar at ETH Zurich, Switzerland, and UC Berkeley, USA. His research interests focus on optimal control with application to green trajectory planning for commercial aircraft in Air Traffic Management (ATM). Dr. Soler has participated in several research projects and has published his work in different international journal and conference papers. He has been awarded with the SESAR young scientist award 2013.

Contents

Preface xvii
Acknowledgments xviii
List of figures xxiii
List of tables xxvi

I Introduction   1

1 The Scope   3
1.1 Engineering 4
1.2 Aerospace activity 5
1.3 Aviation research agenda 12
    1.3.1 Challenges 12
    1.3.2 Clean Sky 13
    1.3.3 SESAR 15
References 17

2 Generalities   19
2.1 Classification of aerospace vehicles 20
    2.1.1 Fixed wing aircraft 21
    2.1.2 Rotorcraft 23
    2.1.3 Missiles 24
    2.1.4 Space vehicles 25
2.2 Parts of the aircraft 27
    2.2.1 Fuselage 27
    2.2.2 Wing 28
    2.2.3 Empennage 30
    2.2.4 Main control surfaces 31
    2.2.5 Propulsion plant 32

	2.3	Standard atmosphere	33
		2.3.1 Hypotheses	33
		2.3.2 Fluid-static equation	34
		2.3.3 ISA equations	34
		2.3.4 Warm and cold atmospheres	36
		2.3.5 Barometric altitude	37
	2.4	System references	37
	2.5	Problems	39
	References		43

II The aircraft 45

3 Aerodynamics 47

3.1	Fundamentals of fluid mechanics	48
	3.1.1 Generalities	48
	3.1.2 Continuity equation	49
	3.1.3 Quantity of movement equation	50
	3.1.4 Viscosity	52
	3.1.5 Speed of sound	56
3.2	Airfoils shapes	58
	3.2.1 Airfoil nomenclature	59
	3.2.2 Generation of aerodynamic forces	60
	3.2.3 Aerodynamic dimensionless coefficients	63
	3.2.4 Compressibility and drag-divergence Mach number	66
3.3	Wing aerodynamics	68
	3.3.1 Geometry and nomenclature	68
	3.3.2 Flow over a finite wing	69
	3.3.3 Lift and induced drag in wings	71
	3.3.4 Characteristic curves in wings	72
	3.3.5 Aerodynamics of wings in compressible and supersonic regimes	73
3.4	High-lift devices	74
	3.4.1 Necessity of high-lift devices	74
	3.4.2 Types of high-lift devices	75
	3.4.3 Increase in $C_{L_{max}}$	77
3.5	Problems	79
References		101

4 Aircraft structures 103

4.1	Generalities	104
4.2	Materials	108

		4.2.1	Properties	108
		4.2.2	Materials in aircraft	109
	4.3	Loads		113
		4.3.1	Fuselage loads	113
		4.3.2	Wing and tail loads	114
		4.3.3	Landing gear loads	114
		4.3.4	Other loads	114
	4.4	Structural components of an aircraft		114
		4.4.1	Structural elements and functions of the fuselage	115
		4.4.2	Structural elements and functions of the wing	117
		4.4.3	Tail	118
		4.4.4	Landing gear	118
	References			119
5	**Aircraft instruments and systems**			**121**
	5.1	Aircraft instruments		122
		5.1.1	Sources of data	123
		5.1.2	Instruments requirements	126
		5.1.3	Instruments to be installed in an aircraft	126
		5.1.4	Instruments layout	130
		5.1.5	Aircrafts' cockpits	131
	5.2	Aircraft systems		135
		5.2.1	Electrical system	135
		5.2.2	Fuel system	137
		5.2.3	Hydraulic system	139
		5.2.4	Flight control systems: Fly-By-Wire	140
		5.2.5	Air conditioning & pressurisation system	142
		5.2.6	Other systems	143
	References			145
6	**Aircraft propulsion**			**147**
	6.1	The propeller		148
		6.1.1	Propeller propulsion equations	148
	6.2	The jet engine		150
		6.2.1	Some aspects about thermodynamics	151
		6.2.2	Inlet	153
		6.2.3	Compressor	154
		6.2.4	Combustion chamber	156
		6.2.5	Turbine	158
		6.2.6	Nozzles	160
	6.3	Types of jet engines		162

		6.3.1	Turbojets	162
		6.3.2	Turbofans	164
		6.3.3	Turboprops	165
		6.3.4	After-burning turbojet	166
	References			167

7 Mechanics of flight 169
7.1 Performances 170
		7.1.1	Reference frames	170
		7.1.2	Hypotheses	170
		7.1.3	Aircraft equations of motion	172
		7.1.4	Performances in a steady linear flight	175
		7.1.5	Performances in steady ascent and descent flight	175
		7.1.6	Performances in gliding	176
		7.1.7	Performances in turn maneuvers	177
		7.1.8	Performances in the runway	179
		7.1.9	Range and endurance	182
		7.1.10	Payload-range diagram	184
	7.2	Stability and control		187
		7.2.1	Fundamentals of stability	187
		7.2.2	Fundamentals of control	189
		7.2.3	Longitudinal balancing	191
		7.2.4	Longitudinal stability and control	192
		7.2.5	Lateral-directional stability and control	194
	7.3	Problems		196
	References			229

III Air Transportation, Airports, and Air Navigation 231

8 Air transportation 233
	8.1	Regulatory framework		234
		8.1.1	ICAO	234
		8.1.2	IATA	239
	8.2	The market of aircraft for commercial air transportation		240
		8.2.1	Manufacturers in the current market of aircraft	241
		8.2.2	Types of aircraft	243
		8.2.3	Future market of aircraft	245
	8.3	Airlines' cost strucutre		247
		8.3.1	Operational costs	248
	8.4	Environmental impact		256

		8.4.1	Sources of environmental impact	256
		8.4.2	Aircraft operations' environmental fingerprint.	257
	References		. . . .	268

9 Airports — 269

- 9.1 Introduction 270
  - 9.1.1 Airport designation and naming 270
  - 9.1.2 The demand of air transportation 271
  - 9.1.3 The master plan 273
- 9.2 Airport configuration 273
  - 9.2.1 Airport description 273
  - 9.2.2 The runway 276
  - 9.2.3 The terminal 282
  - 9.2.4 Airport services 287
- 9.3 Airport operations 288
  - 9.3.1 Air Traffic Management (ATM) services 288
  - 9.3.2 Airport navigational aids 289
  - 9.3.3 Safety management 295
  - 9.3.4 Environmental concerns 295
- References 297

10 Air navigation — 299

- 10.1 Introduction 300
  - 10.1.1 Definition 300
  - 10.1.2 History 300
- 10.2 Technical and operative framework 306
  - 10.2.1 Communications, Navigation & Surveillance (CNS) 306
  - 10.2.2 Air Traffic Management (ATM) 308
- 10.3 Airspace Management (ASM) 311
  - 10.3.1 ATS routes 311
  - 10.3.2 Airspace organization in regions and control centers 314
  - 10.3.3 Restrictions in the airspace 316
  - 10.3.4 Classification of the airspace according to ICAO 317
  - 10.3.5 Navigation charts 317
  - 10.3.6 Flight plan 321
- 10.4 Technical support: CNS system 321
  - 10.4.1 Communication systems 322
  - 10.4.2 Navigation systems 325
  - 10.4.3 Surveillance systems 343
- 10.5 SESAR concept 348
  - 10.5.1 Single European Sky 348

      10.5.2   SESAR 348
  References 349

IV Appendixes   351

A 6-DOF Equations of Motion   353
  A.1   Reference frames 354
  A.2   Orientation between reference frames 355
       A.2.1   Wind axes-Local horizon orientation 357
       A.2.2   Body axed-Wind axes orientation 358
  A.3   General equations of motion 358
       A.3.1   Dynamic relations 358
       A.3.2   Forces acting on an aircraft 360
  A.4   Point mass model 361
       A.4.1   Dynamic relations 361
       A.4.2   Mass relations 362
       A.4.3   Kinematic relations 362
       A.4.4   Angular kinematic relations 364
       A.4.5   General differential equations system 364
  References 367

Index   **374**

PREFACE

FUNDAMENTALS OF AEROSPACE ENGINEERING covers an undergraduate, introductory course to aeronautical engineering and aims at combining theory and practice to provide a comprehensive, thorough introduction to the fascinating, yet complex discipline of aerospace engineering. This book is the ulterior result of three year of teaching a course called *Aerospace Engineering* in the first year of a degree in aerospace engineering (with a minor in air navigation) at the Universidad Rey Juan Carlos, in Madrid, Spain.

When I started preparing the course, back in 2010, I realized there was not a suitable text-book reference due to two fundamental reasons:

First, the above mentioned degree was in english, a trend that is becoming more and more popular in Spain now, but it was completely new at those days. Therefore, the classical references used in similar courses in Spain (introductory courses in aeronautical and aerospace engineering) were written in Spanish.

Second, as opposed to most parts of the world, e.g., the USA and most of Europe, where traditionally airports, air transportation, and air navigation are included in the branch of civil or transportation engineering, the studies of aeronautical and aerospace engineering in Spain (due to national legislation) include aspects related to airports, air transportation, and air navigation. As a consequence, the classical references written in english and used as classical references in similar courses did not cover part of the contents of the course.

Therefore, I started writing my own lecture notes: in english and covering issues related to airports, air transportation, and air navigation. After three preliminary, draft versions used as reference lecture notes throughout the past years, they have evolved into the book I'm presenting herein.

The book is divided into three parts, namely: Introduction, The Aircraft, and Air Transportation, Airports, and Air Navigation.

The first part is divided in two chapters in which the student must achieve to understand the basic elements of atmospheric flight (ISA and planetary references) and the technology that apply to the aerospace sector, in particular with a specific comprehension of the elements of an aircraft. The second part focuses on the aircraft and it is divided in five chapters that introduce the student to aircraft aerodynamics (fluid mechanics, airfoils,

Preface

wings, high-lift devices), aircraft materials and structures, aircraft propulsion, aircraft instruments and systems, and atmospheric flight mechanics (performances and stability and control). The third part is devoted to understand the global air transport system (covering both regulatory and economical frameworks), the airports, and the global air navigation system (its history, current status, and future development). The theoretical contents are illustrated with figures and complemented with some problems/exercises. The problems deal, fundamentally, with aerodynamics and flight mechanics, and were proposed in different exams.

The course is complemented by a practical approach. Students should be able to apply theoretical knowledge to solve practical cases using academic (but also industrial) software, such as MATLAB (now we are moving towards open source software such as SciLab). The course also includes a series of assignments to be completed individually or in groups. These tasks comprise an oral presentation, technical reports, scientific papers, problems, etc. The course is supplemented by scientific and industrial seminars, recommended readings, and a visit to an institution or industry related to the study and of interest to the students. All this documentation is not explicitly in the book but can be accessed online at the book's website www.aerospaceengineering.es. The slides of the course are also available at the book's website www.aerospaceengineering.es.

At this point, the reader should have noticed that space engineering is almost totally missing. I'm afraid this is true. The course originally was aimed at providing an introduction to aeronautical engineering with the focus on commercial aircraft, and thus space vehicles, space systems, space materials, space operations, and/or orbital mechanics are not covered in this book. Neither helicopters or unmanned air vehicles are covered. This is certainly something to add in future editions.

FUNDAMENTALS OF AEROSPACE ENGINEERING is licensed under a Creative Commons Attribution-Non Comercial-Share Alike (CC BY-NC-SA) 3.0 License, and it is offered in open access both in "pdf" and "epub" formats. The document can be accessed and downloaded at the book's website www.aerospaceengineering.es This licensing is aligned with a philosophy of sharing and spreading knowledge. Writing and revising over and over this book has been an exhausting, very time consuming activity. To acknowledge author's effort, a donation platform has been activated at the book's website www.aerospaceengineering.es. Also, printed copies can be acquired at low cost price (lower than self printing) via Amazon and/or OMM Campus Libros, which has edited a printed copy of the book at a low price for students.

Manuel Soler.

Acknowledgments

The list of people that has contributed to this book is immense. Unfortunately, I can not cite all of them herein.

First of all, I have to acknowledge the contribution of all the students that I have had the pleasure to teach during these three years. You are the reason of this book. All of them, directly or indirectly, have contributed to the final birth of the manuscript. The initial version and all the improved versions (revision after revision on a daily basis) have been encouraged by a motivation inspired in delivering the best material for their formation. They have also pointed out several grammar errors, typographical errors, structural inconsistencies, passages not properly exposed, and so on and so forth. Thank you guys.

I have to acknowledge all authors by whom the contents of the book are inspired. Special thanks to all my mentors in the School of Aeronautical Engineering at the Polytechnic University of Madrid: the lecture notes that I used as a student almost 15 years ago have been the primary source of material that I consulted when I first started writing this book. Also, special thanks to all contributors to wikipedia and other open source resources: many figures and some passages have been retrieved from wikipedia; also, some CAD figures were downloaded from the open source repository BIBCAD.

I own special thanks to two colleges, Luis Cadarso and Javier Esquillor, who invested part of their busy time in reviewing some of the chapters of the book, motivated me to continue on, and gave me some valuable advises.

For this edition, I have to acknowledge Franscisco Javier García Caro for a preliminary cover design that I used as template for final design. Also the team of Desarrollo Creativo for the book's website design.

Last but not least, this book would have been impossible without the patience and support of my beloved partner Reme and my family.

List of Figures

1.1	Flag companies and low cost companies	8
1.2	International Space Station (ISS)	9
1.3	Contributors to reducing emissions	14
2.1	Classification of air vehicles	20
2.2	Aerostats	21
2.3	Gliders	21
2.4	Military aircraft types	22
2.5	Civil aircraft types	23
2.6	Helicopter	24
2.7	Space shuttle: Discovery	26
2.8	Parts of an aircraft	27
2.9	Types of fuselages	28
2.10	Aircraft's plant-form types	29
2.11	Wing vertical position	29
2.12	Wing and empennage devices	30
2.13	Aircraft's empennage types	31
2.14	Actions on the control surfaces	32
2.15	Propulsion plant	33
2.16	Differential cylinder of air	35
2.17	ISA atmosphere	36
3.1	Stream line	49
3.2	Stream tube	49
3.3	Continuity equation	50
3.4	Quantity of movement	51
3.5	Viscosity	53
3.6	Airfoil with boundary layer	54

List of Figures

3.7	Boundary layer transition	55
3.8	Effects of the speed of sound in airfoils	57
3.9	Aerodynamic forces and moments	58
3.10	Description of an airfoil	59
3.11	Description of an airfoil with angle of attack	60
3.12	Pressure and friction stress over an airfoil	61
3.13	Aerodynamic forces and moments over an airfoil	61
3.14	Aerodynamic forces and moments over an airfoil with angle of attack	61
3.15	Lift generation	62
3.16	Aerodynamic forces and torques over an airfoil	64
3.17	Lift and drag characteristic curves	66
3.18	Divergence Mach	67
3.19	Supercritical airfoils	67
3.20	Wing geometry	69
3.21	Coefficient of lift along a wingspan	70
3.22	Whirlwind trail	70
3.23	Effective angle of attack	71
3.24	Induced drag	71
3.25	Characteristic curves in wings	73
3.26	Types of high-lift devices	76
3.27	Effects of high lift devices in airfoil flow	77
3.28	Distribution of the coefficient of pressures (Problem 3.1)	80
3.29	Coefficient of lift along the wingspan (Problem 3.1)	81
3.30	Characteristic curves of a NACA 4410 airfoil	83
3.31	Plant-form of the wing (Problem 3.3)	87
3.32	Distribution of the coefficient of pressures (Problem 3.4)	93
3.33	Coefficient of lift along the wingspan (Problem 3.4)	94
3.34	Plant-form of the wing (Problem 3.5)	96
4.1	Normal stress	104
4.2	Bending	104
4.3	Torsion	105
4.4	Shear stress due to bending	105
4.5	Shear stress due to torsion	106
4.6	Stresses in a plate	106
4.7	Normal deformation	106
4.8	Tangential deformation	107
4.9	Behavior of an isotropic material	107
4.10	Fibre-reinforced composite materials	111
4.11	Aircraft monocoque skeleton	115
4.12	Aircraft semimonocoque skeleton	116

4.13	Structural wing sketch.	117
4.14	Structural wing torsion box	117
5.1	Barometric altimeter	122
5.2	Pitot tube	123
5.3	Gyroscope and accelerometer	124
5.4	Diagram gimbals with accelerometers and gyroscopes	125
5.5	Flight and navigation instruments I	127
5.6	Flight and navigation instruments II	128
5.7	Navigation instruments	129
5.8	Instruments T layout	130
5.9	Aircraft cockpit	131
5.10	Aircraft glass cockpit displays	132
5.11	EICAS/ECAM cockpit displays	133
5.12	Flight Management System	134
5.13	Aircraft electrical system	136
5.14	A380 power system components	137
5.15	Aircraft fuel system	138
5.16	Aircraft hydraulic system	140
5.17	Flight control system	141
6.1	Propeller schematic.	149
6.2	Core elements and station numbers in a jet engine	151
6.3	Adiabatic process	152
6.4	Types of inlets	154
6.5	Types of jet compressors	155
6.6	Axial compressor	155
6.7	Combustion chamber or combustor	157
6.8	Turbine	159
6.9	Variable extension nozzle	160
6.10	Convergent–divergent nozzle	161
6.11	Relative suitability of different types of jets	163
6.12	Turbojet with centrifugal compressor	163
6.13	Turbojet with axial compressor	164
6.14	Turbofan	165
6.15	Turboprop engines	166
6.16	Afterburner	166
7.1	Wind axes reference frame.	171
7.2	Aircraft forces.	173
7.3	Aircraft forces in a horizontal loop.	177
7.4	Aircraft forces in a vertical loop.	179

List of Figures

7.5	Take off distances and velocities.	180
7.6	Forces during taking off.	181
7.7	Landing distances and velocities.	182
7.8	Take-off weight components	185
7.9	Payload-range diagram.	187
7.10	Aircraft static stability	188
7.11	Aircraft dynamic stability	189
7.12	Feedback loop control	190
7.13	Longitudinal equilibrium	191
7.14	Longitudinal stability	192
7.15	Effects of elevator on moments coefficient	195
7.16	Forces during taking off (Problem 7.2).	200
7.17	Payload-range diagram (Problem 7.4).	217
7.18	Longitudinal equilibrium (Problem 7.5)	223
8.1	Aircraft manufacturers.	242
8.2	Airbus A320 family.	245
8.3	European percentage share of airline operational costs in 2008	250
8.4	Evolution of the price of petroleum 1987-2012.	251
8.5	CO_2 and global warming emissions	259
8.6	Aircraft emissions contributing to global warming.	260
8.7	Contrails	261
8.8	Favorable regions of contrail formation	264
8.9	Favorable regions of contrail formation over USA	265
9.1	Schematic configuration of an airport	274
9.2	Typical airport infrastructure	275
9.3	Madrid Barajas layout	279
9.4	FAA airport diagram of O'Hare International Airport.	280
9.5	Runway designators	281
9.6	Finger	283
9.7	Terminal layout	284
9.8	Departure terminal layout	285
9.9	Terminal configurations	286
9.10	Airport visual aids	290
9.11	Runway pavement signs	291
9.12	Runway lighting	292
9.13	PAPI	292
9.14	ILS modulation	294
9.15	ILS: Localizer array and approach lighting	294
10.1	Triangle of velocities	301

10.2	Astronomic navigation	303
10.3	ATM levels.	308
10.4	Air Traffic Control	310
10.5	Air navigation chart	313
10.6	FIR/UIR structure	314
10.7	Volumes of responsibility	315
10.8	Phases in a flight.	318
10.9	En-route lower navigation chart.	319
10.10	Instrumental approximation chart.	320
10.11	Doppler effect.	326
10.12	Inertial Navigation System (INS).	327
10.13	Accuracy of navigation systems.	328
10.14	Scanning beam radiation.	331
10.15	VOR-DME	333
10.16	GNSS systems.	334
10.17	SBAS Augmentation Systems.	335
10.18	LORAN	336
10.19	NDB	337
10.20	VOR	339
10.21	Animation that demonstrates the spatial modulation principle of VORs	339
10.22	VOR displays interpretation.	340
10.23	MLS	342
10.24	Radar.	344
10.25	TCAS	345
10.26	ADS-B	347
A.1	Euler angles	357

List of Tables

1.1	Flag companies and low cost companies	7
3.1	Increase in $c_{l_{max}}$ of airfoils with high lift devices	78
3.2	Typical values for $C_{L_{max}}$ in wings with high-lift devices	78
3.3	Chap 3. Prob. 2 Data $c_l - \alpha$	84
3.4	Chap 3. Prob. 2 Data $c_l - c_d$	84
8.1	Long-haul aircraft specifications	243
8.2	Medium-haul aircraft specifications	244
8.3	Regional aircraft specifications	244
8.4	Airbus 2012 medium-haul aircraft prices	244
8.5	Airbus 2012 long-haul aircraft prices	244
8.6	A320neo family specifications	246
8.7	B737 MAX family specifications	246
8.8	Cost structure of a typical airline	248
8.9	Evolution of airlines' operational costs 2001-2008	250
8.10	Route's waypoints, navaids, and fixes	263
9.1	Busiest airports by passengers in 2012	271
9.2	Busiest airports by passengers in 2011	272
9.3	Runway ICAO categories	278
9.4	Runway's minimum width according to ICAO	278
9.5	ICAO minimum distance in airport operations	282
9.6	ILS categories	295
10.1	Navigation charts	318
10.2	Navigational systems	325
10.3	Navigation aids based on situation surface and the technique	331
10.4	Classification of the navigation aids based on the flight phase	331

Part I

Introduction

1

THE SCOPE

Contents

1.1	Engineering	4
1.2	Aerospace activity	5
1.3	Aviation research agenda	12
	1.3.1 Challenges	12
	1.3.2 Clean Sky	13
	1.3.3 SESAR	15
References		17

The aim of this chapter is to give a broad overview of the activities related to the field referred to as aerospace engineering. More precisely, it aims at summarizing briefly the main scope in which the student will develop his or her professional career in the future as an aerospace engineer. First, a rough overview of what engineering is and what engineers do is given, with particular focus on aerospace engineering. Also, a rough taxonomy of the capabilities that an engineer is supposed to have is provided. Second, the focus is on describing the different aerospace activities, i.e., the industry, the airlines, the military air forces, the infrastructures on earth, the research institutions, the space agencies, and the international organizations. Last but not least, in the believe that research, development, and innovation is the key element towards the future, an overview of the current aviation research agenda is presented.

1.1 Engineering

Following WIKIPEDIA [7], engineering can be defined as:

> The application of scientific, economic, social, and practical knowledge in order to design, build, and maintain structures, machines, devices, systems, materials, and processes. It may encompass using insights to conceive, model, and scale an appropriate solution to a problem or objective. The discipline of engineering is extremely broad, and encompasses a range of more specialized fields of engineering, each with a more specific emphasis on particular areas of technology and types of application.

The foundations of engineering lays on mathematics and physics, but more important, it is reinforced with additional study in the natural sciences and the humanities. Therefore, attending to the previously given definition, engineering might be briefly summarized with the following six statements:

- to adapt scientific discovery for useful purposes;
- to create useful devices for the service of society;
- to invent solutions to meet society's needs;
- to come up with solutions to technical problems;
- to utilize forces of nature for society's purposes;
- to convert energetic resources into useful work.

On top of this, according to current social sensitivities, one should add: in an environmentally friendly manner.

Following WIKIPEDIA [6], aerospace engineering can be defined as:

> a primary branch of engineering concerned with the research, design, development, construction, testing, and science and technology of aircraft and spacecraft. It is divided into two major and overlapping branches: aeronautical engineering and astronautical engineering. The former deals with aircraft that operate in Earth's atmosphere, and the latter with spacecraft that operate outside it.

Therefore, an aerospace engineering education attempts to introduce the following capabilities NEWMAN [3, Chap. 2]:

- Engineering fundamentals (maths and physics), innovative ideas conception and problem solving skills, the vision of high-technology approaches to engineering complex systems, the idea of technical system integration and operation.

- knowledge in the technical areas of aerospace engineering including mechanics and physics of fluids, aerodynamics, structures and materials, instrumentation, control and estimation, humans and automation, propulsion and energy conversion, aeronautical and astronautical systems, infrastructures on earth, the air navigation system, legislation, air transportation, etc.
- The methodology and experience of analysis, modeling, and synthesis.
- Finally, an engineering goal of addressing socio-humanistic problems.

As a corollary, an aerospace engineering education should produce engineers capable of the following NEWMAN [3, Chap. 2]:

- **Conceive**: conceptualize technical problems and solutions.
- **Design**: study and comprehend processes that lead to solutions to a particular problem including verbal, written, and visual communications.
- **Development**: extend the outputs of research.
- **Testing**: determine performance of the output of research, development, or design.
- **Research**: solve new problems and gain new knowledge.
- **Manufacturing**: produce a safe, effective, economic final product.
- **Operation and maintenance**: keep the products working effectively.
- **Marketing and sales**: look for good ideas for new products or improving current products in order to sell.
- **Administration (management)**: coordinate all the above.

Thus, the student as a future aerospace engineer, will develop his or her professional career accomplishing some of the above listed capabilities in any of the activities that arise within the aerospace industry.

1.2 Aerospace activity

It seems to be under common agreement that the aerospace activities (in which aerospace engineers work) can be divided into seven groups FRANCHINI *et al.* [2]:

- the **industry**, manufacturer of products;
- the **airlines**, transporters of goods and people;
- the **military air forces**, demanders of high-level technologies;
- the **space agencies**, explorers of the space;
- the **infrastructures on earth**, supporter of air operations;
- the **research institutions**, guarantors of technological progress;
- the **international organizations**, providers of jurisprudence.

The aerospace industry

The aerospace industry is considered as an strategic activity given that it is a high technology sector with an important economic impact. The Aerospace sector is an important contributor to economic growth everywhere in the world. The european aerospace sector represents a pinnacle of manufacturing which employed almost half a million highly skilled people directly (20.000 in Spain) in 2010 and it continuously spins-out technology to other sectors. About 2.6 million indirect jobs can be attributed to air transport related activities and a contribution of around €250 billion[1] (around 2.5%) to european gross domestic product in 2010. Therefore, the aerospace industry is an important asset for Europe economically, being a sector that invests heavily in Research and Development (R & D) compared with other industrial sectors. The aerospace sector is also an important pole for innovation.

The aerospace industry accomplish three kinds of activities: aeronautics (integrated by airships, propulsion systems, and infrastructures and equipments); space; and missiles. *Grosso modo*, the aeronautical industry constitutes around the 80-90% of the total activity.

The fundamental characteristics of the aerospace industry are:

- Great dynamism in the cycle research-project-manufacture-commercialization.
- Specific technologies in the vanguard which spin-out to other sectors.
- High-skilled people.
- Limited series (non mass production) and difficult automation of manufacturing processes.
- Long term development of new projects.
- Need for huge amount of capital funding.
- Governmental intervention and international cooperation.

The linkage between research and project-manufacture is essential because the market is very competitive and the product must fulfill severe safety and reliability requirements in order to be certified. Thus, it is necessary to continuously promote the technological advance to take advantage in such a competitive market.

The quantity of units produced a year is rather small if we compare it with other manufacture sectors (automobile manufacturing, for instance). An airship factory only produces tens of units a year; in the case of space vehicles the common practice is to produce a unique unit. These facts give a qualitative measure of the difficulties in automating manufacturing processes in order to reduce variable costs.

The governmental intervention comes from different sources. First, directly participating from the capital of the companies (many of the industries in Spain and Europe are state

---

[1] one billion herein refers to 100.000.000.000 monetary units.

Flag companies	Low Cost companies
Operate hubs and spoke	Operate point to point
Hubs in primary international airports	Mostly regional airports
Long rotation times (50 min)	Short rotation times (25 min)
Short and long haul routes	Short haul routes
Mixed fleets	Standardized fleets
Low density seats layout	High density seats layout
Selling: agencies and internet	Selling: internet
Extras included (Business, VIP lounges, catering)	No extras included in the tickets

Table 1.1: Comparison between flag companies and low cost companies.

owned). Indirectly, throughout research subsides. Also, as a direct client, as it is the case for military aviation. The fact that many companies do not have the critical size to absorb the costs and the risks of such projects makes common the creation of long-term alliances for determined aircrafts (Airbus) or jet engines (International Aero Engines or Eurojet).

Airlines

Among the diverse elements that conform the air transportation industry, airlines represent the most visible ones and the most interactive with the consumer, i.e., the passenger. An airline provides air transport services for traveling passengers and/or freight. Airlines lease or own their aircraft with which to supply these services and may form partnerships or alliances with other airlines for mutual benefit, e.g., Oneworld, Skyteam, and Star alliance. Airlines vary from those with a single aircraft carrying mail or cargo, through full-service international airlines operating hundreds of aircraft. Airline services can be categorized as being intercontinental, intra-continental, domestic, regional, or international, and may be operated as scheduled services or charters.

The first airlines were based on dirigibles. DELAG (Deutsche Luftschiffahrts-Aktiengesellschaft) was the world's first airline. It was founded on November 16, 1909, and operated airships manufactured by the zeppelin corporation. The four oldest non-dirigible airlines that still exist are Netherlands' KLM, Colombia's Avianca, Australia's Qantas, and the Czech Republic's Czech Airlines. From those first years, going on to the elite passenger of the fifties and ultimately to the current mass use of air transport, the world airline companies have evolved significantly.

Traditional airlines were state-owned. They were called *flag companies* and used to have a strong strategic influence. It was not until 1978, with the United States Deregulation Act, when the market started to be liberalized. The main purpose of the act was to remove government control over fares, routes, and market entry of new airlines in the commercial

(a) Iberia's A340. © Javier Bravo Muñoz / Wikimedia Commons / GNU FDL.

(b) Iberia Express's A320. © Curimedia / Wikimedia Commons / CC-BY-SA-2.0.

(c) Ryanair's B737-800. © AlejandroDiRa / Wikimedia Commons / CC-BY-SA-3.0.

Figure 1.1: Flag companies (e.g., Iberia) and low cost companies (e.g., Iberia Express and Ryanair).

aviation sector. Up on that law, private companies started to emerge in the 80's and 90's, specially in USA. Very recently, a new phenomena have arisen within the last 10-15 years: the so called *low cost companies*, which have favored the mass transportation of people. A comparison between low cost companies and traditional flag companies is presented in Table 1.1. It provides a first understanding of the main issues involved in the direct operating costs of an airline, which will be studied in Chapter 8. The competition has been so fierce that many traditional companies have been pushed to create their own low cost filial companies, as it the case of Iberia and its filial Iberia Express. See Figure 1.1.

Military air forces

The military air forces are linked to the defense of each country. In that sense, they play a strategic role in security, heavily depending on the economical potential of the country and its geopolitical situation. Historically, it has been an encouraging sector for technology and innovation towards military supremacy. Think for instance in the advances due to World War II and the Cold War. Nowadays, it is mostly based on cooperation and alliances. However, inherent threats in nations still make this sector a strategic sector whose demand in high technology will be maintained. An instance of this is the encouraging trend of the USA towards the development of Unmanned Air Vehicles (UAV) in the last 20 years in order to maintain the supremacy in the middle east minimizing the risk of soldiers life.

Space agencies

There are many government agencies engaged in activities related to outer space exploration. Just to mention a few, the China National Space Administration (CNSA), the Indian Space Research Organization (ISRO), the Russian Federal Space Agency (RFSA)

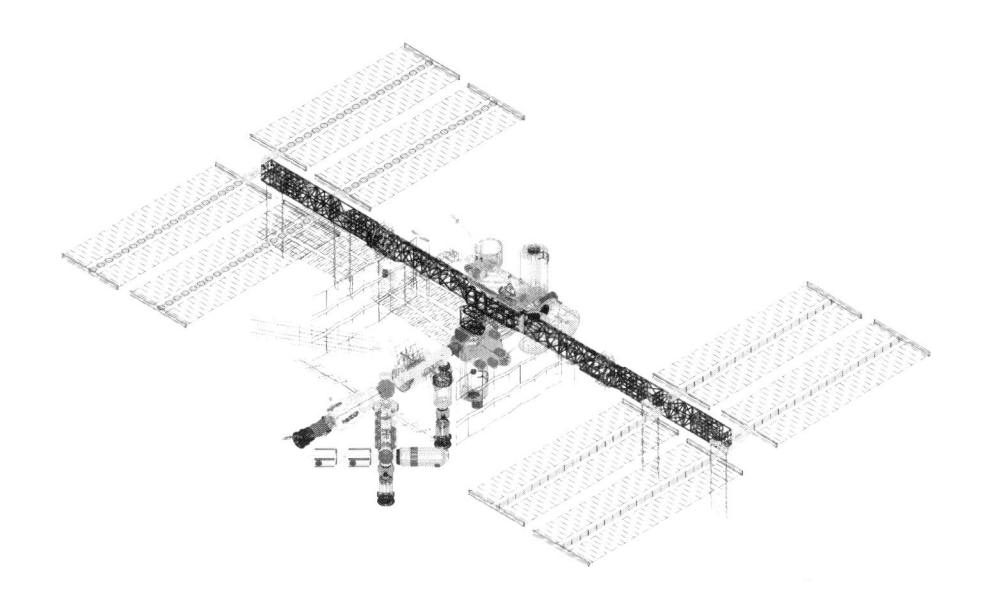

Figure 1.2: International Space Station (ISS).

(successor of the Soviet space program), the European Space Agency (ESA), and the National Aeronautics and Space Administration (NASA). For their interest, the focus will be on these last two.

The European Space Agency (ESA) was established in 1975, it is an intergovernmental organization dedicated to the exploration of space. It counts currently with 20 member states: Austria, Belgium, Check Republic, Denmark, Finland, France, Germany, Greece, Ireland, Italy, Luxembourg, the Nederland, Norway, Poland, Portugal, Romania, Spain, Sweden, Switzerland, and United Kingdom. Moreover, Hungary and Canada have a special status and cooperate in certain projects.

In addition to coordinating the individual activities of the member states, ESA's space flight program includes human spaceflight, mainly through the participation in the International Space Station (ISS) program (Columbus lab, Node-3, Cupola), the launch and operation of unmanned exploration missions to other planets and the Moon (probe Giotto to observe Halley's comet; Cassine-Hyugens, joint mission with NASA, to observe Saturn and its moons; Mars Express, to explore mars), Earth observation (Meteosat), science (Spacelab), telecommunication (Eutelsat), as well as maintaining a major spaceport, the Guiana Space Centre at Kourou, French Guiana, and designing launch vehicles (Ariane).

The National Aeronautics and Space Administration (NASA) is the agency of the United States government that is responsible for the civilian space program and for aeronautics and aerospace research. NASA was established by the National Aeronautics and Space Act on July 29, 1958, replacing its predecessor, the National Advisory Committee for Aeronautics (NACA). NASA science is focused on better understanding of Earth through the Earth Observing System, advancing heliophysics through the efforts of the Science Mission Directorate's Heliophysics Research Program, exploring bodies throughout the Solar System with advanced robotic missions such as New Horizons, and researching astrophysics topics, such as the Big Bang, through the Great Observatories and associated programs.

United States space exploration efforts have since 1958 been led by NASA, including the Apollo moon-landing missions, the Skylab space station, the Space Shuttle, a reusable space vehicles program whose last mission took place in 2011 (see Figure 2.7), the probes (Pioner, Viking, etc.) which explore the outer space. Currently, NASA is supporting the ISS, and the Mars Science Laboratory unmanned mission known as *curiosity*. NASA not only focuses on space, but conducts fundamental research in aeronautics, such in aerodynamics, propulsion, materials, or air navigation.

Infrastructures on earth

In order to perform safe operations either for airliners, military aircraft, or space missions, a set of infrastructures and human resources is needed. The necessary infrastructures on earth to assist flight operations and space missions are: airports and air navigation services on the one hand (referring to atmospheric flights); launch bases and control and surveillance centers on the other (referring to space missions).

The airport is the localized infrastructure where flights depart and land, and it is also a multi-modal node where interaction between flight transportation and other transportation modes (rail and road) takes place. It consists of a number of conjoined buildings, flight field installations, and equipments that enable: the safe landing, take-off, and ground movements of aircrafts, together with the provision of hangars for parking, service, and maintenance; the multi-modal (earth-air) transition of passengers, baggage, and cargo.

The air navigation is the process of steering an aircraft in flight from an initial position to a final position, following a determined route and fulfilling certain requirements of safety and efficiency. The navigation is performed by each aircraft independently, using diverse external sources of information and proper on-board equipment. The fundamental goals are to avoid getting lost, to avoid collisions with other aircraft or obstacles, and to minimize the influence of adverse meteorological conditions. Air navigation demands juridic, organizational, operative, and technical framework to assist aircraft on air fulfilling safe operations. The different Air Navigation Service Providers (ANSP) (AENA in Spain, FAA in USA, Eurocontrol in Europe, etc.) provide these frameworks, comprising three main components:

- Communication, Navigation, and Surveillance (CNS).
- Meteorological services.

- Air Traffic Management (ATM).
    - Air Space Management (ASM).
    - Air Traffic Services (ATS) such traffic control and information.
    - Air Traffic Flow Management (ATFM).

A detailed insight on these concepts will be given in Chapter 10.

A launch base is an earth-based infrastructure from where space vehicles are launched to outer space. The situation of launch bases depends up on different factors, including latitudes close to the ecuador, proximity to areas inhabited or to the sea to avoid danger in the first stages of the launch, etc. The most well known bases are: Cape Kennedy in Florida (NASA); Kourou in the French Guyana (ESA); Baikonur en Kazakhstan (ex Soviet Union space program). Together with the launch base, the different space agencies have control centers to monitor the evolution of the space vehicles, control their evolution, and communicate with the crew (in case there is crew).

Aerospace research institutions

The research institutions fulfill a key role within the aerospace activities because the development of aviation and space missions is based on a continuos technological progress affecting a variety of disciplines such as aerodynamics, propulsion, materials, avionics, communication, airports, air navigation, etc. The research activity is fundamentally fulfilled at universities, aerospace companies, and public institutions.

Spain counts with the Instituto Nacional de Técnica Aeroespacial (INTA), which is the spanish public research organization specialized in aerospace research and technology development. It pursues the acquisition, maintenance, and continuous improvement of all those technologies that can be applied to the aerospace field. There exist "sister" institutions such the French Aerospace Lab (ONERA), the German Aerospace Center (DLR), or the National Aeronautics and Space Administration (NASA) in the United States of America (USA), just to mention a few significative ones.

International organizations

In order to promote a reliable, efficient, and safe air transportation, many regulations are needed. This regulatory framework arises individually in each country but always under the regulatory core of two fundamental supranational organizations: The International Civil Aviation Organization (ICAO) and the International Air Transport Association (IATA) .

ICAO was created as a result of the Chicago Convention. ICAO was created as a specialized agency of the United Nations charged with coordinating and regulating international air travel. The Convention establishes rules of airspace, aircraft registration and safety, and details the rights of the signatories in relation to air travel. In the successive revisions ICAO has agreed certain criteria about the freedom of overflying and landing in countries, to develop the safe and ordered development of civil aviation world wide, to encourage the design and use techniques of airships, to stimulate the development of the necessary infrastructures for air navigation. Overall, ICAO has encourage the evolution of civil aviation.

The modern IATA is the successor to the International Air Traffic Association founded in the Hague in 1919. IATA was founded in Havana, Cuba, in April 1945. It is the prime vehicle for inter-airline cooperation in promoting safe, reliable, secure, and economical air services. IATA seeks to improve understanding of the industry among decision makers and increase awareness of the benefits that aviation brings to national and global economies. IATA ensures that people and goods can move around the global airline network as easily as if they were on a single airline in a single country.

In addition to the cited organizations, it is convenient to mention the two most important organization with responsibility in safety laws and regulations, including the airship project and airship certification, maintenance labour, crew training, etc.: The European Aviation Safety Agency (EASA) in the European Union and the Federal Aviation Administration (FAA). Spain counts with the Agencia Estatal de Seguridad Aérea (AESA), dependent on the ministry of infrastructures (*fomento*). AESA is also responsible for safety legislation in civil aviation, airships, airports, air navigation, passengers rights, general aviation, etc.

1.3 Aviation research agenda

Aviation has dramatically transformed society over the past 100 years. The economic and social benefits throughout the world have been immense in shrinking the planet with the efficient and fast transportation of people and goods. However, encouraging challenges must be faced to cope with the expected demand, but also to meet social sensitivities.

These challenges have led to the formation of the Advisory Council for Aeronautics Research in Europe (ACARE) to define a Strategic Research Agenda (SRA) for aeronautics and air transport in Europe ACARE [1]. The goals set by the SRA have had a clear influence on current aeronautical research, delivering important initiatives and benefits for the aviation industry, including among others the Clean Sky Joint Technology Initiative, which pursues a greener aviation, and the Single European Sky ATM Research (SESAR) Joint Undertaking, which pursues a more efficient ATM system. Initiatives in the same direction have been also driven in USA within the Next Generation of air transportation system (NextGen).

1.3.1 Challenges

The growth of air traffic in the past 20 years has been spectacular, and forecasts indicate that there will continue in the future. According to Eurocontrol, in 2010 there was 9.493

million IFR[2] flights in Europe, around 26000 flights a day. Traffic demand will nearly triple, and airlines will more than double their fleets of passenger aircraft within 20 years time. This continuous growth in demand will bring increased challenges for dealing with mass transportation and congestion of ATM and airport infrastructure.

Aviation is directly impacted by energy trends. The oil price peaks in the last period (2008-2011) due to diverse geopolitical crisis are not isolated events, the cost of oil will continue to increase. Dependence on fuel availability will continue to be a risk for air transport, specially if energy sources are held in a few hands. Aviation will have to develop long-term strategies for energy supply, such as alternative fuels, that will be technically suitable and commercially scaleable as well as environmentally sustainable.

Climate change is a major societal and political issue and is becoming more. Globally civil aviation is responsible for 2% of CO_2 of man made global emissions according to the United Nations' Intergovernmental Panel on Climate Change (IPCC). As aviation grows to meet increasing demand, the IPCC forecasts that its share of global man made CO_2 emissions will increase to around 3% to 5% in 2050 PENNER [4]. Non-CO_2 emissions including oxides of nitrogen and condensation trails which may lead to the formation of cirrus clouds, also have impacts but require better scientific understanding. Thus, reducing emissions represents a major challenge, maybe the biggest ever. Reducing disturbance around airports is also a challenge with the need to ensure that noise levels and air quality around airports remain acceptable.

1.3.2 CLEAN SKY

Clean Sky is a Joint Technology Initiative (JTI) that aims at developing a mature breakthrough clean technologies for air transport. By accelerating their deployment, the JTI will contribute to Europe's strategic environmental and social priorities, and simultaneously promote competitiveness and sustainable economic growth.

Joint Technology Initiatives are specific large scale research projects created by the European Commission within the 7th Framework Programme (FP7) in order to allow the achievement of ambitious and complex research goals, set up as a public-private partnership between the European Commission and the European aeronautical industry.

Clean Sky will speed up technological breakthrough developments and shorten the time to market for new and cleaner solutions tested on full scale demonstrators, thus contributing significantly to reducing the environmental footprint of aviation (i.e. emissions and noise reduction but also green life cycle) for our future generations. The purpose of Clean Sky is to demonstrate and validate the technology breakthroughs that are necessary to make major steps towards the environmental goals set by ACARE and to be reached in 2020 when compared to 2000 levels:

---

[2]IFR stands for Instrumental Flight Rules and refers to instrumental flights

The Scope

CO_2 emissions

Aircraft design and Flight management — 35-45%

ATM and Aispace design — 5-10%

Bio fuels — 45-60%

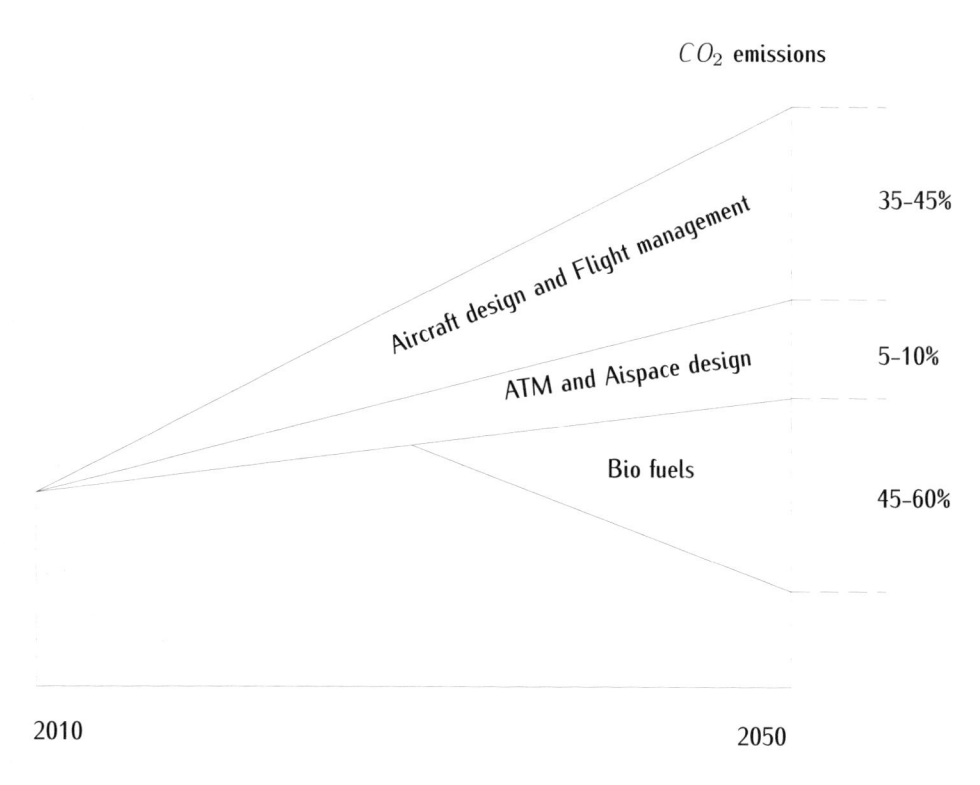

2010   2050

Figure 1.3: Contributors to reducing emissions. Adapted from Clean Sky JTI.

- 50% reduction of CO_2 emissions;
- 80% reduction of NOx emissions;
- 50% reduction of external noise; and
- a green product life cycle.

Clean Sky JTI is articulated around a series of the integrated technology demonstrators:

- Eco Design.
- Smart Fixed Wing Aircraft.
- Green Regional Aircraft.
- Green Rotorcraft.
- Systems for Green Operations.
- Sustainable and Green Engines.

Therefore, the reduction in fuel burn and CO_2 will require contributions from new technologies in aircraft design (engines, airframe materials, and aerodynamics), alternative fuels (bio fuels), and improved ATM and operational efficiency (mission and trajectory management). See Figure 1.3. ACARE has identified the main contributors to achieving the above targets. The predicted contributions to the 50% CO_2 emissions reduction target are: efficient aircraft: 20-25%; efficient engines: 15-20%; improved air traffic management: 5-10%; bio fuels: 45-60%.

1.3.3 SESAR

ATM, which is responsible for sustainable, efficient, and safe operations in civil aviation, is still nowadays a very complex and highly regulated system. A substantial change in the current ATM paradigm is needed because this system is reaching the limit of its capabilities. Its capacity, efficiency, environmental impact, and flexibility should be improved to accommodate airspace users' requirements. The Single European Sky ATM Research (SESAR) Program aims at developing a new generation of ATM system.

The SESAR program is one of the most ambitious research and development projects ever launched by the European Community. The program is the technological and operational dimension of the Single European Sky (SES) initiative to meet future capacity and air safety needs. Contrary to the United States, Europe does not have a single sky, one in which air navigation is managed at the European level. Furthermore, European airspace is among the busiest in the world with over 33,000 flights on busy days and high airport density. This makes air traffic control even more complex. The Single European Sky is the only way to provide an uniform and high level of safety and efficiency over Europe's skies. The major elements of this new institutional and organizational framework for ATM in Europe consist of: separating regulatory activities from service provision and the possibility of cross-border ATM services; reorganizing European airspace that is no longer constrained by national borders; setting common rules and standards, covering a wide range of issues, such as flight data exchanges and telecommunications.

The mission of the SESAR Joint Undertaking is to develop a modernized air traffic management system for Europe. This future system will ensure the safety and fluidity of air transport over the next thirty years, will make flying more environmentally friendly, and reduce the costs of air traffic management system. Indeed, the main goals of SESAR are SESAR CONSORTIUM [5]:

- 3-fold increase the air traffic movements whilst reducing delays;
- improvement the safety performance by a factor of 10;
- 10% reduction in the effects aircraft have on the environment;
- provide ATM services at a cost to airspace users with at least 50% less.

References

[1] ACARE (2010). Beyond Vision 2020 (Towards 2050). Technical report, European Commission. The Advisory Council for Aeronautics Research in Europe.

[2] FRANCHINI, S., LÓPEZ, O., ANTOÍN, J., BEZDENEJNYKH, N., and CUERVA, A. (2011). *Apuntes de Tecnología Aeroespacial*. Escuela de Ingeniería Aeronáutica y del Espacio. Universidad Politécnica de Madrid.

[3] NEWMAN, D. (2002). *Interactive aerospace engineering and design*. McGraw-Hill.

[4] PENNER, J. (1999). Aviation and the global atmosphere: a special report of IPCC working groups I and III in collaboration with the scientific assessment panel to the Montreal protocol on substances that deplete the ozone layer. Technical report, International Panel of Climate Change (IPCC).

[5] SESAR CONSORTIUM (April 2008). SESAR Master Plan, SESAR Definition Phase Milestone Deliverable 5.

[6] WIKIPEDIA. Aerospace Engineering. http://en.wikipedia.org/wiki/Aerospace_engineering. Last accesed 30 sept. 2013.

[7] WIKIPEDIA. Engineering. http://en.wikipedia.org/wiki/Engineering. Last accesed 30 sept. 2013.

2
GENERALITIES

Contents

2.1	Classification of aerospace vehicles		20
	2.1.1	Fixed wing aircraft	21
	2.1.2	Rotorcraft	23
	2.1.3	Missiles	24
	2.1.4	Space vehicles	25
2.2	Parts of the aircraft		27
	2.2.1	Fuselage	27
	2.2.2	Wing	28
	2.2.3	Empennage	30
	2.2.4	Main control surfaces	31
	2.2.5	Propulsion plant	32
2.3	Standard atmosphere		33
	2.3.1	Hypotheses	33
	2.3.2	Fluid-static equation	34
	2.3.3	ISA equations	34
	2.3.4	Warm and cold atmospheres	36
	2.3.5	Barometric altitude	37
2.4	System references		37
2.5	Problems		39
References			43

The aim of this chapter is to present the student some generalities focusing on the atmospheric flight of airplanes. First, a classification of aerospace vehicles is given. Then, focusing on airplanes (which will be herein also referred to as aircraft), the main parts of an aircraft will be described. Third, the focus is on characterizing the atmosphere, in which atmospheric flight takes place. Finally, in order to be able to describe the movement of an aircraft, different system references will be presented. For a more detailed description of an aircraft, please refer for instance to any of the following books in aircraft design: TORENBEEK [8], HOWE [5], JENKINSON et al. [6], and RAYMER et al. [7].

GENERALITIES

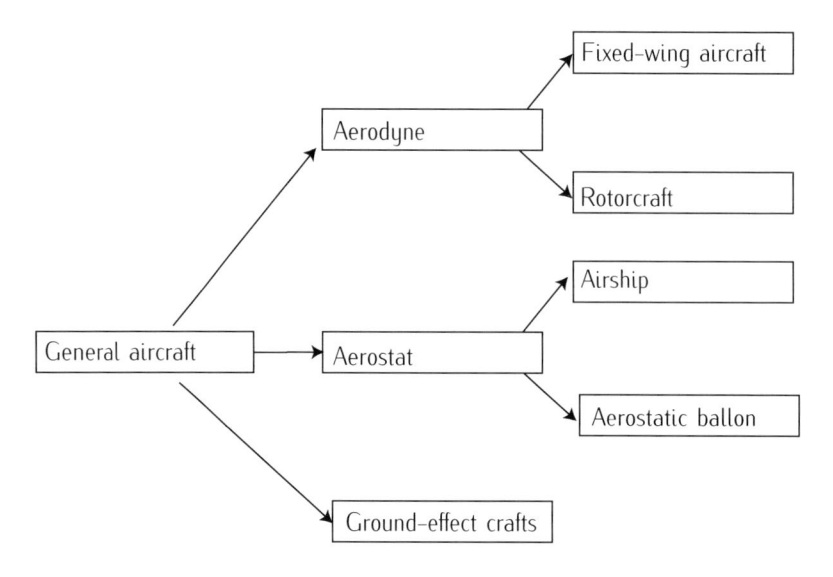

Figure 2.1: Classification of air vehicles. Adapted from FRANCHINI *et al.* [3].

2.1 CLASSIFICATION OF AEROSPACE VEHICLES (FRANCHINI *et al.* [3], FRANCHINI AND GARCÍA [2])

An aircraft, in a wide sense, is a vehicle capable to navigate in the air (in general, in the atmosphere of a planet) by means of a lift force. This lift appears due to two different physical phenomena:

- aeroestatic lift, which gives name to the aerostats (lighter than the air vehicles), and
- dynamic effects generating lift forces, which gives name to the aerodynes (heavier than the air vehicles).

An aerostat is a craft that remains aloft primarily through the use of lighter than air gases, which produce lift to the vehicle with nearly the same overall density as air. Aerostats include airships and aeroestatic balloons. Aerostats stay aloft by having a large "envelope" filled with a gas which is less dense than the surrounding atmosphere. See Figure 2.2 as illustration.

Aerodynes produce lift by moving a wing through the air. Aerodynes include fixed-wing aircraft and rotorcraft, and are heavier-than-the-air aircraft. The first group is the one nowadays know as airplanes (also known simply as aircraft). Rotorcraft include helicopters or autogyros (Invented by the Spanish engineer Juan de la Cierva in 1923).

A special category can also be considered: *ground effect* aircraft. Ground effect refers to the increased lift and decreased drag that an aircraft airfoil or wing generates when an aircraft is close the ground or a surface. Missiles and space vehicles will be also analyzed as classes of aerospace vehicles.

(a) Aerostatic ballon.

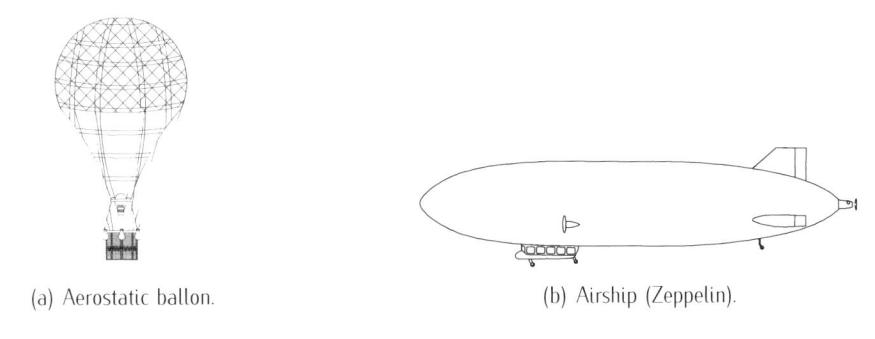

(b) Airship (Zeppelin).

Figure 2.2: Aerostats.

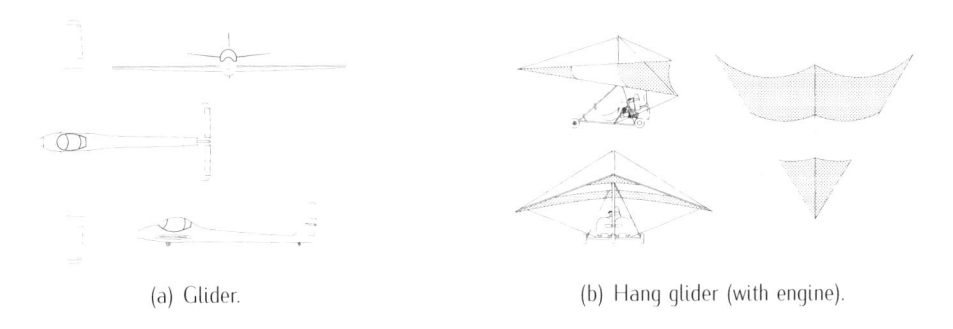

(a) Glider.

(b) Hang glider (with engine).

Figure 2.3: Gliders.

2.1.1 Fixed wing aircraft

A first division arises if we distinguish those fixed-wing aircraft with engines from those without engines.

A glider is an aircraft whose flight does not depend on an engine. The most common varieties use the component of their weight to descent while they exploit meteorological phenomena (such thermal gradients and wind deflections) to maintain or even gain height. Other gliders use a tow powered aircraft to ascent. Gliders are principally used for the air sports of gliding, hang gliding and paragliding, or simply as leisure time for private pilots. See Figure 2.3.

Aerodynes with fixed-wing and provided with a power plant are known as airplanes[1]. An exhaustive taxonomy of airplanes will not be given, since there exist many particularities. Instead, a brief sketch of the fundamentals which determine the design of an aircraft will be drawn. The fundamental variables that must be taken into account for airplane design are: mission, velocity range, and technological solution to satisfy the needs of the mission.

The configuration of the aircraft depends on the aerodynamic properties to fly in

---

[1]Also referred to as aircraft. From now on, when we referred to an *aircraft*, we mean an aerodyne with fixed-wing and provided with a power plant.

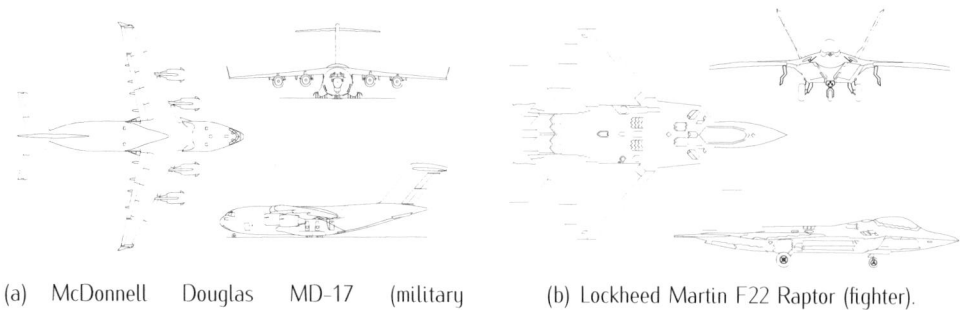

GENERALITIES

(a) McDonnell Douglas MD-17 (military transportation).

(b) Lockheed Martin F22 Raptor (fighter).

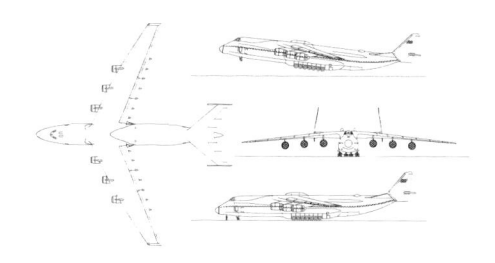

(c) Antonov 225 (military transportation).

Figure 2.4: Military aircraft types.

a determined regime (low subsonic, high subsonic, supersonic). In fact, the general configuration of the aircraft depends upon the layout of the wing, the fuselage, stabilizers, and power plant. This four elements, which are enough to distinguish, *grosso modo*, one configuration from another, are designed according to the aerodynamic properties.

Then one possible classification is according to its configuration. However, due to different technological solutions that might have been adopted, airplanes with the same mission, could have different configurations. This is the reason why it seems more appropriate to classify airplanes attending at its mission.

Two fundamental branches exist: military airplanes and civilian airplanes.

The most usual military missions are: surveillance, recognition, bombing, combat, transportation, or training. For instance, a combat airplane must flight in supersonic regime and perform sharp maneuvers. Figure 2.4 shows some examples of military aircraft.

In the civil framework, the most common airplanes are those dedicated to the transportation of people in different segments (business jets, regional transportation, medium-haul transportation, and long-haul transportation). Other civil uses are also derived to civil aviation such fire extinction, photogrametric activities, etc. Figure 2.5 shows some examples of civilian aircraft.

(a) Cessna 208 (Regional)   (b) Concorde (Supersonic)

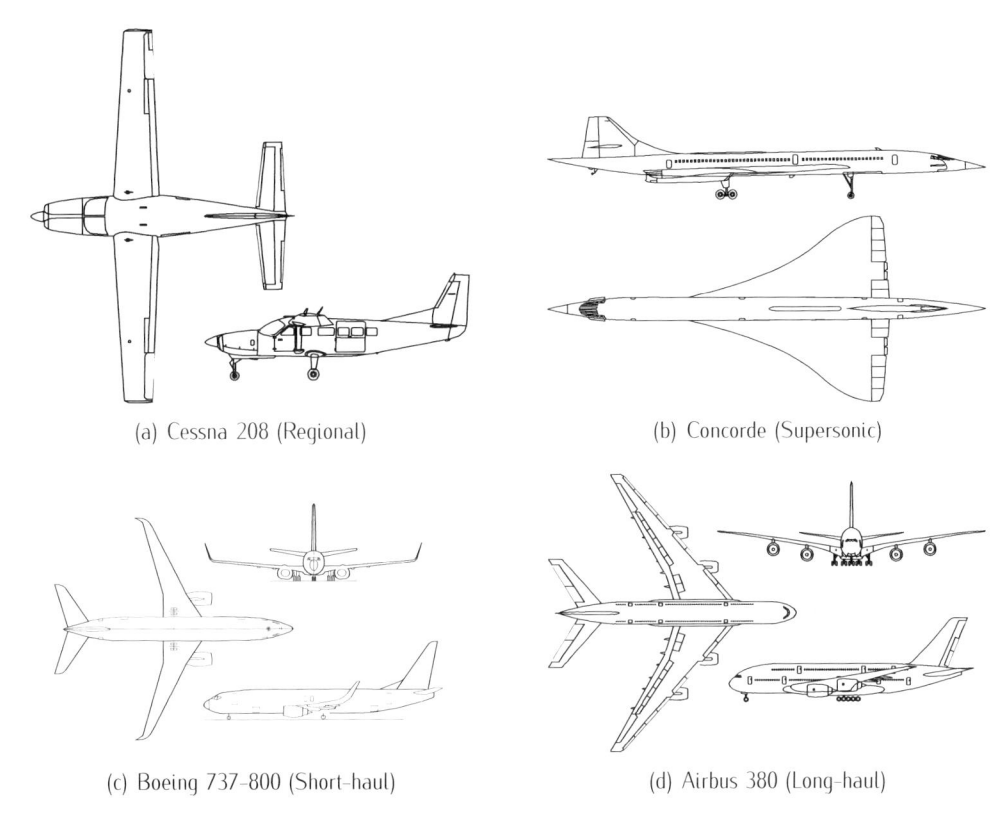

(c) Boeing 737-800 (Short-haul)   (d) Airbus 380 (Long-haul)

Figure 2.5: Types of civilian transportation aircraft.

2.1.2 Rotorcraft

A rotorcraft (or rotary wing aircraft) is a heavier-than-air aircraft that uses lift generated by wings, called rotor blades, that revolve around a mast. Several rotor blades mounted to a single mast are referred to as a rotor. The International Civil Aviation Organization (ICAO) defines a rotorcraft as *supported in flight by the reactions of the air on one or more rotors*. Rotorcraft include:

- Helicopters.
- Autogyros.
- Gyrodinos.
- Combined.
- Convertibles.

A helicopter is a rotorcraft whose rotors are driven by the engine (or engines) during the flight, to allow the helicopter to take off vertically, hover, fly forwards, backwards, and laterally, as well as to land vertically. Helicopters have several different configurations

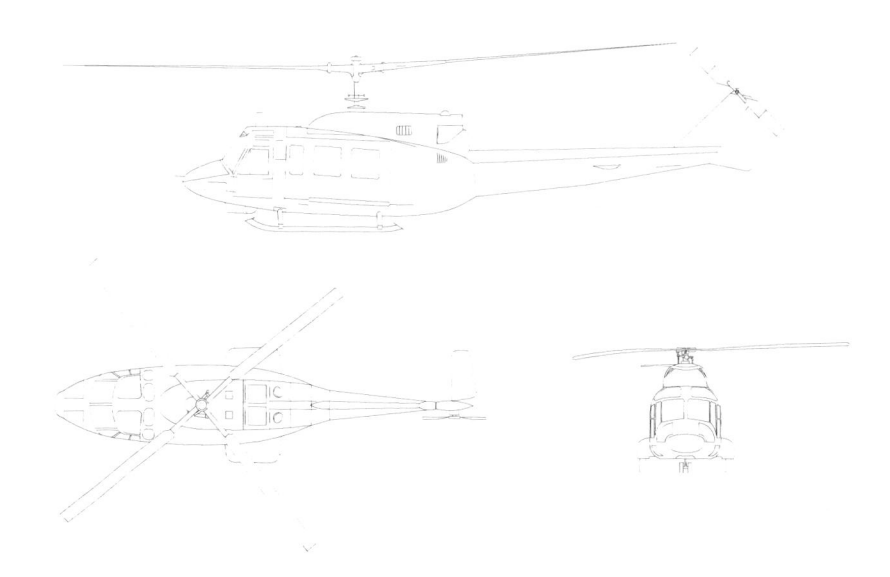

Figure 2.6: Helicopter.

of one or more main rotors. Helicopters with one driven main rotor require some sort of anti-torque device such as a tail rotor. See Figure 2.6 as illustration of an helicopter.

An autogyro uses an unpowered rotor driven by aerodynamic forces in a state of autorotation to generate lift, and an engine-powered propeller, similar to that of a fixed-wing aircraft, to provide thrust and fly forward. While similar to a helicopter rotor in appearance, the autogyro's rotor must have air flowing up and through the rotor disk in order to generate rotation.

The rotor of a gyrodyne is normally driven by its engine for takeoff and landing (hovering like a helicopter) with anti-torque and propulsion for forward flight provided by one or more propellers mounted on short or stub wings.

The combined is an aircraft that can be either helicopter or autogyro. The power of the engine can be applied to the rotor (helicopter mode) or to the propeller (autogyro mode). In helicopter mode, the propeller assumes the function of anti-torque rotor.

The convertible can be either helicopter or airplane. The propoller-rotor (proprotor) changes its attitude 90 [deg] with respect to the fuselage so that the proprotor can act as a rotor (helicopter) or as a propeller with fixed wings (airplane).

2.1.3 Missiles

A missile can be defined as an unmanned self-propelled guided weapon system.

Missiles can be classified attending at different concepts: attending at the trajectory, missiles can be cruise, ballistic, or semi-ballistic. A ballistic missile is a missile that follows a sub-orbital ballistic flightpath with the objective to a predetermined target. The missile is only guided during the relatively brief initial powered phase of flight and its course is subsequently governed by the laws of orbital mechanics and ballistics. Attending at the target, missiles can be classified as anti-submarines, anti-aircraft, anti-missile, anti-tank, anti-radar, etc. If we look at the military function, missiles can be classified as strategic and tactical. However, the most extended criteria is as follows:

- Air-to-air: launched from an airplane against an arial target.
- Surface-to-air: design as defense against enemy airplanes or missiles.
- Air-to-surface: dropped from airplanes.
- Surface-to-surface: supports infantry in surface operations.

The general configuration of a missile consists in a cylindrical body with an ogival warhead and surfaces with aerodynamic control. Missiles also have a guiding system and are powered by an engine, generally either a type of rocket or jet engine.

2.1.4 Space vehicles

A space vehicle (also referred to as spacecraft or spaceship) is a vehicle designed for spaceflight. Space vehicles are used for a variety of purposes, including communications, earth observation, meteorology, navigation, planetary exploration, and transportation of humans and cargo. The main particularity is that such vehicles operate without any atmosphere (or in regions with very low density). However, they must scape the Earth's atmosphere. Therefore, we can identify different kinds of space vehicles:

- Artificial satellites.
- Space probes.
- Manned spacecrafts.
- Space launchers.

A satellite is an object which has been placed into orbit by human endeavor, which goal is to endure for a long time. Such objects are sometimes called artificial satellites to distinguish them from natural satellites such as the Moon. They can carry on board diverse equipment and subsystems to fulfill with the commended mission, generally to transmit data to Earth. A taxonomy can be given attending at the mission (scientific, telecommunications, defense, etc), or attending at the orbit (equatorial, geostationary, etc).

A space probe is a scientific space exploration mission in which a spacecraft leaves Earth and explores space. It may approach the Moon, enter interplanetary, flyby or orbit other bodies, or approach interstellar space. Space probes are a form of robotic spacecraft. Space probes are aimed for research activities.

GENERALITIES

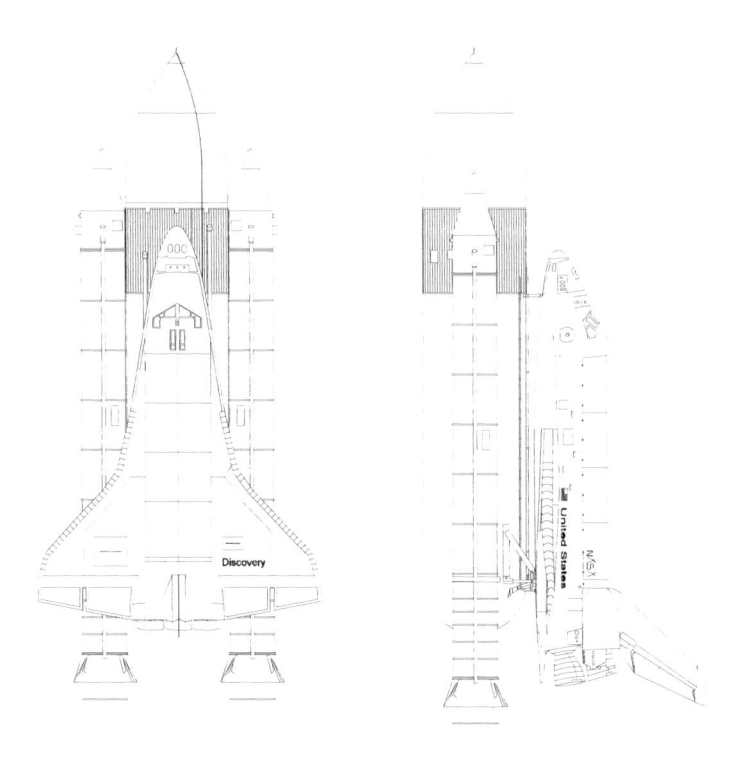

Figure 2.7: Space shuttle: Discovery.

The manned spacecraft are space vehicles with crew (at least one). We can distinguish space flight spacecrafts and orbital stations (such ISS). Those missions are also aimed for research and observation activities.

Space launchers are vehicles which mission is to place another space vehicles, typically satellites, in orbit. Generally, they are not recoverable, with the exception of the the American space shuttles (Columbia, Challenger, Discovery, Atlantis, and Endevour). The space shuttle was a manned orbital rocket and spacecraft system operated by NASA on 135 missions from 1981 to 2011. This system combined rocket launch, orbital spacecraft, and re-entry spaceplane. See Figure 2.7, where the Discovery is sketched. Major missions included launching numerous satellites and interplanetary probes, conducting space science experiments, and 37 missions constructing and servicing the ISS.

The configuration of space vehicles varies depending on the mission and can be unique. As a general characteristic, just mention that launchers have similar configuration as missiles.

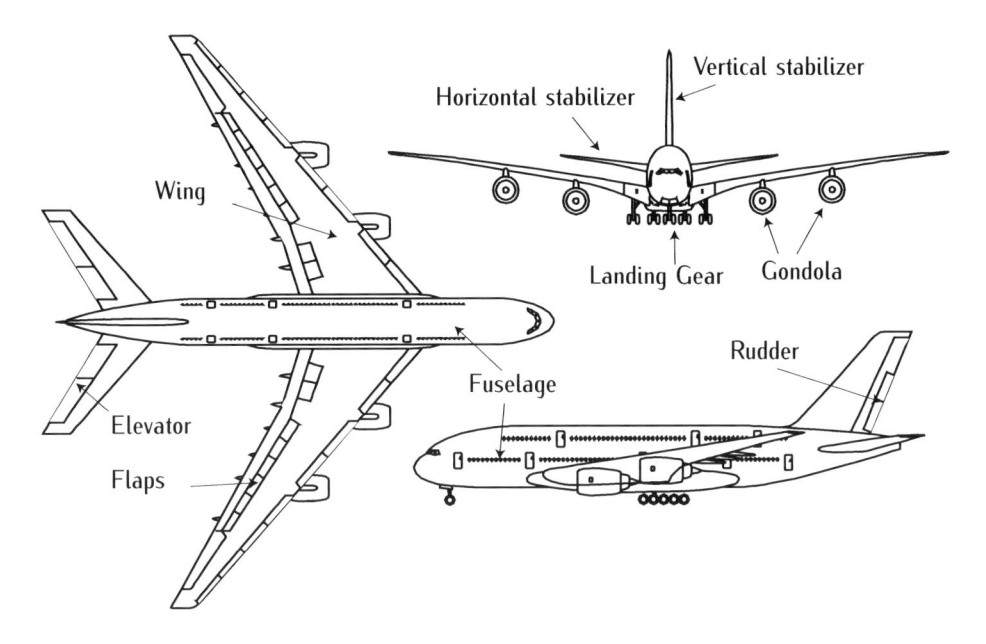

Figure 2.8: Parts of an aircraft.

2.2 Parts of the aircraft (Franchini *et al.* [3])

Before going into the fundamentals of atmospheric flight, it is interesting to identify the fundamental elements of the aircraft[2]. As pointed out before, there are several configurations. The focus will be on commercial airplanes flying in high subsonic regimes, the most common ones. Figure 2.8 shows the main parts of a typical commercial aircraft.

The central body of the airplane, which hosts the crew and the payload (passengers, luggage, and cargo), is the fuselage. The wing is the main contributor to lift force. The surfaces situated at the tail or empennage of the aircraft are referred to as horizontal stabilizer and vertical stabilizer. The engine is typically located under the wing protected by the so-called gondolas (some configurations with three engines locate one engine in the tail).

2.2.1 Fuselage

The fuselage is the aircraft's central body that accommodates the crew and the payload (passengers and cargo) and protect them from the exterior conditions. The fuselage also gives room for the pilot's cabin and its equipments, and serves as main structure to which the rest of structures (wing, stabilizers, etc.) are attached. Its form is a trade off between an aerodynamic geometry (with minimum drag) and enough volume to fulfill its mission.

---

[2]Again, in the sense of a fixed-wing aircraft provided with a power plant.

GENERALITIES

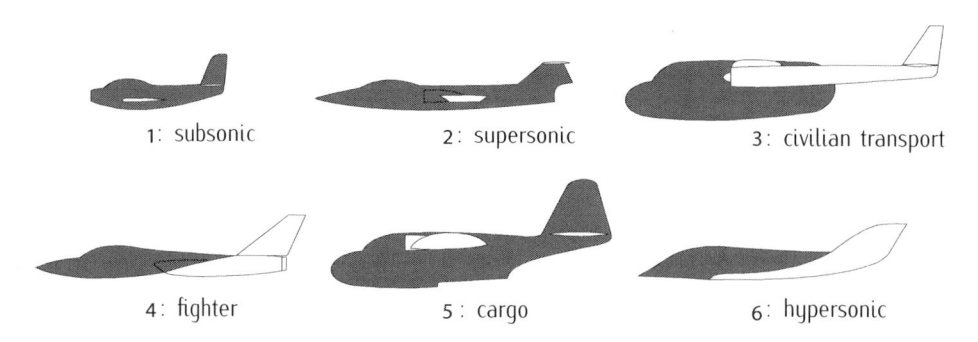

<pre>
      1: subsonic   2: supersonic   3: civilian transport

        4: fighter   5: cargo   6: hypersonic
</pre>

Figure 2.9: Types of fuselages. © Adrián Hermida / Wikimedia Commons / CC-BY-SA-3.0.

Most of the usable volume of the fuselage is derived to passenger transportation in the passenger cabin. The layout of the passenger cabin must fulfill IATA regulations (dimensions of corridors, dimensions of seats, distance between lines, emergency doors), and differs depending on the segment of the aircraft (short and long-haul), the passenger type (economic, business, first class, etc.), or company policies (low cost companies Vs. flag companies). Cargo is transported in the deck (in big commercial transportation aircraft generally situated bellow the passenger cabin). Some standardized types of fuselage are depicted in Figure 2.9.

2.2.2 Wing

A wing is an airfoil that has an aerodynamic cross-sectional shape producing a useful lift to drag ratio. A wing's aerodynamic quality is expressed as its lift-to-drag ratio. The lift that a wing generates at a given speed and angle of attack can be one to two orders of magnitude greater than the total drag on the wing. A high lift-to-drag ratio requires a significantly smaller thrust to propel the wings through the air at sufficient lift.

The wing can be classified attending at the plant-form. The elliptic plant-form is the best in terms of aerodynamic efficiency (lift-to-drag ration), but it is rather complex to manufacture. The rectangular plant-form is much easier to manufacture but the efficiency drops significantly. An intermediate solution is the wing with narrowing (also referred to as trapezoidal wing or tapered wing). As the airspeed increases and gets closer to the speed of sound, it is interesting to design swept wings with the objective of retarding the effects of sharpen increase of aerodynamic drag associated to transonic regimens, the so-called compressibility effects. The delta wing is less common, typical of supersonic flights. An evolution of the delta plant-form is the ogival plant-form. See Figure 2.10.

Attending at the vertical position, the wing can also be classified as high, medium, and low. High wings are typical of cargo aircraft. It allows the fuselage to be nearer the floor, and it is easier to execute load and download tasks. On the contrary, it is difficult to locate space for the retractile landing gear (also referred to as undercarriage). The

2.2 Parts of the aircraft

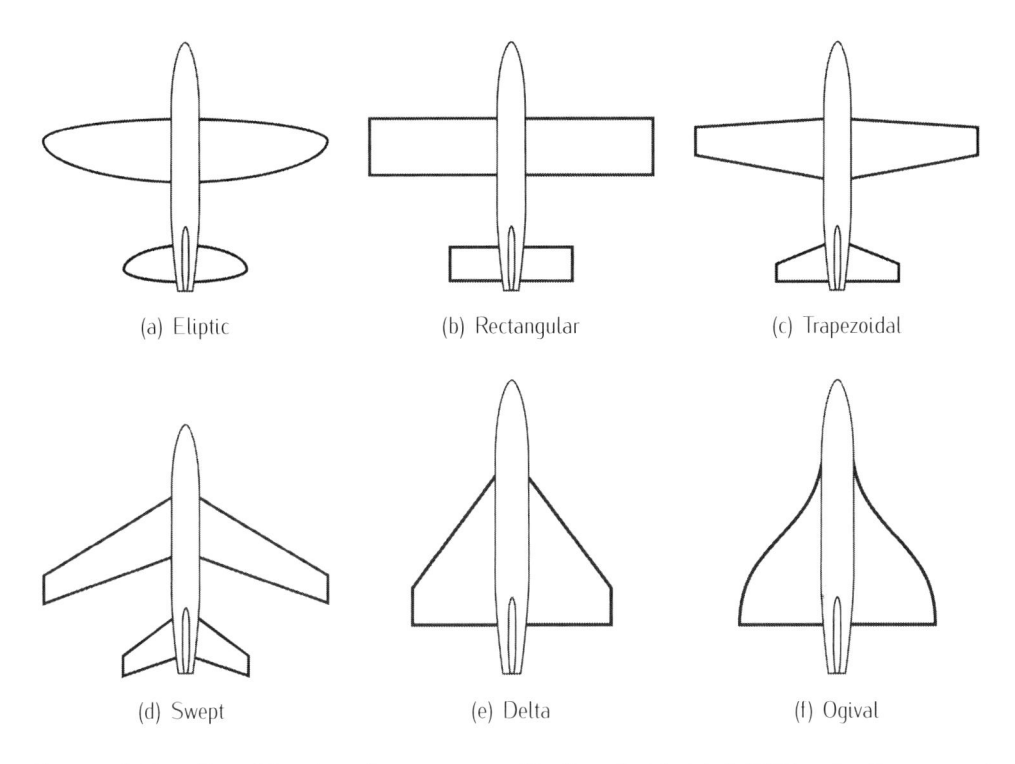

Figure 2.10: Aircraft's plant-form types. © Guy Inchbald / Wikimedia Commons / CC-BY-SA-3.0.

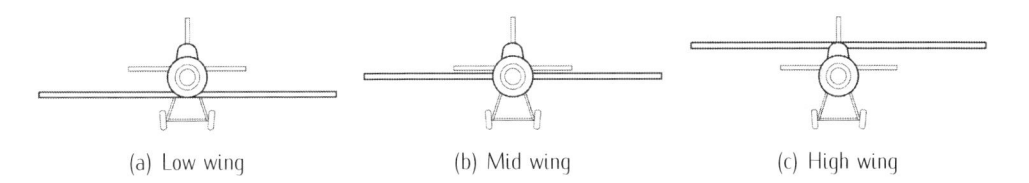

(a) Eliptic (b) Rectangular (c) Trapezoidal
(d) Swept (e) Delta (f) Ogival

Figure 2.11: Wing vertical position. © Guy Inchbald / Wikimedia Commons / CC-BY-SA-3.0.

(a) Low wing (b) Mid wing (c) High wing

low wing is the typical one in commercial aviation. It does not interfere in the passenger cabin, diving the deck into two spaces. It is also useful to locate the retractile landing gear. The medium wing is not typical in commercial aircraft, but it is very common to see it in combat aircraft with the weapons bellow the wing to be dropped. See Figure 2.11.

Usually, aircraft's wings have various devices, such as flaps or slats, that the pilot uses to modify the shape and surface area of the wing to change its aerodynamic characteristics in flight, or ailerons, which are used as control surfaces to make the aircraft roll around its longitudinal axis. Another kind of devices are the spoilers which typically used to help braking the aircraft after touching down. Spoilers are deflected so that the lift gets

GENERALITIES

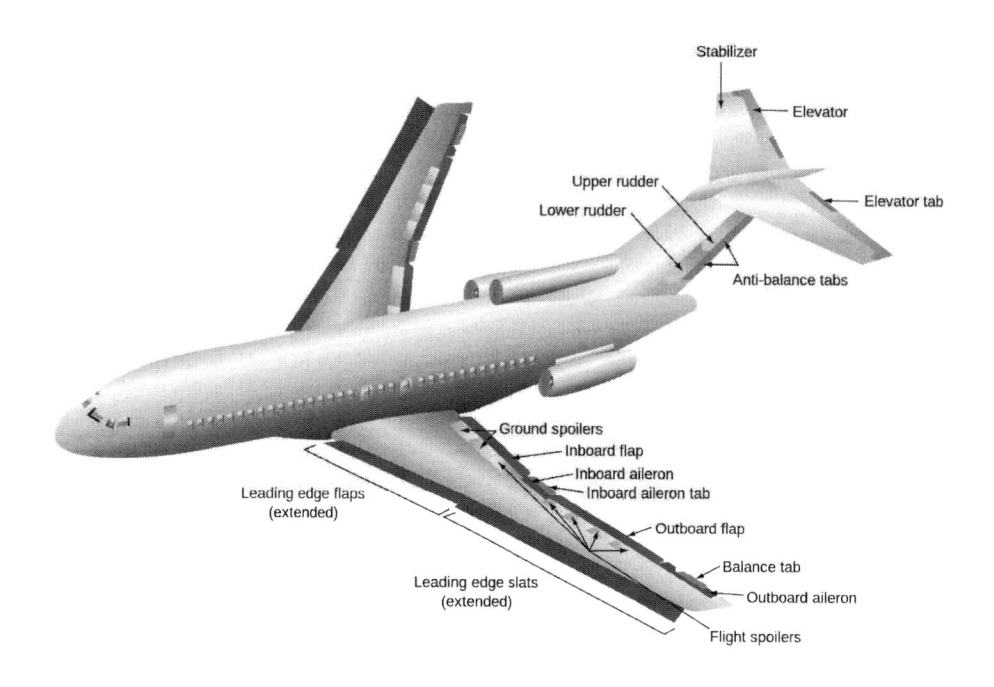

Figure 2.12: Wing and empennage devices. Wikimedia Commons / Public Domain.

reduced in the semi-wing they are acting, and thus they can be also useful to help the aircraft rolling. If both are deflected at the same time, the total lift of the aircraft drops and can be used to descent quickly or to brake after touching down. See Figure 2.12.

2.2.3 EMPENNAGE

The empennage, also referred to as tail or tail assembly, gives stability to the aircraft. Most aircraft feature empennage incorporating vertical and horizontal stabilizing surfaces which stabilize the flight dynamics of pitch and yaw as well as housing control surfaces. Different configurations for the empennage can be identified (See Figure 2.13):

The conventional tail (also referred to as low tail) configuration, in which the horizontal stabilizers are placed in the fuselage. It is the conventional configuration for aircraft with the engines under the wings. It is structurally more compact and aerodynamically more efficient.

The cruciform tail, in which the horizontal stabilizers are placed midway up the vertical stabilizer, giving the appearance of a cross when viewed from the front. Cruciform tails are often used to keep the horizontal stabilizers out of the engine wake, while avoiding many of the disadvantages of a T-tail.

2.2 PARTS OF THE AIRCRAFT

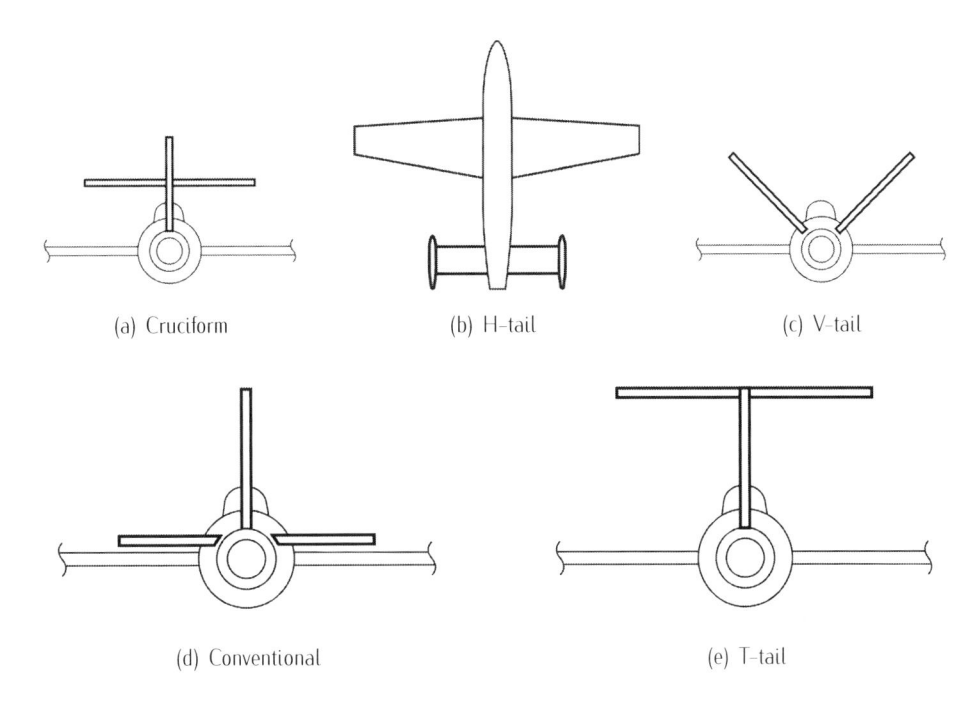

(a) Cruciform

(b) H-tail

(c) V-tail

(d) Conventional

(e) T-tail

Figure 2.13: Aircraft's empennage types. © Guy Inchbald / Wikimedia Commons / CC-BY-SA-3.0.

The T-tail configuration, in which the horizontal stabilizer is mounted on top of the fin, creating a "T" shape when viewed from the front. T-tails keep the stabilizers out of the engine wake, and give better pitch control. T-tails have a good glide ratio, and are more efficient on low speed aircraft. However, T-tails are more likely to enter a deep stall, and is more difficult to recover from a spin. T-tails must be stronger, and therefore heavier than conventional tails. T-tails also have a larger cross section.

Twin tail (also referred to as H-tail) or V-tail are other configuration of interest although much less common.

2.2.4 MAIN CONTROL SURFACES

The main control surfaces of a fixed-wing aircraft are attached to the airframe on hinges or tracks so they may move and thus deflect the air stream passing over them. This redirection of the air stream generates an unbalanced force to rotate the plane about the associated axis.

The main control surfaces are: ailerons, elevator, and rudder.

Ailerons are mounted on the trailing edge of each wing near the wingtips and move in opposite directions. When the pilot moves the stick left, the left aileron goes up and

(a) Roll right (b) Pitch down (c) Yaw left

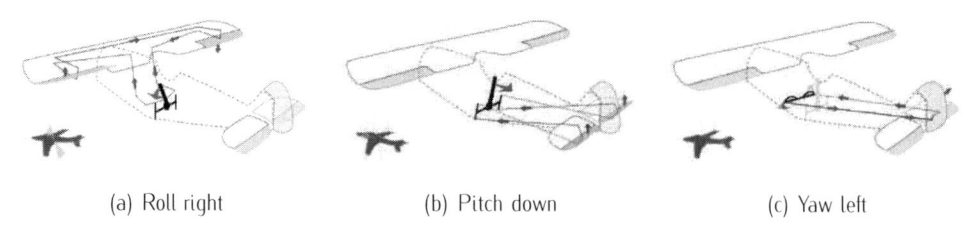

Figure 2.14: Actions on the control surfaces. © Ignacio Icke / Wikimedia Commons / CC-BY-SA-3.0.

the right aileron goes down. A raised aileron reduces lift on that wing and a lowered one increases lift, so moving the stick left causes the left wing to drop and the right wing to rise. This causes the aircraft to roll to the left and begin to turn to the left. Centering the stick returns the ailerons to neutral maintaining the bank angle. The aircraft will continue to turn until opposite aileron motion returns the bank angle to zero to fly straight.

An elevator is mounted on the trailing edge of the horizontal stabilizer on each side of the fin in the tail. They move up and down together. When the pilot pulls the stick backward, the elevators go up. Pushing the stick forward causes the elevators to go down. Raised elevators push down on the tail and cause the nose to pitch up. This makes the wings fly at a higher angle of attack, which generates more lift and more drag. Centering the stick returns the elevators to neutral position and stops the change of pitch.

The rudder is typically mounted on the trailing edge of the fin, part of the empennage. When the pilot pushes the left pedal, the rudder deflects left. Pushing the right pedal causes the rudder to deflect right. Deflecting the rudder right pushes the tail left and causes the nose to yaw to the right. Centering the rudder pedals returns the rudder to neutral position and stops the yaw.

2.2.5 Propulsion plant

The propulsion in aircraft is made by engines that compress air taken from the exterior, mix it with fuel, burn the mixture, and get energy from the resulting high-pressure gases.

There are two main groups: propellers and jets.

A propeller is a type of fan that transmits power by converting rotational motion into thrust. The first aircraft were propelled using a piston engine. Nowadays, piston engines are limited to light aircraft due to its weight and its inefficient performance at high altitudes. Another kind of propelled engine is the turbopropoller (also referred to as turboprop) engine, a type of turbine engine which drives an aircraft propeller using a reduction gear. Turboprop are efficient in low subsonic regimes.

A jet engine is a reaction engine that discharges a fast moving jet to generate thrust by jet propulsion in accordance with the third Newton's laws of motion (action-reaction).

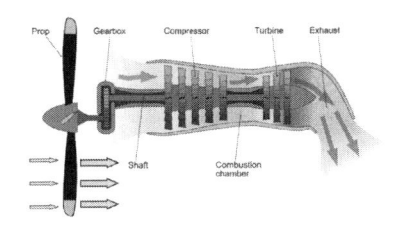

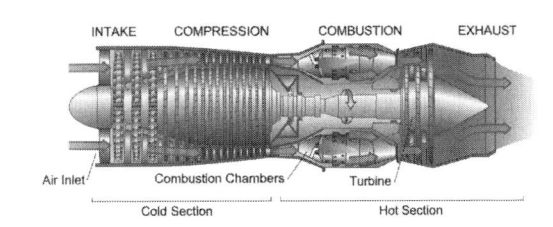

(a) Turbopropeler © Emoscopes / Wikimedia Commons / CC-BY-SA-3.0.

(b) Jet engine © Jeff Dahl / Wikimedia Commons / CC-BY-SA-3.0.

Figure 2.15: Propulsion plant.

It typically consists of an engine with a rotating air compressor powered by a turbine (the so-called *Brayton cycle*), with the leftover power providing thrust via a propelling nozzle. This broad definition of jet engines includes turbojets, turbofans, rockets, ramjets, pulse jets. These types of jet engines are primarily used by jet aircraft for long-haul travel. Early jet aircraft used turbojet engines which were relatively inefficient for subsonic flight. Modern subsonic jet aircraft usually use high-bypass turbofan engines which provide high speeds at a reasonable fuel efficiency (almost as good as turboprops for low subsonic regimes).

2.3 Standard atmosphere

The International Standard Atmosphere (ISA) is an atmospheric model of how the pressure, temperature, density, and viscosity of the Earth's atmosphere change over a wide range of altitudes. It has been established to provide a common reference for the atmosphere consider standard (with an average solar activity and in latitudes around 45N). This model of atmosphere is the standard used in aviation and weather studies. The temperature of air is a function of the altitude, given by the profiles established by the International Standard Atmosphere (ISA) in the different layers of the atmosphere. The reader is referred, for instance, to ANDERSON [1] and FRANCHINI and GARCÍA [2] for a deeper insight.

2.3.1 Hypotheses

Hypothesis 2.1 (*Standard atmosphere*). *The basic hypotheses of ISA are:*

- *Complies with the perfect gas equation:*

$$p = \rho R T, \qquad (2.1)$$

where R is the perfect gas constant for air (R=287.053 [J/kg K]), p is the pressure, ρ is the density, and T the temperature.

GENERALITIES

- In the troposphere the temperature gradient is constant.

$$\text{Troposphere } (0 \leq h < 11000\,[m]):$$
$$T = T_0 - \alpha h, \qquad (2.2)$$

where $T_0 = 288.15[K]$, $\alpha = 6.5[K/km]$.

- In the tropopause and the inferior stratosphere the temperature is constant.

$$\text{Tropopause and inferior stratosphere } (11000\,[m] \leq h < 20000\,[m]):$$
$$T = T_{11} \qquad (2.3)$$

where $T_{11} = 216.65[K]$.

- The air pressure at sea level ($h = 0$) is $p_0 = 101325[Pa]$. In Equation (2.1), the air density at sea level yields $\rho_0 = 1.225[kg/m^3]$.

- The acceleration due to gravity is constant ($g = 9.80665[m/s^2]$).

- The atmosphere is in calm with respect to Earth.

2.3.2 FLUID-STATIC EQUATION

Fluid statics (also called hydrostatics) is the science of fluids at rest, and is a sub-field within fluid mechanics. It embraces the study of the conditions under which fluids are at rest in stable equilibrium.

If we assume the air at rest as in Hypothesis (2.1), we can formulate the equilibrium of a differential cylindrical element where only gravitational volume forces and pressure surface forces act (see Figure 2.16):

$$pdS - (p + dp)dS = \rho g dS dh, \qquad (2.4)$$

which gives rise to the equation of the fluid statics:

$$\frac{dp}{dh} = -\rho g. \qquad (2.5)$$

2.3.3 ISA EQUATIONS

Considering Equation (2.1), Equations (2.2)-(2.3), and Equation (2.5), the variations of ρ and p within altitude can be obtained for the different layers of the atmosphere that affect atmospheric flight:

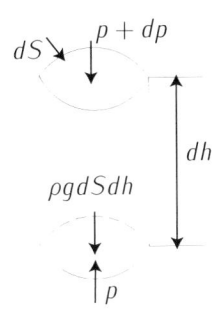

Figure 2.16: Differential cylinder of air. Adapted from FRANCHINI et al. [3].

Troposphere ($0 \leq h < 11000$ [m]): Introducing Equation (2.1) and Equation (2.2) in Equation (2.5), it yields:

$$\frac{dp}{dh} = -\frac{p}{R(T_0 - \alpha h)} g. \tag{2.6}$$

Integrating between a generic value of altitude h and the altitude at sea level ($h = 0$), the variation of pressure with altitude yields:

$$\frac{p}{p_0} = \left(1 - \frac{\alpha}{T_0} h\right)^{\frac{g}{R\alpha}}. \tag{2.7}$$

With the value of pressure given by Equation (2.7), and entering in the equation of perfect gas (2.1), the variation of density with altitude yields:

$$\frac{\rho}{\rho_0} = \left(1 - \frac{\alpha}{T_0} h\right)^{\frac{g}{R\alpha} - 1}. \tag{2.8}$$

Introducing now the numerical values, it yields:

$$T[k] = 288.15 - 0.0065 h[m]; \tag{2.9}$$
$$\rho[kg/m^3] = 1.225(1 - 22.558 \times 10^{-6} \times h[m])^{4.2559}; \tag{2.10}$$
$$p[Pa] = 101325(1 - 22.558 \times 10^{-6} \times h[m])^{5.2559}. \tag{2.11}$$

Tropopause and Inferior part of the stratosphere (11000 [m] $\leq h < 20000$ [m]): Introducing Equation (2.1) and Equation (2.3) in Equation (2.5), and integrating between a generic altitude ($h > 11000$ [m]) and the altitude at the tropopause ($h_{11} = 11000$ [m]):

$$\frac{p}{p_{11}} = \frac{\rho}{\rho_{11}} = e^{-\frac{g}{RT_{11}}(h - h_{11})}. \tag{2.12}$$

GENERALITIES

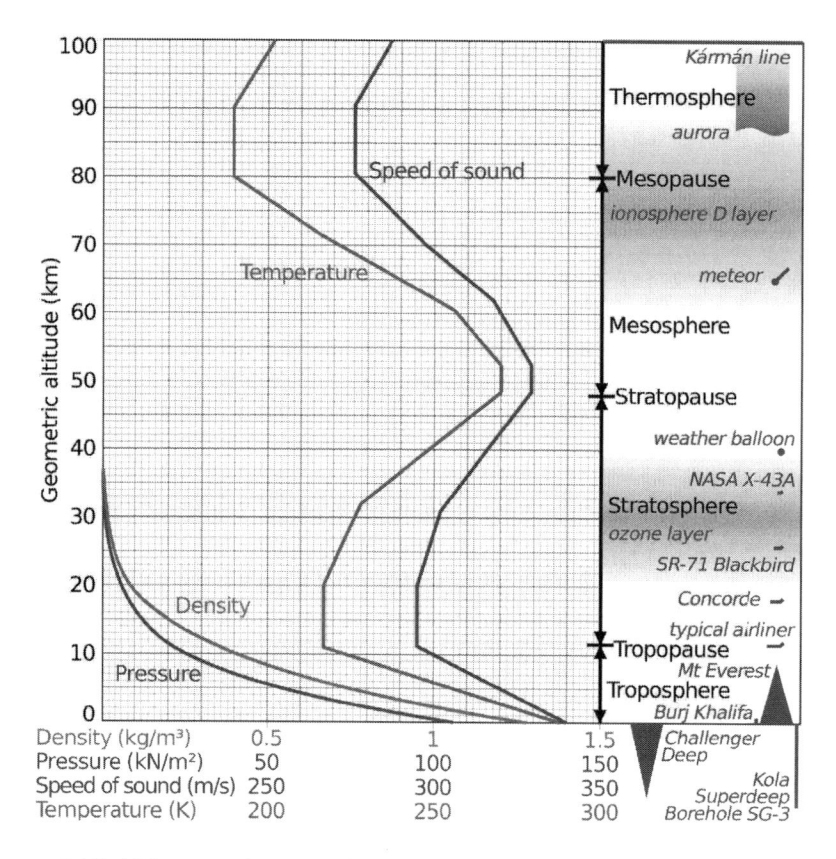

Figure 2.17: ISA atmosphere. © Cmglee / Wikimedia Commons / CC-BY-SA-3.0.

Introducing now the numerical values, it yields:

$$T[k] = 216.65; \tag{2.13}$$

$$\rho[kg/m^3] = 0.3639 e^{-157.69 \cdot 10^{-6}(h[m]-11000)}; \tag{2.14}$$

$$p[Pa] = 22632 e^{-157.69 \cdot 10^{-6}(h[m]-11000)}. \tag{2.15}$$

2.3.4 WARM AND COLD ATMOSPHERES

For warm and cold days, it is used the so called warm (ISA+5, ISA+10, ISA+15, etc.) and cold (ISA-5, ISA-10, ISA-15, etc.), where the increments (decrements) represent the difference with respect to the 288.15 [K] of an average day.

Given the new T_0, and given the same pressure at sea level (p_0 does not change), the new density at sea level can be calculated. Then the ISA equation are obtained proceeding in the same manner.

2.3.5 Barometric altitude

Altitude can be determined based on the measurement of atmospheric pressure. Still nowadays, most of the aircraft use the barometric altimeters to determine the altitude of the aircraft. An altimeter cannot, however, be adjusted for variations in air temperature. Differences in temperature from the ISA model will, therefore, cause errors in the indicated altitude.

Typically, an aneroid barometer or mercury barometer measures the atmospheric pressure from a static port outside the aircraft and based on a reference pressure. According to Equation (2.11) one has that:

$$\frac{p}{p_0} = (1 - \frac{\alpha}{T_0}h)^{\frac{g}{R\alpha}}; \quad \text{and} \tag{2.16}$$

$$\frac{p_{ref}}{p_0} = (1 - \frac{\alpha}{T_0}h_{ref})^{\frac{g}{R\alpha}}. \tag{2.17}$$

Isolating h and h_{ref}, respectively, and subtracting, it yields:

$$h - h_{ref}[m] = \frac{T_0}{\alpha}\left[(\frac{p_{ref}}{p_0})^{\frac{R\alpha}{g}} - (\frac{p}{p_0})^{\frac{R\alpha}{g}}\right]. \tag{2.18}$$

The reference values can be adjusted, and there exist three main standards:

QNE setting: the baseline pressure is 101325 Pa. This setting is equivalent to the air pressure at mean sea level (MSL) in the International Standard Atmosphere (ISA).

QNH setting: the baseline pressure is the real pressure at sea level (not necessarily 101325 [Pa]). In order to estimate the real pressure at sea level, the pressure is measured at the airfield and then, using equation (2.8), the real pressure at mean sea level is estimated (notice that now $p_0 \neq 101325$). It captures better the deviations from the ISA.

QFE setting: where p_{ref} is the pressure in the airport, so that $h - h_{ref}$ reflects the altitude above the airport.

2.4 System references (Gómez-Tierno *et al.* [4])

The atmospheric flight mechanics uses different coordinates references to express the positions, velocities, accelerations, forces, and torques. Therefore, before going into the fundamentals of flight mechanics, it is useful to define some of the most important ones:

Definition 2.1 (*Inertial Reference Frame*). *According to classical mechanics, an inertial reference frame $F_I(O_I, x_I, y_I, z_I)$ is either a non accelerated frame with respect to a quasi-fixed reference star, or either a system which for a punctual mass is possible to apply the second Newton's law:*

$$\sum \vec{F_I} = \frac{d(m \cdot \vec{V_I})}{dt}$$

GENERALITIES

Definition 2.2 (*Earth Reference Frame*). *An earth reference frame $F_e(O_e, x_e, y_e, z_e)$ is a rotating topocentric (measured from the surface of the earth) system. The origin O_e is any point on the surface of earth defined by its latitude θ_e and longitude λ_e. Axis z_e points to the center of earth; x_e lays in the horizontal plane and points to a fixed direction (typically north); y_e forms a right-handed thrihedral (typically east).*

Such system is sometimes referred to as *navigational system* since it is very useful to represent the trajectory of an aircraft from the departure airport.

Hypothesis 2.2. Flat earth: *The earth can be considered flat, non rotating, and approximate inertial reference frame. Consider F_I and F_e. Consider the center of mass of the aircraft denoted by CG. The acceleration of CG with respect to F_I can be written using the well-known formula of acceleration composition from the classical mechanics:*

$$\vec{a}_I^{CG} = \vec{a}_e^{CG} + \vec{\Omega} \wedge (\vec{\Omega} \wedge \vec{r}_{O_ICG}) + 2\vec{\Omega} \wedge \vec{V}_e^{CG}, \tag{2.19}$$

where the centripetal acceleration $(\vec{\Omega} \wedge (\vec{\Omega} \wedge \vec{r}_{O_ICG}))$ and the Coriolis acceleration $(2\vec{\Omega} \wedge \vec{V}_e^{CG})$ are neglectable if we consider typical values: $\vec{\Omega}$ (the earth angular velocity) is one revolution per day; $\vec{r}$ is the radius of earth plus the altitude (around 6380 [km]); V_e^{CG} is the velocity of the aircraft in flight (200-300 [m/s]). This means $\vec{a}_I^{CG} \approx \vec{a}_e^{CG}$ and therefore F_e can be considered inertial reference frame.

Definition 2.3 (*Local Horizon Frame*). *A local horizon frame $F_h(O_h, x_h, y_h, z_h)$ is a system of axes centered in any point of the symmetry plane (assuming there is one) of the aircraft, typically the center of gravity. Axes (x_h, y_h, z_h) are defined parallel to axes (x_e, y_e, z_e).*

In atmospheric flight, this system can be considered as quasi-inertial.

Definition 2.4 (*Body Axes Frame*). *A body axes frame $F_b(O_b, x_b, y_b, z_b)$ represents the aircraft as a rigid solid model. It is a system of axes centered in any point of the symmetry plane (assuming there is one) of the aircraft, typically the center of gravity. Axis x_b lays in to the plane of symmetry and it is parallel to a reference line in the aircraft (for instance, the zero-lift line), pointing forwards according to the movement of the aircraft. Axis z_b also lays in to the plane of symmetry, perpendicular to x_b and pointing down according to regular aircraft performance. Axis y_b is perpendicular to the plane of symmetry forming a right-handed thrihedral (y_b points then to the right wing side of the aircraft).*

Definition 2.5 (*Wind Axes Frame*). *A wind axes frame $F_w(O_w, x_w, y_w, z_w)$ is linked to the instantaneous aerodynamic velocity of the aircraft. It is a system of axes centered in any point of the symmetry plane (assuming there is one) of the aircraft, typically the center of gravity. Axis x_w points at each instant to the direction of the aerodynamic velocity of the aircraft $\vec{V}$. Axis z_w lays in to the plane of symmetry, perpendicular to x_w and pointing down according to regular aircraft performance. Axis y_b forms a right-handed thrihedral.*

Notice that if the aerodynamic velocity lays in to the plane of symmetry, $y_w \equiv y_b$.

2.5 Problems

Problem 2.1: International Standard Atmosphere

After the launch of a spatial probe into a planetary atmosphere, data about the temperature of the atmosphere have been collected. Its variation with altitude (h) can be approximated as follows:

$$T = \frac{A}{1 + e^{\frac{h}{B}}}, \qquad (2.20)$$

where A and B are constants to be determined.

Assuming the gas behaves as a perfect gas and the atmosphere is at rest, using the following data:

- *Temperature at $h = 1000$, $T_{1000} = 250$ K;*
- *$p_0 = 100000 \frac{N}{m^2}$;*
- *$\rho_0 = 1 \frac{Kg}{m^3}$;*
- *$T_0 = 300$ K;*
- *$g = 10 \frac{m}{s^2}$.*

determine:

1. *The values of A and B, including their unities.*
2. *Variation law of density and pressure with altitude, respectively $\rho(h)$ and $p(h)$ (do not substitute any value).*
3. *The value of density and pressure at $h = 1000$ m.*

GENERALITIES

Solution to Problem 2.1:

We assume the following hypotheses:

a) The gas is a perfect gas.

b) It fulfills the fluidostatic equation.

Based on hypothesis a):

$$P = \rho RT. \tag{2.21}$$

Based on hypothesis b):

$$dP = -\rho g dh. \tag{2.22}$$

Based on the data given in the statement, and using Equation (2.21):

$$R = \frac{P_0}{\rho_0 T_0} = 333.3 \frac{J}{(Kg \cdot K)}. \tag{2.23}$$

1. The values of A and B:

   Using the given temperature at an altitude $h = 0$ ($T_0 = 300$ K), and Equation (2.20):

   $$300 = \frac{A}{1+e^0} = \frac{A}{2} \rightarrow A = 600 \; K. \tag{2.24}$$

   Using the given temperature at an altitude $h = 1000$ ($T_{1000} = 250$ K), and Equation (2.20):

   $$250 = \frac{A}{1+e^{\frac{1000}{B}}} = \frac{600}{1+e^{\frac{1000}{B}}} \rightarrow B = 2972 \; m. \tag{2.25}$$

2. Variation law of density and pressure with altitude:

   Using Equation (2.21) and Equation (2.22):

   $$dP = -\frac{P}{RT}g dh. \tag{2.26}$$

Integrating the differential Equation (2.26) between $P(h = 0)$ and P; $h = 0$ and h:

$$\int_{P_0}^{P} \frac{dP}{P} = \int_{h=0}^{h} -\frac{g}{RT} dh. \tag{2.27}$$

40

Introducing Equation (2.20) in Equation (2.27):

$$\int_{P_0}^{P} \frac{dP}{P} = \int_{h=0}^{h} -\frac{g(1+e^{\frac{h}{B}})}{RA} dh. \qquad (2.28)$$

Integrating Equation (2.28):

$$Ln\frac{P}{P_0} = -\frac{g}{RA}(h + Be^{\frac{h}{B}} - B) \to P = P_0 e^{-\frac{g}{RA}(h+Be^{\frac{h}{B}}-B)}. \qquad (2.29)$$

Using Equation (2.21), Equation (2.20), and Equation (2.29):

$$\rho = \frac{P}{RT} = \frac{P_0 e^{-\frac{g}{RA}(h+Be^{\frac{h}{B}}-B)}}{R\frac{A}{1+e^{\frac{h}{B}}}}. \qquad (2.30)$$

3. *Pressure and density at an altitude of 1000 m:*

   Using Equation (2.29) and Equation (2.30), the given data for P_0 and g, and the values obtained for R, A, and B:

   - $\rho(h = 1000) = 1.0756 \frac{kg}{m^3}$.
   - $P(h = 1000) = 89632.5 \, Pa$.

References

[1] Anderson, J. (2012). *Introduction to flight, seventh edition.* McGraw-Hill.

[2] Franchini, S. and García, O. (2008). *Introducción a la ingeniería aeroespacial.* Escuela Universitaria de Ingeniería Técnica Aeronáutica, Universidad Politécnica de Madrid.

[3] Franchini, S., López, O., Antoín, J., Bezdenejnykh, N., and Cuerva, A. (2011). *Apuntes de Tecnología Aeroespacial.* Escuela de Ingeniería Aeronáutica y del Espacio. Universidad Politécnica de Madrid.

[4] Gómez-Tierno, M., Pérez-Cortés, M., and Puentes-Márquez, C. (2009). *Mecánica de vuelo.* Escuela Técnica Superior de Ingenieros Aeronáuticos, Universidad Politécnica de Madrid.

[5] Howe, D. (2000). *Aircraft conceptual design synthesis*, volume 5. Wiley.

[6] Jenkinson, L. R., Simpkin, P., Rhodes, D., Jenkison, L. R., and Royce, R. (1999). *Civil jet aircraft design*, volume 7. Arnold London.

[7] Raymer, D. P. *et al.* (1999). *Aircraft design: a conceptual approach*, volume 3. American Institute of Aeronautics and Astronautics.

[8] Torenbeek, E. (1982). *Synthesis of subsonic airplane design: an introduction to the preliminary design of subsonic general aviation and transport aircraft, with emphasis on layout, aerodynamic design, propulsion and performance.* Springer.

Part II

The aircraft

3
AERODYNAMICS

Contents

3.1	Fundamentals of fluid mechanics		48
	3.1.1	Generalities	48
	3.1.2	Continuity equation	49
	3.1.3	Quantity of movement equation	50
	3.1.4	Viscosity	52
	3.1.5	Speed of sound	56
3.2	Airfoils shapes		58
	3.2.1	Airfoil nomenclature	59
	3.2.2	Generation of aerodynamic forces	60
	3.2.3	Aerodynamic dimensionless coefficients	63
	3.2.4	Compressibility and drag-divergence Mach number	66
3.3	Wing aerodynamics		68
	3.3.1	Geometry and nomenclature	68
	3.3.2	Flow over a finite wing	69
	3.3.3	Lift and induced drag in wings	71
	3.3.4	Characteristic curves in wings	72
	3.3.5	Aerodynamics of wings in compressible and supersonic regimes	73
3.4	High-lift devices		74
	3.4.1	Necessity of high-lift devices	74
	3.4.2	Types of high-lift devices	75
	3.4.3	Increase in $C_{L_{max}}$	77
3.5	Problems		79
References			101

Aerodynamics is the discipline that studies the forces and the resulting motion of objects in the air. Therefore, the basis of atmospheric flight is found on the study of aerodynamics (a branch of fluid mechanics). We start then giving the fundamentals of fluid mechanics in Section 3.1. Section 3.2 and Section 3.3 are devoted to the study of aerodynamics of airfoils and wings, respectively. Finally, Section 3.4 analyzes high-lift devices. Thorough references are ANDERSON [2] and FRANCHINI and GARCÍA [3, Chap. 3].

3.1 Fundamentals of fluid mechanics (Franchini et al. [4])

A fluid is a substance, such a liquid or a gas, that changes its shape rapidly and continuously when acted on by forces. Fluid mechanics is the science that study how the fluid qualities respond to such forces and what forces the fluid applies to solids in contact with the fluid. Readers are referred to Franchini et al. [4]. More detailed information can be found in Franchini and García [3, Chap. 3].

3.1.1 Generalities

Many of the fundamental laws of fluid mechanics apply to both liquid and gases. Liquids are nearly incompressible. Unlike a gas, the volume of a given mass of the liquid remains almost constant when a pressure is applied to the fluid. The interest herein is centered in gases since the atmosphere in which operate is a gas commonly know as air. Air is a viscous, compressible fluid composed mostly by nitrogen (78%) and oxygen (21%). Under some conditions (for instance, at low flight velocities), it can be considered incompressible.

Gas: A gas consists of a large number of molecules in random motion, each molecule having a particular velocity, position, and energy, varying because of collisions between molecules. The force per unit created on a surface by the time rate of change of momentum of the rebounding molecules is called the pressure. As long as the molecules are sufficiently apart so that the intermolecular magnetic forces are negligible, the gas acts as a continuos material in which the properties are determined by a statistical average of the particle effects. Such a gas is called a perfect gas.

Stream line: The air as a continuos fluid flows under determined patterns[1] confined into a finite space (the atmosphere). The curves tangent to the velocity vector of the flow at each point of the fluid in an instant of time are referred to as streamlines. Two streamlines can not cross each other except in points with null velocity, otherwise will mean that one point has two different velocities.

Stream tube: A stream tube is the locus[2] of the streamlines which pass through a closed curve in a given instant. The stream tube can be though as a pipe inside the fluid, through its walls there is no flow.

---

[1] the movement of a fluid is governed by the Navier-Stokes partial differential equations. The scope of this course does not cover the study of Navier-Stokes.

[2] In geometry, a locus is a collection of points that share a property.

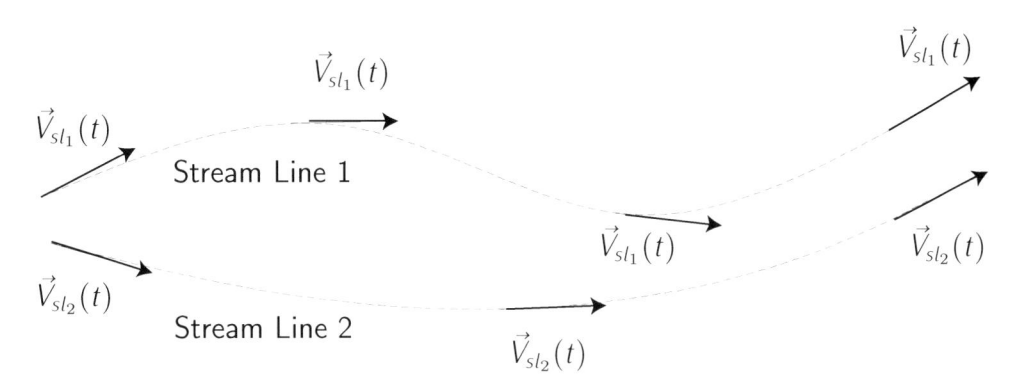

Figure 3.1: Stream line.

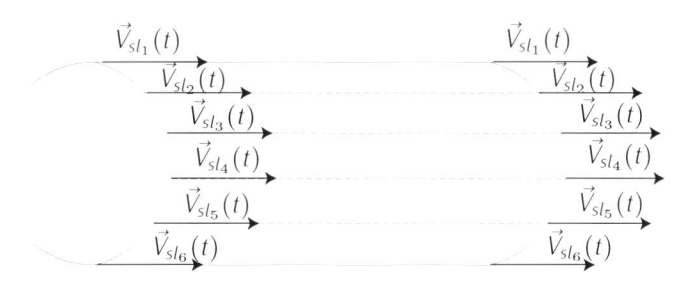

Figure 3.2: Stream tube.

3.1.2 Continuity equation

One of the fundamentals of physics stays that the matter in the interior of an isolated system is not created nor destroyed, it is only transformed. If one thinks in open systems (not isolated), such as human beings or airplanes in flight, its mass is constantly varying.

In a fluid is not easy to identify particles or fluid volumes since they are moving and deforming constantly within time. That is way the conservation of mass must be understood in a different way:

Recall the concept of stream tube. Assuming through its walls there is no flow, and that the flow is stationary across any section area (the velocity is constant), the mass that enters per unit of time in Section A_1 ($\rho_1 V_1 A_1$) will be equal to the mass that exits Section A_2 ($\rho_2 V_2 A_2$), where ρ is the density, V is the velocity, and A is the area. Therefore, the continuity of mass stays:

$$\rho_1 V_1 A_1 = \rho_2 V_2 A_2. \tag{3.1}$$

Since Section A_1 and Section A_2 are generic, one can claim that the product $\rho V A$ is constant along the stream tube. The product $\rho V A$ is referred to as mass flow $\dot{m}$ (with dimensions [kg/s]).

$$\dot{m}_1 = \dot{m}_2$$

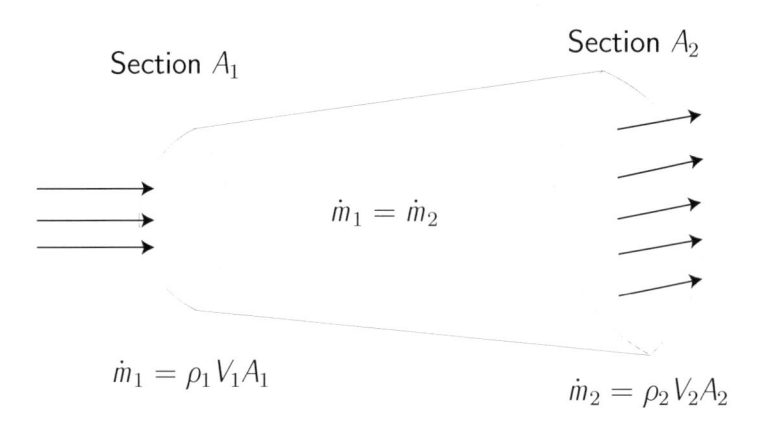

$$\dot{m}_1 = \rho_1 V_1 A_1 \qquad \dot{m}_2 = \rho_2 V_2 A_2$$

Figure 3.3: Continuity equation.

Compressible and incompressible flow

In many occasions occurs that the density of a fluid does not change due to the fact that it is moving. This happens in liquids and, in some circumstances, in gases (think in the air confined in a room). Notice that one can not say that the air is incompressible, but an air flow is incompressible.

The movement of air in which the velocity is inferior to 100 [m/s] can be considered incompressible. When the air moves faster, as is the case in a jet airplane, the flow is compressible and the studies become more complicated as it will be seen in posterior courses.

3.1.3 Quantity of movement equation

The quantity of movement equation in a fluid (also referred to as momentum equation) is the second Newton law expression applied to a fluid:

$$\sum \vec{F} = m \frac{d(\vec{V})}{dt}. \qquad (3.2)$$

The forces exerted over the matter, solid or fluid, can be of two types:

- Distance-exerted forces, as gravitational, electric, and magnetic, related to mass or volume.

- Contact-exerted forces, pressure or friction, related to the surface of contact.

3.1 FUNDAMENTALS OF FLUID MECHANICS

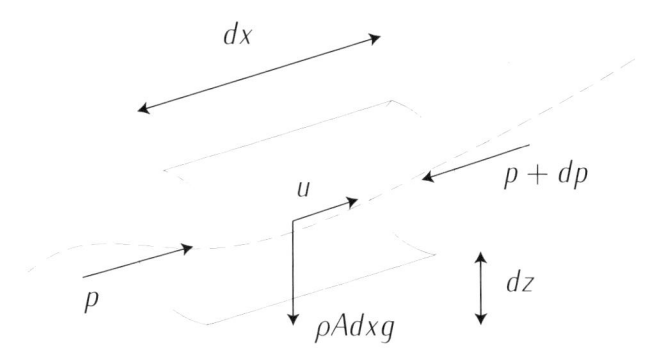

Figure 3.4: Quantity of movement. Adapted from FRANCHINI et al. [4].

Euler equation

We assume herein that the fluid is inviscous and, therefore, we do not consider frictional forces. The only sources of forces are pressure and gravity. We consider one-dimensional flow along the longitudinal axis.

Assume there is a fluid particle with circular section A and longitude dx moving with velocity u along direction x. Apply second Newton law ($m\dot{u} = \sum F$; $m = \rho A dx$):

$$(\rho A dx)\frac{du}{dt} = -A dp - (\rho A dx) g \frac{dz}{dx}. \qquad (3.3)$$

The force due to gravity only affects in the direction of axis x.

Considering a reference frame attached to the particle in which $dt = \frac{dx}{u}$ (stationary flow), the acceleration can be written as $\frac{du}{dt} = \frac{du}{dx/u}$ and Equation (3.3), dividing by Adx, yields:

$$\rho u \frac{du}{dx} = -\frac{dp}{dx} - \rho g \frac{dz}{dx}. \qquad (3.4)$$

Equation (3.4) is referred to as Euler equation. This is the unidimensional case. The Euler equation is more complex, considering the tridimensional motion. The complete equation will be seen is posterior courses of fluid mechanics.

Bernoulli equation

Consider Equation (3.4) and notice that the particle moves on the direction of the streamline. If the flow is incompressible (or there exist a relation between pressure and density, relation called barotropy), the equation can be integrated along the streamline:

$$\rho\frac{u^2}{2} + p + \rho g z = C. \tag{3.5}$$

The value of the constant of integration, C, should be calculated with the known conditions of a point. Bernouilli equation expresses that the sum of the dynamic pressure ($\rho\frac{u^2}{2}$), the static pressure (p), and the piezometric pressure ($\rho g z$) is constant along a stream tube.

In the case of the air moving around an airplane, the term $\rho g z$ does not vary significantly between the different points of the streamline and can be neglected. Equation (3.5) is then simplified to:

$$\rho\frac{u^2}{2} + p = C. \tag{3.6}$$

This does not occur in liquids, where the density is an order of thousands higher than in gases and the piezometric term is always as important as the rest of terms (except if the movement is horizontal).

If at any point of the streamline the velocity is null, the point will be referred to as stagnation point. At that point, the pressure takes a value known as stagnation pressure (p_T), so that at any other point holds:

$$u = \sqrt{2\frac{p_T - p}{\rho}}. \tag{3.7}$$

3.1.4 Viscosity

Viscosity is a measure of the resistance of a fluid which is being deformed by either shear or tensile stress.

Every single fluid inherently has viscosity. The less viscous the fluid is, the easier is the movement (fluidity). For instance, it is well known that the honey has low fluidity (high viscosity), while water presents, compared to honey, high fluidity (low viscosity).

Viscosity is a property of the fluid, but affects only if the fluid is under movement. Viscosity describes a fluid's internal resistance to flow and may be thought as a measure of fluid friction. In general, in any flow, layers move at different velocities and the fluid's viscosity arises from the shear stress between the layers that ultimately opposes any applied force.

Viscosity stress

The viscosity force is a friction force and therefore the shear stress τ (viscosity stress) is a force over a unity of surface.

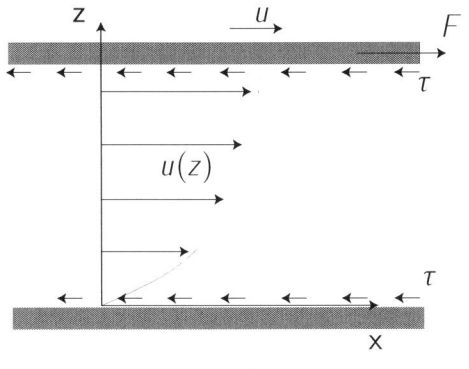

(a) Field of velocities of a fluid inside two plates.

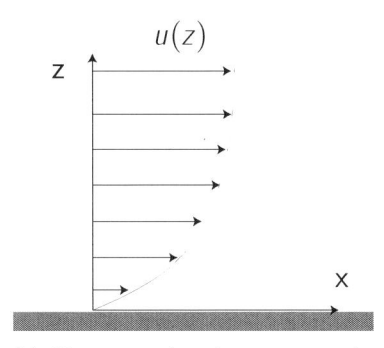

(b) Variation of velocity near the boundary.

Figure 3.5: Viscosity. Adapted from FRANCHINI *et al.* [4].

The relationship between the shear stress and the velocity gradient can be obtained considering two plates closely spaced at a distance z, and separated by a homogeneous fluid, e.g., water or oil. Assuming that the plates are very large, with a large area A, and that the lower plate is fixed, let a force F be applied to the upper plate. Thus, the force causes the substance between the plates to undergo shear flow with a velocity gradient du/dz. The applied force is proportional to the area and velocity gradient in the fluid:

$$F = \mu A \frac{du}{dz}, \qquad (3.8)$$

where coefficient μ is the dynamic viscosity.

This equation can be expressed in terms of shear stress, $\tau = F/A$. Thus expressed in differential form for straight, parallel, and uniform flow, the shear stress between layers is proportional to the velocity gradient in the direction perpendicular to the layers:

$$\tau = \mu \frac{du}{dz}. \qquad (3.9)$$

Some typical values of μ in regular conditions are: 0.26 $[N \cdot s/m^2]$ for oil; 0.001 $[N \cdot s/m^2]$ for water; and 0.000018 $[N \cdot s/m^2]$ for air.

Boundary layer

If one observes the flow around an airfoil, it will be seen that fluid particles in contact with the airfoil have null relative velocity. However, the velocity of the particles at a (relatively low) distance is approximately the velocity of the exterior stream. This thin layer, in which the velocity perpendicular to the airfoil varies dramatically, is known as boundary layer

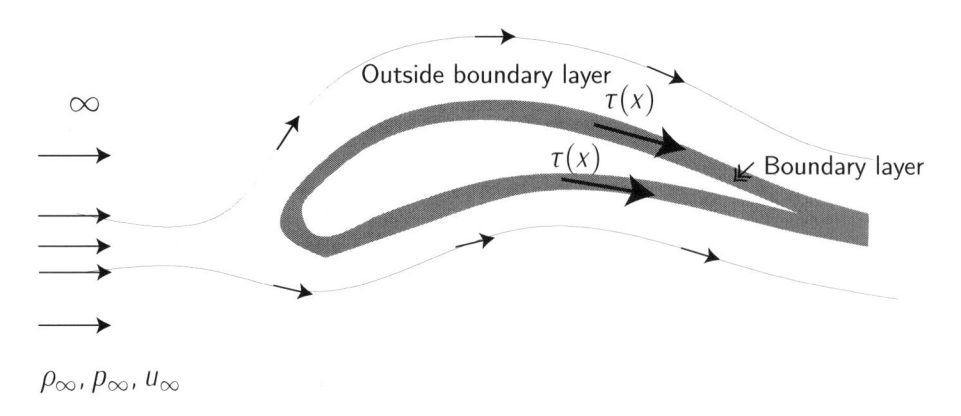

Figure 3.6: Airfoil with boundary layer. The boundary layer has been overemphasized for clarity. Adapted from FRANCHINI *et al.* [4].

and plays a very important role.

The aerodynamic boundary layer was first defined by Ludwig Prandtl in 1904 PRANDTL [5]. It allows to simplify the equations of fluid flow by dividing the flow field into two areas: one inside the boundary layer, where viscosity is dominant and the majority of the drag experienced by a body immersed in a fluid is created, and one outside the boundary layer where viscosity can be neglected without significant effects on the solution. This allows a closed-form solution for the flow in both areas, which is a significant simplification over the solution of the full Navier-Stokes equations. The majority of the heat transfer to and from a body also takes place within the boundary layer, again allowing the equations to be simplified in the flow field outside the boundary layer.

In high-performance designs, such as commercial transport aircraft, much attention is paid to controlling the behavior of the boundary layer to minimize drag. Two effects have to be considered. First, the boundary layer adds to the effective thickness of the body through the displacement thickness, hence increasing the pressure drag. Secondly, the shear forces at the surface of the wing create skin friction drag.

Reynolds number

The dimensionless Reynolds number is due to the studies of Professor Osborne Reynolds (1842-1912) about the conditions in which the flow of fluid in pipes transitioned from laminar flow to turbulent flow. Reynolds used a fluid made of a mixture of water and glycerin, so that varying the mixture the viscosity of the fluid could be modified. When the proportion of glycerin was high, the flow was smooth; by injecting a thread of ink in a pipe with the fluid the thread of ink was flowing smoothly. When the proportion of water was high, by injecting the ink in the pipe a spinning movement was noticed (vortexes) and soon the ink was blurred into the fluid. Reynolds called laminar flow the smooth flow and

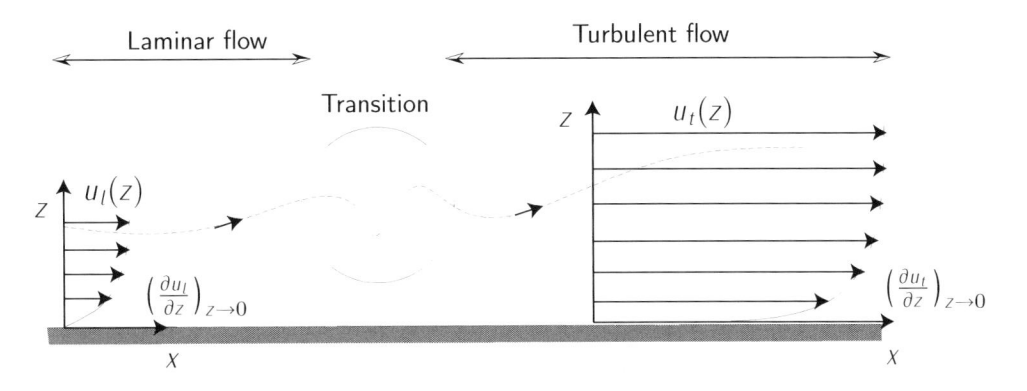

Figure 3.7: Boundary layer transition. Adapted from FRANCHINI et al. [4].

turbulent flow the chaotic one. He also proved that the character of the flow depended on an dimensionless parameter:

$$Re = \rho V D / \mu, \quad (3.10)$$

where V is the mean velocity of the fluid and D is the diameter of the pipe. Posterior research named this number the Reynolds number.

More precisely, the Reynolds number Re is a dimensionless number that gives a measure of the ratio of inertial forces to viscous forces and consequently quantifies the relative importance of these two types of forces for given flow conditions:

$$Re = \frac{\rho V^2 D^2}{\mu V D}. \quad (3.11)$$

Laminar flow occurs at low Reynolds numbers, where viscous forces are dominant, and is characterized by smooth, constant fluid motion; turbulent flow occurs at high Reynolds numbers and is dominated by inertial forces, which tend to produce chaotic eddies, vortices, and other flow instabilities.

In the movement of air around the wing, instead of D it is used the chord of the airfoil, c, the most adequate characteristic longitude. In flight, the Reynolds number is high, of an order of millions, which means the viscosity effects are low and the boundary layer is thin. In this case, the Euler Equation (3.4) can be used to determine the exterior flow around the airfoil. However, the boundary layer can thicken and the boundary layer drops off along the body, resulting in turbulent flow which increases drag.

Therefore, at high Reynolds numbers, such as typical full-sized aircraft, it is desirable to have a laminar boundary layer. This results in a lower skin friction due to the characteristic velocity profile of laminar flow. However, the boundary layer inevitably thickens and becomes less stable as the flow drops off along the body, and eventually becomes turbulent, the process known as boundary layer transition. One way of dealing with this problem is to suck the boundary layer away through a porous surface (Boundary layer suction). This can result in a reduction in drag, but is usually impractical due to the mechanical complexity involved and the power required to move the air and dispose of it. Natural laminar flow is the name for techniques pushing the boundary layer transition aft by shaping of an airfoil or a fuselage so that their thickest point is aft and less thick. This reduces the velocities in the leading part and the same Reynolds number is achieved with a greater length.

3.1.5 Speed of sound

The speed of sound in a perfect gas is:

$$a = \sqrt{\gamma R T}, \qquad (3.12)$$

where R is the constant of the gas, T the absolute temperature, and γ the adiabatic coefficient which depends on the gas. In the air $\gamma = 1.4$ and $R = 287.05$ [J/KgK]. Therefore, the speed of sound in the air is 340.3 [m/s] at sea level in regular conditions.

Mach number

Mach number is the quotient between the speed of an object moving in the air (or any other fluid substance), typically an aircraft or a fluid particle, and the speed of sound of the air (or substance) for its particular physical conditions, that is:

$$M = \frac{V}{a}. \qquad (3.13)$$

Depending on the Mach number of an air vehicle (airplane, space vehicle, or missile, for instance), five different regimes can be considered:

1. Incompressible: $M < 0.3$, approximately. In this case, the variation of the density with respect to the density at rest can be neglected.
2. Subsonic (compressible subsonic): $0.3 \leq M < 0.8$, approximately. The variations in density must be included due to compressibility effects. Two different regimes can be distinguished: low subsonic ($0.3 \leq M < 0.6$, approximately) and high subsonic ($0.6 \leq M < 0.8$, approximately). While regional aircraft typically fly in low subsonic regimes, commercial jet aircraft typically fly in high subsonic regimes (trying to be the closest to transonic regimes while avoiding its negative effects in terms of aerodynamic drag).
3. Transonic: $0.8 \leq M < 1$, approximately. This is complex situation since around the aircraft coexist both subsonic flows and supersonic flows (for instance, in the

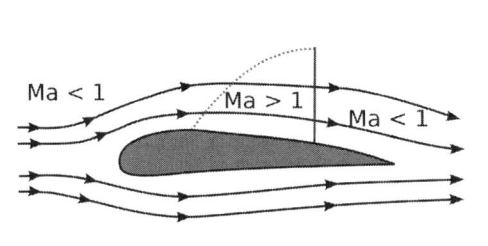

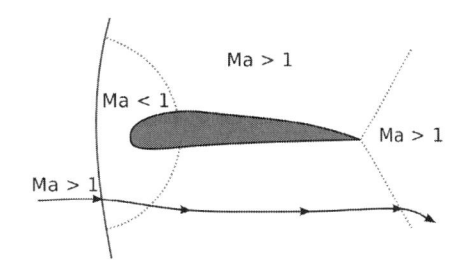

(a) Compressible subsonic Wikimedia Commons / Public Domain.

(b) Supersonic Wikimedia Commons / Public Domain.

Figure 3.8: Effect of the speed of sound in airfoils (M_a corresponds to Mach number).

extrados of the airfoil the flow accelerates and can be supersonic while the flow entering through the leading edge was subsonic).

4. Supersonic: $M \geq 1$, and then the flow around the aircraft is also at $M \geq 1$. Notice that the flow at $M = 1$ is known as sonic.

5. Hipersonic: $M \gg 1$ (in practice, $M > 5$). In these cases phenomena such as the kinetic heat or molecules dissociation appears.

In order to understand the importance of the Mach number it is important to notice that the speed of sound is the velocity at which the pressure waves or perturbations are transmitted in the fluid.

Imagine a compressible air flow with no obstacles. In this case, the pressure will be constant along the whole flow, there are no perturbations. If we introduce an airplane moving in the air, immediately appears a perturbation in the field of pressures near the airplane. Moreover, this perturbation will travel in the form of a wave at the speed of sound throughout the whole fluid field. This wave represents some kind of information emitted to the rest of fluid particles, so that the fluid adapts its physical conditions (trajectory, pressure, temperature) to the upcoming object.

If the airplane flies very slow ($M = 0.2$), the waves will travel fast relative to the airplane ($M = 1$ versus $M = 0.2$) in all directions. In this form the particles approaching the airplane are well *informed* of what is coming and can modify smoothly its conditions. If the velocity is higher, however still below $M = 1$, the modification of the fluid field is not so smooth. If the airplane flies above the speed of sound (say $M = 2$), then in this case the airplane flies twice faster than the perturbation waves, so that waves can not progress forwards to *inform* the fluid field. The consequence is that the fluid particles must adapt its velocity and position in a sudden way, resulting in a phenomena called shock wave.

Therefore, when an aircraft exceeds the sound barrier, a large pressure difference is created just in front of the aircraft resulting in a shock wave. The shock wave spreads backwards and outwards from the aircraft in a cone shape (a so-called Mach cone). It

Aerodynamics

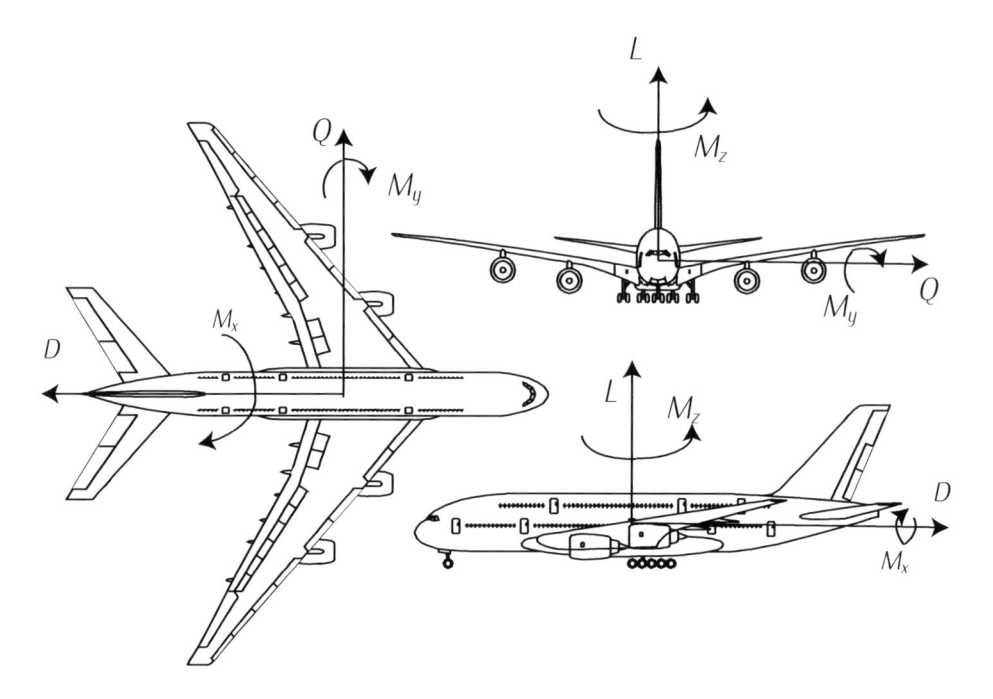

Figure 3.9: Aerodynamic forces and moments.

is this shock wave that causes the sonic boom heard as a fast moving aircraft travels overhead. At fully supersonic speed, the shock wave starts to take its cone shape and the flow is either completely supersonic, or only a very small subsonic flow area remains between the object's nose and the shock wave. As the Mach number increases, so does the strength of the shock wave and the Mach cone becomes increasingly narrow. As the fluid flow crosses the shock wave, its speed is reduced and temperature, pressure, and density increase. The stronger the shock, the greater the changes. At high enough Mach numbers the temperature increases so much over the shock that ionization and dissociation of gas molecules behind the shock wave begins.

3.2 Airfoils shapes (Franchini *et al.* [4])

The geometric figure obtained as a cross section of an airplane wing is referred to as airfoil. An airfoil-shaped body moved through a fluid produces an aerodynamic force. The component of this force perpendicular to the direction of motion is called lift. The component parallel to the direction of motion is called drag. In order to be able to calculate the movement of an airplane, an important issue is to determine the forces and torques around the center of gravity produced by the effects of air. Readers are referred to Franchini *et al.* [4]. Other introductory references on airfoil aerodynamics are Franchini and García [3, Chap. 3] and Anderson [1, Chap. 4–5].

3.2 AIRFOILS SHAPES

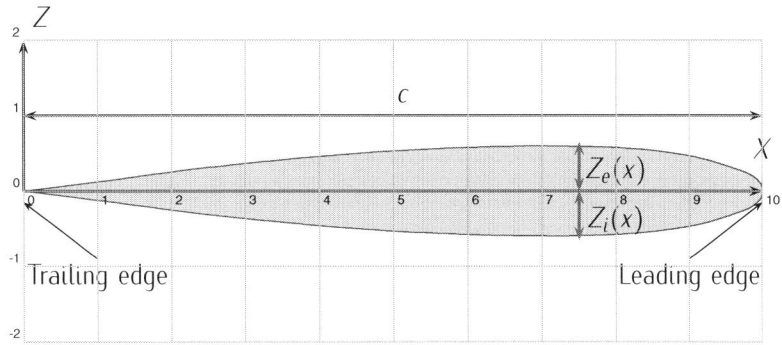

Figure 3.10: Description of an airfoil.

The aerodynamic forces are:

- L: Lift force.
- D: Drag force.
- Q: Lateral force.

The aerodynamic torques are:

- M_x: Roll torque.
- M_y: Pitch torque.
- M_z: Yaw torque.

3.2.1 AIRFOIL NOMENCLATURE

The main parts of an aerodynamic airfoil are: the chord, c, which is the segment joining the leading edge of the airfoil, x_{le}, and the trailing edge, x_{te}. To describe the airfoil, one only needs to know the functions $z_e(x)$ and $z_i(x)$, the extrados and the intrados of the airfoil, respectively. See Figure 3.10.

Another form of describing airfoils is to consider them as the result of two different contributors:

- $C(x)$, representing the camber of the airfoil:

$$C(x) = \frac{z_e(x) + z_i(x)}{2}. \tag{3.14}$$

- $E(x)$, which gives the thickness of the airfoil:

$$E(x) = z_e(x) - z_i(x). \tag{3.15}$$

According to Equation (3.14) and Equation (3.15), it yields:

$$z_e(x) = C(x) + \frac{1}{2}E(x). \tag{3.16}$$

$$z_i(x) = C(x) - \frac{1}{2}E(x). \tag{3.17}$$

AERODYNAMICS

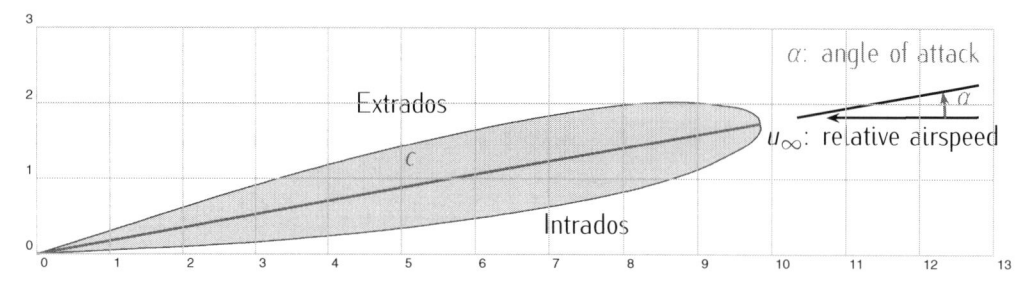

Figure 3.11: Description of an airfoil with angle of attack.

A symmetric airfoil is that in which $C(x) = 0 \ \forall \ x$.

The thickness of the airfoil is the maximum of $E(x)$. The relative thickness of the airfoil is the quotient between thickness and camber, and it is usually expressed in percentage.

The angle of attack α is the angle formed by the direction of the velocity (u_∞) of the air current with no perturbation, that is, sufficiently far away, with a reference line in the airfoil, typically the chord line. See Figure 3.11. Therefore, to give an angle of attack to the airfoil it is only necessary to rotate it around an arbitrary point x_0. In the hypothesis of a low relative thickness ($\leq 15-18\%$), a mathematical approximation of this rotation is obtained adding to both $z_e(x)$ and $z_i(x)$ the straight line $z = (x_0 - x)\alpha$, so that it yields:

$$z_e(x) = C(x) + \frac{1}{2}E(x) + (x_0 - x)\alpha. \tag{3.18}$$

$$z_i(x) = C(x) - \frac{1}{2}E(x) + (x_0 - x)\alpha. \tag{3.19}$$

This breaking down analysis into different contributors will be very useful in the future to analyze the contributors of an airfoil to aerodynamic lift: camber, thickness, and angle of attack.

3.2.2 GENERATION OF AERODYNAMIC FORCES

The actions of the air over a body which moves with respect to it give rise, at each point of the body's surface, to a shear stress tangent to the surface due to viscosity and a perpendicular stress due to the pressure. Thus, a pressure distribution and a shear stress distribution are obtained over the surface of the body.

Integrating the distribution over the surface of the body, one obtains the aerodynamic forces:

$$f_{aero} = \int (p(x) - p_\infty) \cdot dx + \int \tau(x) \cdot dx. \tag{3.20}$$

Taking the resultant of the distribution and multiplying by the distance to a fixed

3.2 Airfoils shapes

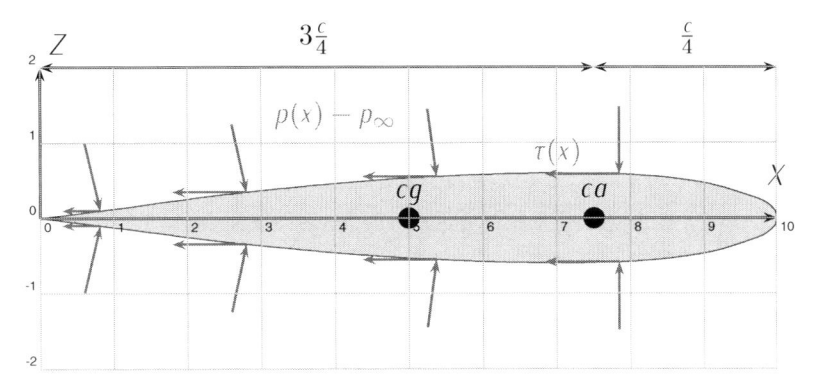

Figure 3.12: Pressure and friction stress over an airfoil.

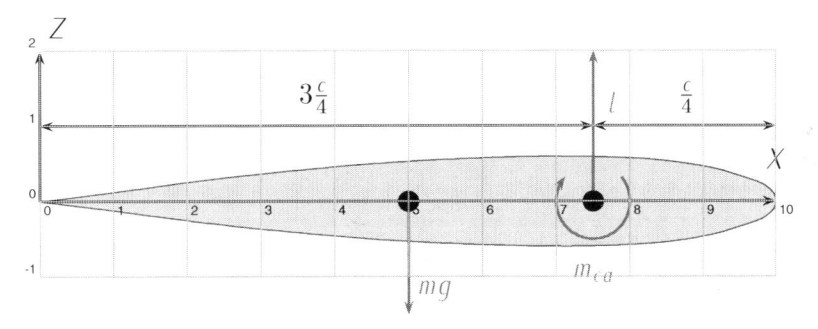

Figure 3.13: Aerodynamic forces and torques over an airfoil.

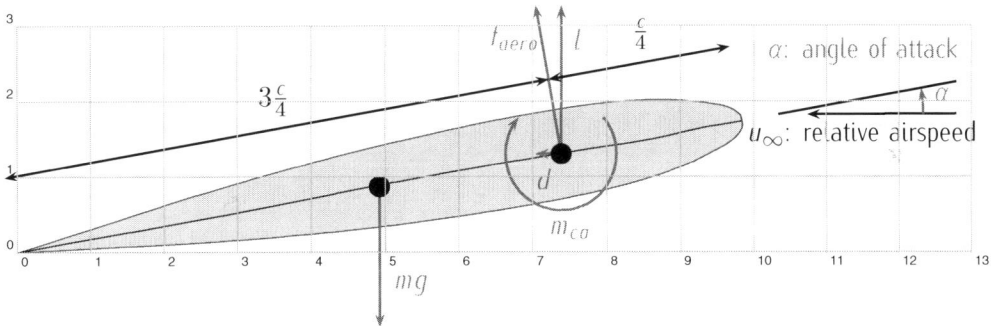

Figure 3.14: Aerodynamic forces and torques over an airfoil with angle of attack.

point (typically the aerodynamic center, located approximately at $c/4$), one obtains the aerodynamic torques:

$$m_{ca} = \int (p(x) - p_\infty)(x - x_{c/4}) \cdot dx + \int \tau(x - x_{c/4}) \cdot dx. \quad (3.21)$$

Figure 3.12 , Figure 3.13, and Figure 3.14 illustrate it.

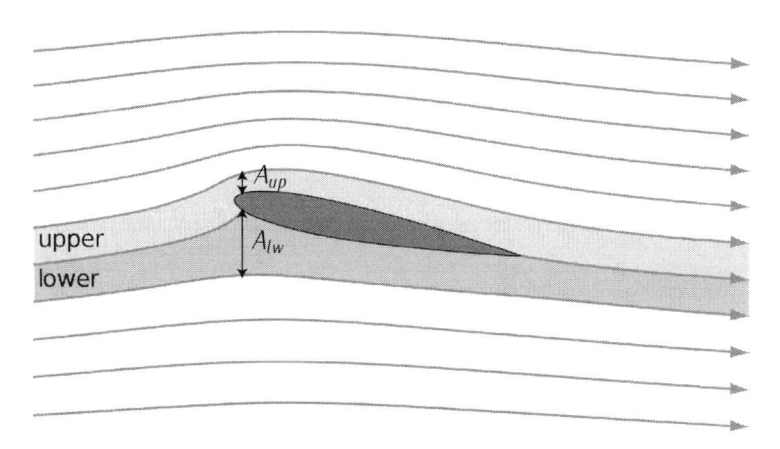

Figure 3.15: Lift generation. Modified from Wikimedia Commons / Public Domain.

The lift in an airfoil comes, basically, from the pressure forces. The drag in an airfoil comes from both the friction forces (shear stress) and the pressure forces.

Regarding drag forces, it is important to remember the already mentioned about the boundary layer. The thicker the boundary layer is, the greater the drag due to pressure effects is. In particular, when the flow drops off along the airfoil and becomes turbulent (boundary layer transition), the drag due to pressure effects increases dramatically. Furthermore, friction forces exist, which are greater in turbulent flow rather than in laminar flow. Typically, the contribution to drag of friction forces is lower than the contribution of pressure forces. Therefore, a smart design of an airfoil regarding the behavior of the boundary layer is key to minimize drag forces.

The lift forces are due to the camber, the angle of attack, and the thickness of the airfoil, which conform an airfoil shape so that the pressures in the extrados are lower that pressures in the intrados. The generation of lift can be summarized as follows:

- Because of the law of mass continuity (Equation (3.1)) the flow velocity increases over the top surface of the airfoil more than it does over the bottom surface. This is illustrated in Figure 3.15.
- As a consequence of Bernoulli effect[3] (Equation (3.6)), the pressure over the top surface of the airfoil is less than the pressure over the bottom surface.
- Because of the lower pressure over the top surface of the airfoil is less than the pressure over the bottom surface, the airfoil experiences a lift force upwards.

Therefore, this simplified statement of the equations of fluid mechanics give a qualitative idea of the aerodynamic forces. However, the resolution of the equations of fluid mechanics

---

[3]For an incompressible flow, from Bernoulli Equation (3.6), where the velocity increases, the static pressure decreases.

(Navier-Stokes equations) is extremely difficult, even though counting with the most powerful numerical tools. From the theoretical point of view they are studied using simplifications. From the experimental point of view, it is common practice to test scale-models in wind tunnels. The wind tunnels is an experimental equipment able to produce a controlled air flow into a testing chamber.

3.2.3 Aerodynamic dimensionless coefficients

The fundamental curves of an aerodynamic airfoil are: lift curve, drag curve, and momentum curve. These curves represent certain dimensionless coefficients related to lift, drag, and momentum.

The interest first focuses on determining the pressure distribution over airfoil's intrados and extrados so that, integrating such distributions, the global loads can be calculated. Again, instead of using the distribution of pressures $p(x)$, the distribution of the coefficient of pressures $c_p(x)$ will be used.

The coefficient of pressures is defined as the pressure in the considered point minus the reference pressure, typically the static pressure of the incoming current p_∞, over the dynamic pressure of the incoming current, $q = \rho_\infty u_\infty^2/2$, that is:

$$c_p = \frac{p - p_\infty}{\frac{1}{2}\rho_\infty u_\infty^2}. \tag{3.22}$$

Using Equation (3.6) and considering constant density, it yields:

$$c_p = 1 - \left(\frac{V}{u_\infty}\right)^2, \tag{3.23}$$

being V the velocity of the air flow at the considered point.

Figure 3.16 shows a typical distribution of coefficient of pressures over an airfoil. Notice that z-axis shows negative c_p and the direction of arrows means the sign of c_p. An arrow which exits the airfoil implies c_p is negative, which means the air current accelerates in that area (airfoil's extrados) and the pressure decreases (suction). On the other hand, where arrows enter the airfoil there exist overpressure, that is, decelerated current and positive c_p. Notice that if there exist a stagnation point ($V = 0$), which is by the way typical, $c_p = 1$.

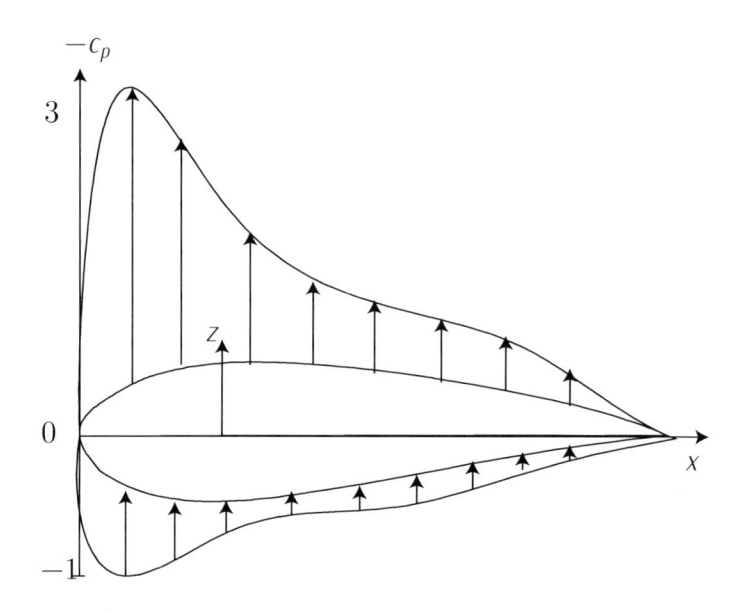

Figure 3.16: Aerodynamic forces and torques over an airfoil. Adapted from FRANCHINI et al. [4].

The dimensionless coefficients are:

$$c_l = \frac{l}{\frac{1}{2}\rho_\infty u_\infty^2 c}; \qquad (3.24)$$

$$c_d = \frac{d}{\frac{1}{2}\rho_\infty u_\infty^2 c}; \qquad (3.25)$$

$$c_m = \frac{m}{\frac{1}{2}\rho_\infty u_\infty^2 c^2}. \qquad (3.26)$$

The criteria of signs is as follows: for c_l, positive if lift goes upwards; for c_d, positive if drag goes backwards; for c_m, positive if the moment makes the aircraft pitch up.

The equation that allow c_l and c_m to be obtained from the distributions of coefficients of pressure in the extrados, $c_{pe}(x)$, and the intrados, $c_{pe}(x)$, are:

$$c_l = \frac{1}{c}\int_{x_{le}}^{x_{te}}(c_{pi}(x) - c_{pe}(x))dx = \frac{1}{c}\int_{x_{le}}^{x_{te}} c_l(x)dx, \qquad (3.27)$$

$$c_m = \frac{1}{c^2}\int_{x_{le}}^{x_{te}}(x_0 - x)(c_{pi}(x) - c_{pe}(x))dx = \ldots \qquad (3.28)$$

$$\ldots = \frac{1}{c^2}\int_{x_{le}}^{x_{te}}(x_0 - x)c_l(x)dx.$$

3.2 Airfoils shapes

If x_0 is chosen so that the moment is null, x_0 will coincide with the center of pressures of the airfoil (also referred to as aerodynamic center), x_{cp}, which is the point of application of the vector lift:

$$x_{cp} = \frac{\int_{x_{le}}^{x_{te}} x c_l(x) dx}{\int_{x_{le}}^{x_{te}} c_l(x) dx}. \tag{3.29}$$

Notice that in the case of a wing or a full aircraft, lift and drag forces have unities of force [N], and the pitch torque has unities of momentum [Nm]. To represent them it is common agreement to use L, D, and M. In the case of airfoils, due to its bi-dimensional character, typically one talks about force and momentum per unity of distance. In order to notice the difference, they are represented as l, d, and m.

Characteristic curves

The characteristic curves of an airfoil are expressed as a function of the dimensionless coefficients. These characteristic curves are, given a Mach number, a Reynolds number, and the geometry of the airfoil, as follows:

- The lift curve given by $c_l(\alpha)$.
- The drag curve, $c_d(c_l)$, also referred to as polar curve.
- The momentum curve, $c_m(\alpha)$.

Moreover, there is another typical curve which represents the aerodynamic efficiency as a function of c_l given Re and M. The aerodynamic efficiency is $E = \frac{c_l}{c_d}$ ($E = \frac{L}{D}$), and measures the ratio between lift generated and drag generated. The designer aims at maximizing this ratio.

Lets focus on the curve of lift. Typically this curve presents a linear zone, which can be approximated by:

$$c_l(\alpha) = c_{l0} + c_{l\alpha} \alpha = c_{l\alpha}(\alpha - \alpha_0), \tag{3.30}$$

where $c_{l\alpha} = dc_l/d\alpha$ is the slope of the lift curve, c_{l0} is the value of c_l for $\alpha = 0$ and α_0 is the value of α for $c_l = 0$. The linear theory of airfoils in incompressible regime gives a value to $c_{l\alpha} = 2\pi$, while c_{l0}, which depends on the airfoil's camber, is null for symmetric airfoils. There is an angle of attack, referred to as stall angle, at which this linear behavior does not hold anymore. At this point the curve presents a maximum. One this point is past lift decreases dramatically. This effect is due to the boundary layer dropping of the airfoil when we increase too much the angle of attack, reducing dramatically lift and increasing drag due to pressure effects. Figure 3.17.b illustrates it.

Aerodynamics

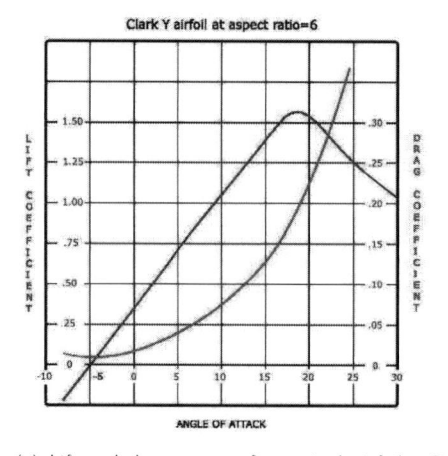

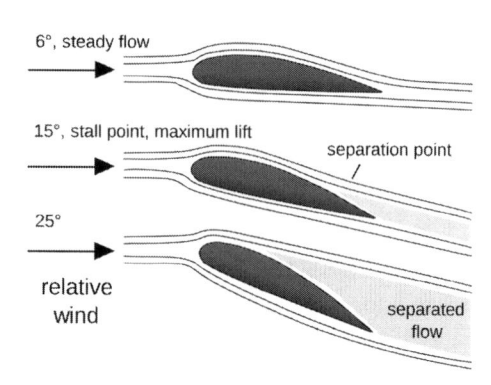

(a) Lift and drag curves of a typical airfoil. © Meggar / Wikimedia Commons / CC-BY-SA-3.0.

(b) Stall formation. Wikimedia Commons / Public Domain.

Figure 3.17: Lift and drag characteristic curves.

The drag polar can be approximated (under the same hypothesis of incompressible flow) to a parabolic curve of the form:

$$c_d(c_l) = c_{d_0} + c_{d_i} c_l^2, \qquad (3.31)$$

where c_{d_0} is the parasite drag coefficient (the one that exist when $c_l = 0$) and c_{d_i} is the induced coefficient (drag induced by lift). This curve is referred to as parabolic drag polar.

The momentum curve can be approximately constant (under the same hypothesis of incompressible flow) if one choses adequately the point x_0. This point is the aerodynamic center of the airfoil. Under incompressible regime, this point is near $0.25c$.

Lift and drag curves are illustrated in Figure 3.17.a for a typical airfoil.

3.2.4 Compressibility and drag-divergence Mach number

Given an airfoil with a specific angle of attack, if the speed of flight increases, the velocities of the air flow over the airfoil also increase. In that case, the coefficient of pressures increases and also the coefficient of lift does so. For Mach number close to $M = 1$, c_p and c_l can be approximated by the Prandtl-Glauert transformation:

$$c_p = \frac{c_{p,inc}}{\sqrt{1 - M^2}}; \qquad (3.32)$$

$$c_l = \frac{c_{l,inc}}{\sqrt{1 - M^2}}; \qquad (3.33)$$

where subindex *inc* refers to the value of the coefficient in incompressible flow.

3.2 AIRFOILS SHAPES

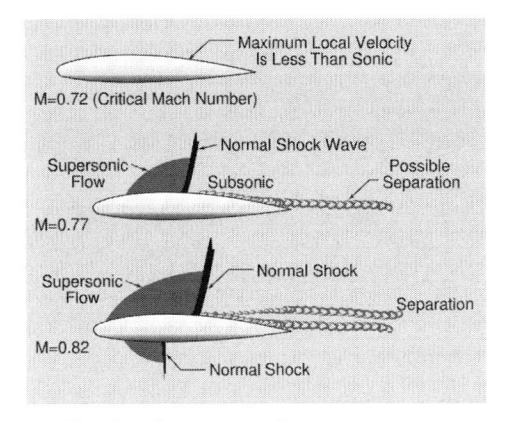

(a) Sketch of transonic flow patterns on an aircraft wing showing the effects at critical Mach. Wikimedia Commons / Public Domain.

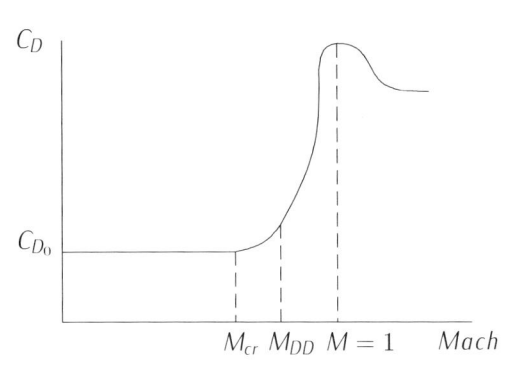

(b) Curve showing the evolution of the coefficient of drag with the Mach number. Critical Mach, divergency Mach, and the sonic barrier are depicted.

Figure 3.18: Divergence Mach.

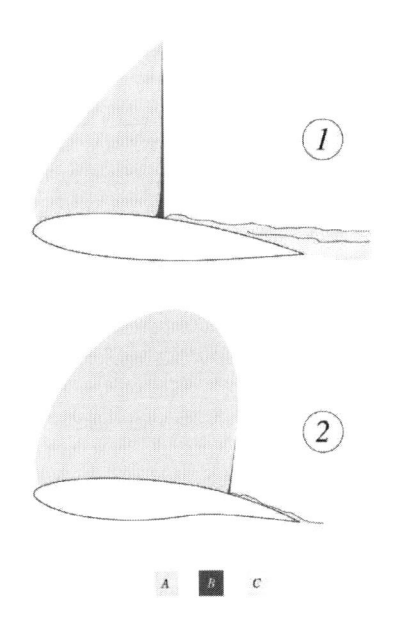

(a) The supercritical airfoil (2) maintains a lower Mach number over its upper surface than the conventional airfoil (1); this induces a weaker shock. © Olivier Cleynen / Wikimedia Commons / CC-BY-SA-3.0.

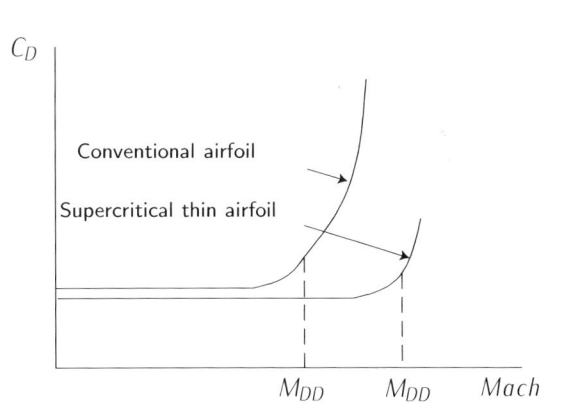

(b) Supercritical, thin airfoils retard the divergency Mach number, M_{DD}.

Figure 3.19: Supercritical airfoils.

On the other hand, the coefficient of drag remains practically constant until the airplane reaches the so called critic velocity, a subsonic velocity for which a point of extrados reaches the sonic velocity. It appears a supersonic region and waves shocks are created, giving rise to an important increase of drag. This phenomena is referred to as drag divergence.

The velocity at which this phenomena appears is refereed to as drag divergence Mach number, M_{DD}. Commercial aircraft can not typically overpass this velocity. There is not a unique definition on how to calculate this velocity. Two of the most used conditions are:

$$\frac{\partial c_d}{\partial M} = 0.1; \text{ and} \qquad (3.34)$$
$$\Delta c_d = 0.002. \qquad (3.35)$$

Airlines seek to fly faster if the consumption does not raise too much. For that reason, it is interesting to increase M_{DD}. In transonic regimes, airfoils can be designed with thin relative thickness. Another design is the so-called supercritical airfoils, whose shape permits reducing the intensity of the shock wave. In supersonic regimes, it appears another contributor to drag, the wave drag. Supersonic airfoils are designed very thin with very sharp leading edges.

3.3 Wing aerodynamics (Franchini *et al.* [4])

After having studied the aerodynamics fundamentals in bi-dimensional airfoils, we proceed on studying the aerodynamic fundamentals in three-dimensional wings. Readers are referred to Franchini *et al.* [4]. Other introductory references on wing aerodynamics are Franchini and García [3, Chap. 3] and Anderson [1, Chap. 4-5].

3.3.1 Geometry and nomenclature

In order to characterize the geometry and nomenclature of a typical commercial aircraft wing, the following wing elements are illustrated in Figure 3.20:

- Wingspan b.
- Chords: root chord c_r and tip chord c_t.
- Leading and trailing edges, and the line corresponding to the locus of $c/4$ points.
- $c/4$ swept $\Lambda_{c/4}$.

The area enclosed into the leading and trailing edge and the marginal borders (the section with c_r) view in a plant-form is referred to as wet wing surface S_w. The quotient between the wet wing surface and the wingspan is referred to as the geometric mean chord $\bar{c}$, which represents the mean chord that a rectangular wing with the same b and S_w would have.

3.3 WING AERODYNAMICS

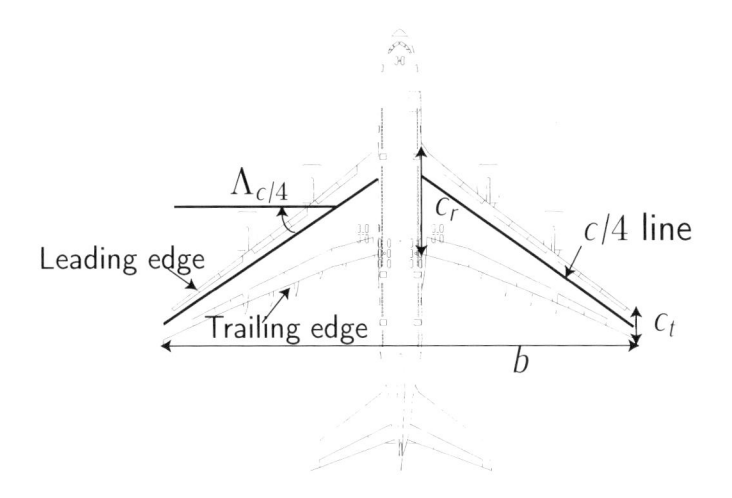

Figure 3.20: Wing geometry

The enlargement, A, is defined as:

$$A = \frac{b}{\bar{c}} = \frac{b^2}{S_w}. \tag{3.36}$$

There is also a parameter measuring the narrowing of the wing: $\lambda = c_t/c_r$.

3.3.2 FLOW OVER A FINITE WING

Figure 3.21 shows the distribution of coefficient of lift along the wingspan of four rectangular wings flying in incompressible flow with an attack angle of 10 [deg]. The four wings use the same aerodynamic airfoil and differ in the enlargement (8,10,12, and infinity). It can be observed that if the enlargement is infinity the wing behaves as the bi-dimensional airfoil y ($c_l(y)$ constant). On the other hand, if the enlargement is finite, $c_l(y)$ shows a maximum in the root of the wing ($y = 0$) and goes to zero in the tip of the wing ($y/c = A/2$). As the enlargement decreases, the maximum $c_l(y)$ also decreases.

The explanation behind this behavior is due to the difference of pressures between extrados and intrados. In particular, in the region close to the marginal border, there is an air current surrounding the marginal border which passes from the intrados, where the pressure is higher, to extrados, where the pressure is lower, giving rise to two vortexes, one in each border rotating clockwise and counterclockwise. This phenomena produces downstream a whirlwind trail. Figure 3.22 illustrates it.

The presence of this trail modifies the fluid field and, in particular, modifies the velocity each wing airfoil *sees*. In addition to the freestream velocity, u_∞, a vertical induced velocity, u_i, must be added (See Figure 3.23). The closer to the marginal border, the higher the

Aerodynamics

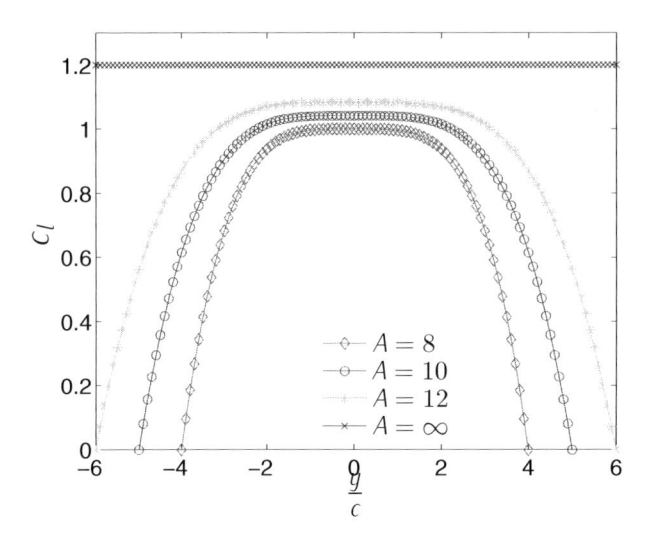

Figure 3.21: Sketch of the coefficient of lift along a wingspan.

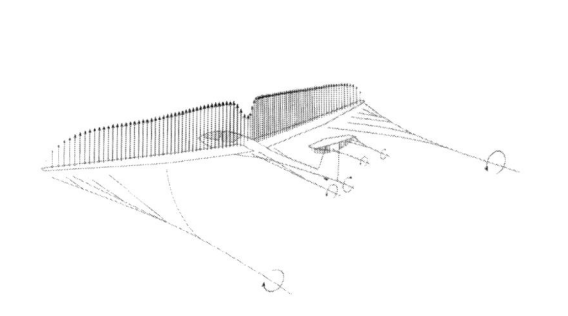

(a) A realistic lift distribution over the wing of an aircraft. The continuously changing lift distribution causes the shedding of a vortex sheet whose strength varies span-wise.. © Olivier Cleynen / Wikimedia Commons / CC-BY-SA-3.0.

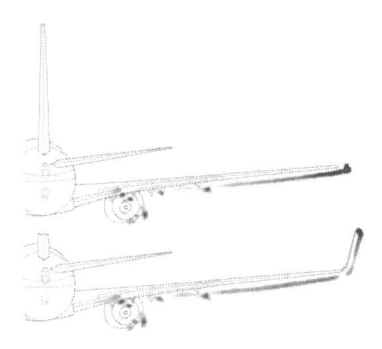

(b) The effect of the installation of winglets on the wing of a Boeing 737. The grey-scale color represents the amount of rotation present in the air immediately behind the trailing edge. © Olivier Cleynen / Wikimedia Commons / CC-BY-SA-3.0.

Figure 3.22: Whirlwind trail

induced velocity is. Therefore, the effective angle of attack of the airfoil is lower that the geometric angle, which explains both the reduction in the coefficient of lift (with respect to the bi-dimensional coefficient) and the fact that this reduction is higher when one gets closer to the marginal border.

3.3 Wing aerodynamics

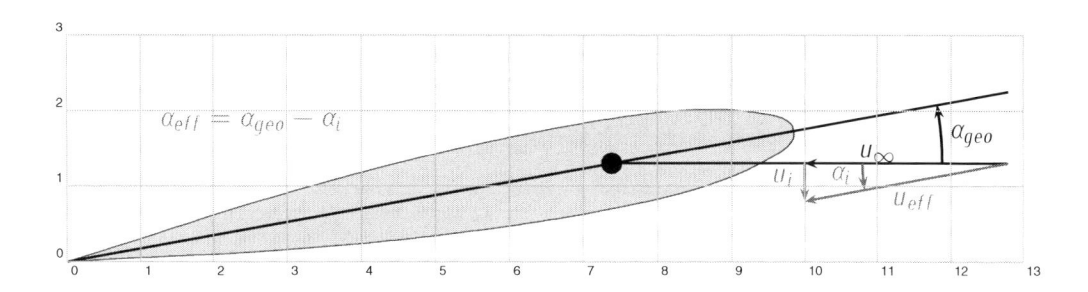

Figure 3.23: Effective angle of attack.

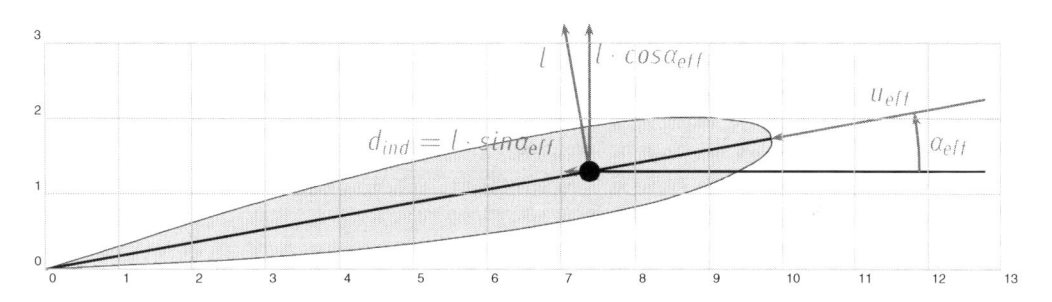

Figure 3.24: Induced drag.

3.3.3 Lift and induced drag in wings

In order to represent the lift curve, a dimensionless coefficient (C_L) will be used. C_L is defined as:

$$C_L = \frac{L}{\frac{1}{2}\rho_\infty u_\infty^2 S_w}, \qquad (3.37)$$

which can also be expressed as:

$$C_L = \frac{L}{\frac{1}{2}\rho_\infty u_\infty^2 S_w} = \frac{1}{\frac{1}{2}\rho_\infty u_\infty^2 S_w} \int_{-b/2}^{b/2} \frac{1}{2}\rho_\infty u_\infty^2 c(y) c_l(y) dy =$$

$$= \frac{1}{S_w} \int_{-b/2}^{b/2} c(y) c_l(y) dy. \qquad (3.38)$$

Another consequence of the induced velocity is the appearance of a new component of drag (see Figure 3.24), the induced drag. This occurs because the lift is perpendicular to the effective velocity and therefore it has a component in the direction of the freestream (the direction used to measure the aerodynamic drag).

3.3.4 Characteristic curves in wings

The curve of lift and the drag polar permit knowing the aerodynamic characteristics of the aircraft.

Lift curve

The coefficient of lift depends, in general, on the angle of attack, Mach and Reynolds number, and the aircraft configuration (flaps, see Section 3.4). The most general expression is:

$$C_L = f(\alpha, M, Re, configuration). \tag{3.39}$$

As in airfoils (under the same hypothesis of incompressible flow), in wings typically the lift curve presents a linear zone, which can be approximated by:

$$C_L(\alpha) = C_{L0} + C_{L\alpha}\alpha = C_{L\alpha}(\alpha - \alpha_0), \tag{3.40}$$

where $C_{L\alpha} = dC_L/d\alpha$ is the slope of the lift curve, C_{L0} is the value of C_L for $\alpha = 0$ and α_0 is the value of α for $C_L = 0$. There is a point at which the linear behavior does not hold anymore, whose angle is referred to as stall angle. At this angle the curve presents a maximum. Once this angle is past, lift decreases dramatically.

According to Prandtl theory of large wings, the slope of the curve is:

$$C_{L\alpha} = \frac{dC_L}{d\alpha} = \frac{c_{l\alpha} e}{1 + \frac{c_{l\alpha}}{\pi A}}, \tag{3.41}$$

where $e \leq 1$ is an efficiency form factor of the wing, also referred to as Oswald factor. In elliptic plantform $e = 1$.

Drag polar

The aircraft's drag polar is the function relating the coefficient of drag with the coefficient of lift, as mentioned for airfoils.

The coefficient of drag depends, in general, on the coefficient of lift, Mach, and Reynolds number, and the aircraft configuration (flaps, see Section 3.4). The most general expression is:

$$C_D = f(C_L, M, Re, configuration). \tag{3.42}$$

3.3 Wing aerodynamics

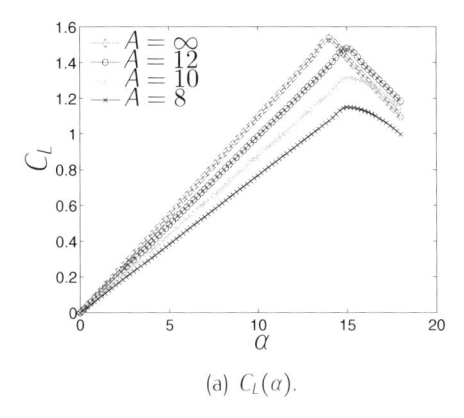

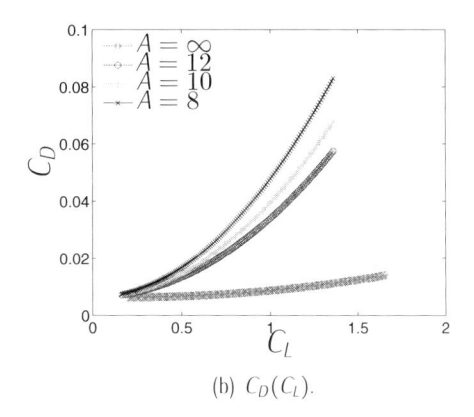

(a) $C_L(\alpha)$.

(b) $C_D(C_L)$.

Figure 3.25: Characteristic curves in wings.

The polar can be approximated to a parabolic curve of the form:

$$C_D(C_L) = C_{D_0} + C_{D_i} C_L^2, \qquad (3.43)$$

where C_{D_0} is the parasite drag coefficient (the one that exists when $C_L = 0$) due to friction and pressure effects in the wing, fuselage, etc., and $C_{D_i} = \frac{1}{\pi A e}$ is the induced coefficient (drag induced by lift) fundamentally due to the induced velocity and the whirlwind trail. This curve is referred to as parabolic drag polar. The typical values of C_{D_0} depend on the aircraft but are approximately $0.015 - 0.030$ and the parameter of aerodynamic efficiency e can be approximately $0.75 - 0.85$.

The lift curve (C_L-α) and the drag ploar ($C_D - C_L$) are represented in Figure 3.25 for a wing with four different enlargements. Both the slope and the maximum value of the lift curves increase when the enlargement increases. For the polar case, it can be observed how drag reduces as the enlargement increases.

3.3.5 Aerodynamics of wings in compressible and supersonic regimes

The evolution of the coefficients of lift and drag for a wing presents similarities with which has already been exposed for airfoils. However, instead of reducing the relative thickness and the use of supercritical airfoil to aft the divergency, in the case of wings there exist an additional resource: wing swept $\Delta_{c/4}$.

The use of swept for the design of the wing permits reducing the effective Mach number (the reduction factor is approximately $\cos \Delta_{c/4}$). Then the behavior of the airfoils is as they were flying slower and consequently the divergence Mach can be seen as higher. However, the use of swept makes the aircraft structurally complicated and, moreover, both C_{L_α} and $C_{L_{max}}$ decrease. That is why commercial aviation tends to develop supercritical airfoils to minimize the swept.

For supersonic flight, the wings are typically designed with great swept and small enlargement. The extreme case is the delta wing.

3.4 High-lift devices (Franchini *et al.* [4])

High-lift devices are designed to increase the maximum coefficient of lift. A first classification differences active and passive devices. Active high-lift devices require energy to be applied directly to the air (typically provided by the engine). Their use has been limited to experimental applications. The passive high-lift devices are, on the other hand, extensively used. Passive high-lift devices are normally hinged surfaces mounted on trailing edges and leading edges of the wing. By their deployment, they increase the aerodynamic chord and the camber of the airfoil, modifying thus the geometry of the airfoil so that the stall speed during specific phases of flight such as landing or take-off is reduced significantly, allowing to fly slower than in cruise. There also exist other types of high-lift devices that are not explicitly flaps, but devices to control the boundary layer. Readers are referred to Franchini *et al.* [4].

3.4.1 Necessity of high-lift devices

As shown in previous sections, there is a maximum coefficient of lift, $C_{L_{max}}$, that can not be exceeded by increasing the angle of attack. Consider the uniform horizontal flight, where the weight of the aircraft ($W = mg$) must be balanced by the lift force, i.e.:

$$W = L = \frac{1}{2}\rho V^2 S_w C_L. \qquad (3.44)$$

Therefore, the existence of the maximum coefficient of lift, $C_{L_{max}}$, implies that the aircraft can not fly below a minimum velocity, the stall speed, V_S:

$$V_S = \sqrt{\frac{W}{\frac{1}{2}\rho S_w C_{L_{max}}}}. \qquad (3.45)$$

Looking at equation (3.45), it can be deduced that increasing the area (S_w) and the maximum coefficient of lift ($C_{L_{max}}$) allows to fly at a lower airspeed since the minimum speed (V_S) decreases.

Deploying high-lift devices also increases the drag coefficient of the aircraft. Therefore, for any given weight and airspeed, deflected flaps increase the drag force. Flaps increase the drag coefficient of an aircraft because of higher induced drag caused by the distorted span-wise lift distribution on the wing with flaps extended. Some devices increase the planform area of the wing and, for any given speed, this also increases the parasitic drag component of total drag.

By decreasing operating speed and increasing drag, high-lift device shorten takeoff and landing distances as well as improve climb rate. Therefore, these devices are fundamental during take-off (reduce the velocity at which the aircraft's lifts equals aircraft's weight), during the initial phase of climb (increases the rate of climb so that obstacles can be avoided) and landing (decrease the impact velocity and help braking the aircraft).

3.4.2 Types of high-lift devices

The passive high-lift devices, commonly referred to as flaps, are based on the following three principles:

- Increase of camber.
- Increase of wet surface (typically by increasing the chord).
- Control of the boundary layer.

There are many different types of flaps depending on the size, speed, and complexity of the aircraft they are to be used on, as well as the era in which the aircraft was designed. Plain flaps, slotted flaps, and Fowler flaps are the most common trailing edge flaps. Flaps used on the leading edge of the wings of many jet airliners are Krueger flaps, slats, and slots (Notice that slots are not explicitly flaps, but more precisely boundary layer control devices).

The plain flap is the simplest flap and it is used in light . The basic idea is to design the airfoil so that the trailing edge can rotate around an axis. The angle of that deflexion is the flap deflexion δ_f. The effect is an increase in the camber of the airfoil, resulting in an increase in the coefficient of lift.

Another kind of trailing edge high-lift device is the slotted flap. The only difference with the plain flap is that it includes a slot which allows the extrados and intrados to be communicated. By this mean, the flap deflexion is higher without the boundary layer dropping off.

The last basic trailing edge high-lift device is the flap Fowler. This kind of flap combines the increase of camber with the increase in the chord of the airfoil (and therefore the wet surface). This fact increases also the slope of the lift curve. Combining the different types, there exist double and triple slotted Fowler flaps, combining also the control of the boundary layer. The Fairey-Youngman, Gouge, and Junkers flaps combine some of the exposed properties.

The last trailing edge high-lift device is the split flap (also refereed to as intrados flap). This flap provides, for the same increase of lift coefficient, more drag but with less torque.

AERODYNAMICS

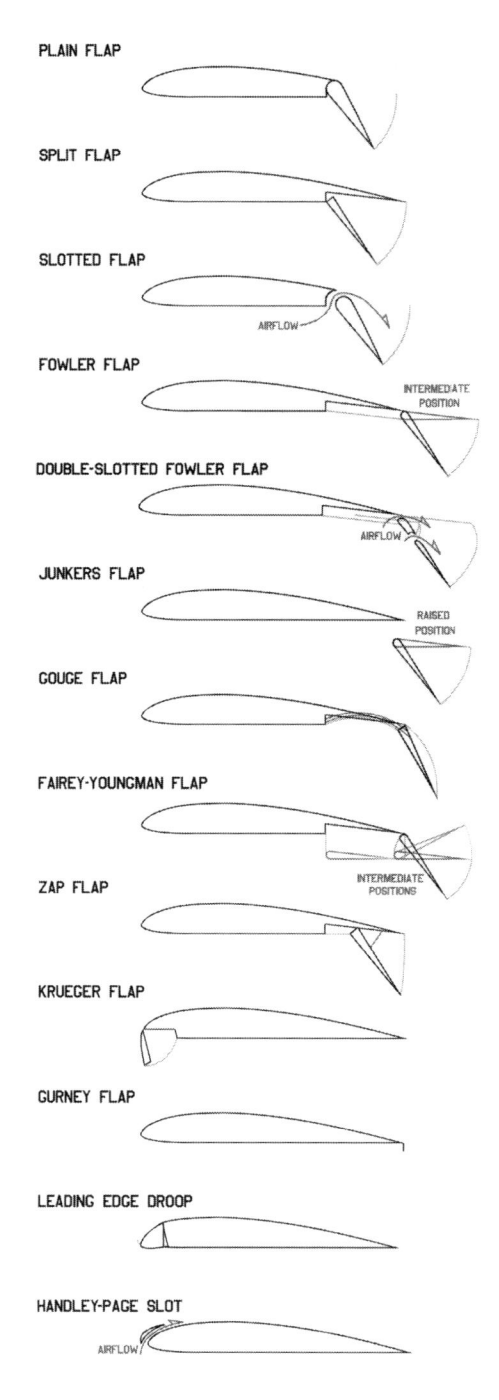

Figure 3.26: Types of high-lift devices. © NiD.29 / Wikimedia Commons / CC-BY-SA-3.0.

3.4 HIGH-LIFT DEVICES

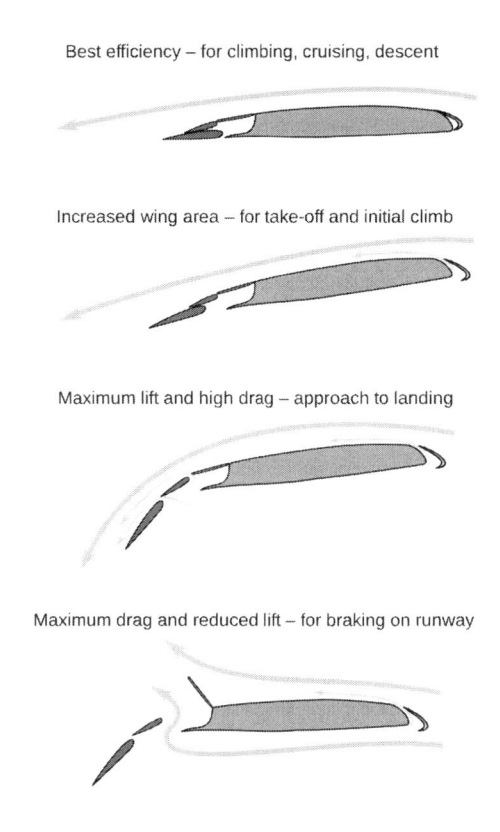

Figure 3.27: Effects of high lift devices in airfoil flow, showing configurations for normal, take-off, landing, and braking. © Andrew Fry / Wikimedia Commons / CC-BY-SA-3.0.

The most important leading edge high devices are: slot, the leading edge drop flap, and the flap Krueger.

The *slot* is a slot in the leading edge. It avoid the dropping off of the boundary layer by communicating extrados and intrados. The leading edge drop has the same philosophy as the plain flap, but applied in the leading edge instead of the trailing edge. The Kruger flaps works modifying the camber of the airfoil but also acting in the control of the boundary layer.

See Figure 3.26 and Figure 3.27.

3.4.3 INCREASE IN $C_{L_{max}}$

Table 3.1 shows the typical values for the increase of coefficient of lift in airfoils.

The increase in the maximum coefficient of lift of the wing ($\Delta C_{L_{max}}$) can be related with the increase of the maximum coefficient of lift of an airfoil ($\Delta c_{l_{max}}$). For slotted and

High-lift devices	$\Delta c_{l_{max}}$
Trailing edge devices	
Plain flap and intrados flap	0.9
Slotted flap	1.3
Fowler flap	$1.3 c'/c^*$
Doble slotted Fowler flap	$1.6 c'/c$
Tripple slotted Fowler flap	$1.9 c'/c$
Leading edge devices	
Slot	0.2
Krueger and drop flap	0.3
Slat	$0.4 c'/c$

* c' is the extended chord and c to the nominal chord.

Table 3.1: Increase in $c_{l_{max}}$ of airfoils with high lift devices. Data retrieved from FRANCHINI et al. [4].

High-lift device	δ_f TO*	δ_f LD	$\frac{C_{L_{max}}}{\cos \Lambda_{1/4}}$ TO	$\frac{C_{L_{max}}}{\cos \Lambda_{1/4}}$ LD
Plain flap	20°	60°	1.4–1.6	1.7–2
Slotted flap	20°	40°	1.5–1.7	1.8–2.2
Fowler flap	15°	40°	2–2.2	2.5–2.9
Doble slotted** flap	20°	50°	1.7–1.95	2.3–2.7
Tripple slotted flap and slat	20°	40°	2.4–2.7	3.2–3.5

* TO and LD refers to take off and landing, respectively.
** Double and triple slotted flaps have always Fowler effects increasing the chord.

Table 3.2: Typical values for $C_{L_{max}}$ in wings with high-lift devices. Data retrieved from FRANCHINI et al. [4].

Fowler flaps, the expression is:

$$\Delta C_{L_{max}} = 0.92 \Delta c_{l_{max}} \frac{S_{fw}}{S_w} \cos \Lambda_{1/4}, \qquad (3.46)$$

where $\Lambda_{1/4}$ refers to the swept measured from the locus of the c/4 of all airfoils and S_{fw} refers to the surface of the wing between the two extremes of the flap. If the flap is a plain flap, the expression is:

$$\Delta C_{L_{max}} = 0.92 \Delta c_{l_{max}} \frac{S_{fw}}{S_w} \cos^3 \Lambda_{1/4}. \qquad (3.47)$$

In the Table 3.2 the typical values of $C_{L_{max}}$ and flap deflections in different configurations are given.

3.5 Problems

Problem 3.1: Airfoils

1. In a wind tunnel experiment it has been measured the distribution of pressures over a symmetric airfoil for an angle of attack of 14°. The distribution of coefficient of pressures at the intrados, C_{pI}, and extrados, C_{pE}, of the airfoil can be respectively approximated by the following functions:

$$C_{pI}(x) = \begin{cases} 1 - 2\frac{x}{c}, & 0 \leq x \leq \frac{c}{4}, \\ \frac{2}{3}(1 - \frac{x}{c}) & \frac{c}{4} \leq x \leq c; \end{cases}$$

$$C_{pE}(x) = \begin{cases} -12\frac{x}{c}, & 0 \leq x \leq \frac{c}{4}, \\ 4(-1 + \frac{x}{c}) & \frac{c}{4} \leq x \leq c. \end{cases}$$

  (a) Draw the curve that represents the distribution of pressures.
  (b) Considering a chord $c = 1$ m, obtain the coefficient of lift of the airfoil.
  (c) Calculate the slope of the characteristic curve $c_l(\alpha)$.

2. Based on such airfoil as cross section, we build a rectangular wing with a wingspan of $b = 20$ m and constant chord $c = 1$ m. The distribution of the coefficient of lift along the wingspan of the wing (y axis) for an angle of attack $\alpha = 14°$ is approximated by the following parabolic function:

$$c_l(y) = 1.25 - 5(\frac{y}{b})^2, \quad -\frac{b}{2} \leq y \leq \frac{b}{2}.$$

  (a) Draw the curve $c_l(y)$.
  (b) Calculate the coefficient of lift of the wing.

Aerodynamics

Solution to Problem 3.1:

1. **Airfoil:**

   a) The curve is as follows:

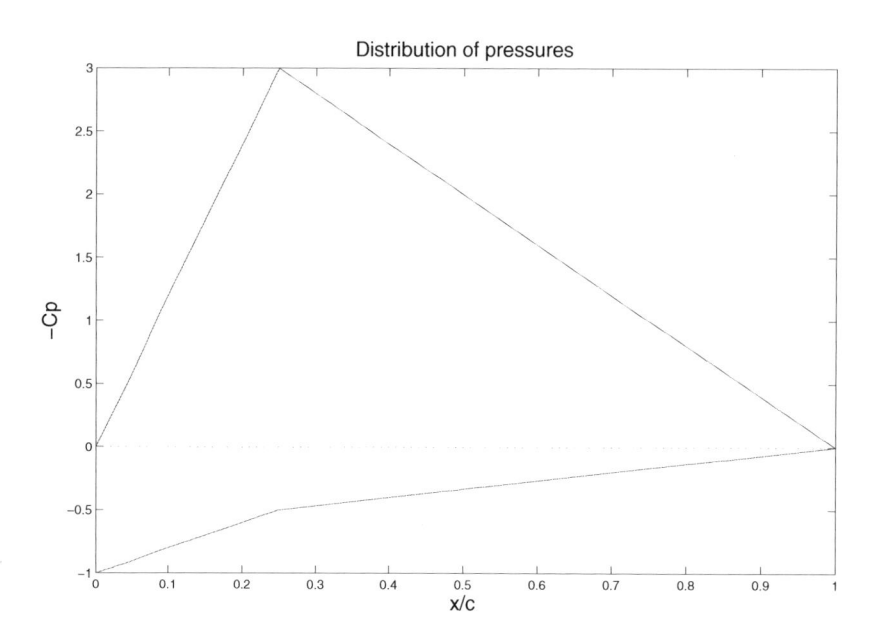

   Figure 3.28: Distribution of the coefficient of pressures (Problem 3.1).

   b) The coefficient of lift of the airfoil for $\alpha = 14°$ can be calculated as follows:

   $$c_l = \frac{1}{c} \int_{x_{le}}^{x_{te}} (c_{pI}(x) - c_{pE}(x)) dx, \qquad (3.48)$$

   In this case, with $c = 1$ and the given distributions of pressures of Intrados and extrados, Equation (3.48) becomes:

   $$c_l = \frac{1}{c} \left[\int_0^{1/4} \left((1-2x) - (-12x)\right) dx + \int_{1/4}^1 \left(2/3(1-x) - 4(-1+x)\right) dx \right] = 1.875.$$

   c) The characteristic curve is given by:

   $$c_l = c_{l_0} + c_{l_\alpha} \alpha.$$

Since the airfoil is symmetric: $c_{l_0} = 0$. *Therefore* $c_{l_\alpha} = \frac{c_l}{\alpha} = \frac{1.875 \cdot 360}{14 \cdot 2\pi} = 7.16 \ 1/rad$.

2. **Wing:**

   a) *The curve is as follows:*

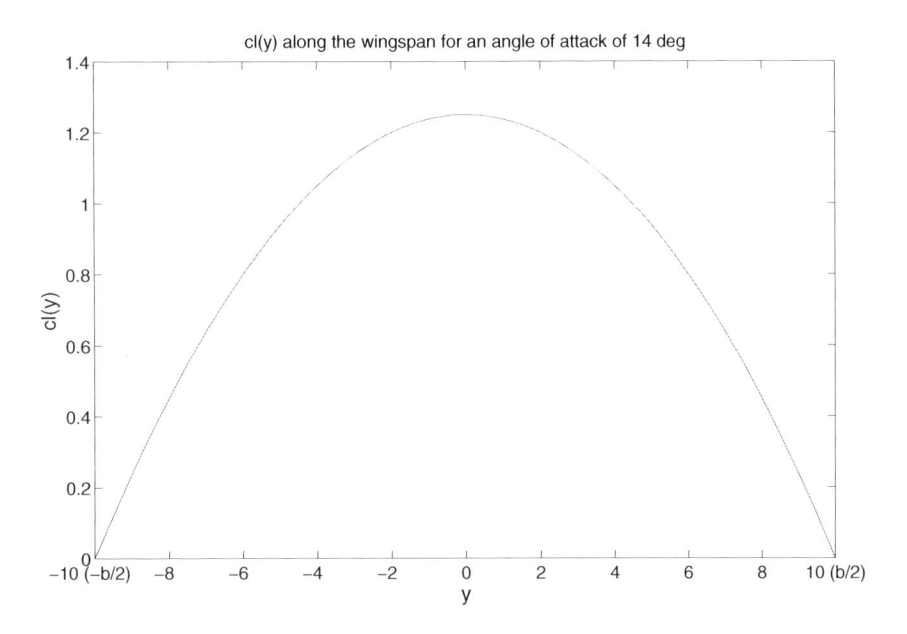

   Figure 3.29: Coefficient of lift along the wingspan (Problem 3.1).

   b) *The coefficient of lift for the wing for* $\alpha = 14°$ *can be calculated as follows:*

   $$C_L = \frac{1}{S_w} \int_{-b/2}^{b/2} c(y) c_l(y) dy. \qquad (3.49)$$

   Substituting in Equation (3.49) considering c(y)=1 and b=20:

   $$C_L = \frac{1}{20} \int_{-10}^{10} \left(1.25 - 5(\frac{y}{20})^2\right) dy = 0.83.$$

Problem 3.2: Airfoils

We want to know the aerodynamic characteristics of a NACA-4410 airfoil for a Reynolds number Re=100000. Experimental results gave the characteristic curves shown in Figure 3.30.

Calculate:

1. *The expression of the lift curve in the linear range in the form:* $c_l = c_{l0} + c_{l\alpha}\alpha$.

2. *The expression of the parabolic polar of the airfoil in the form:* $c_d = c_{d0} + bc_l + kc_l^2$.

3. *The angle of attack and the coefficient of lift corresponding to the minimum coefficient of drag.*

4. *The angle of attack, the coefficient of lift and the coefficient of drag corresponding to the maximum aerodynamic efficiency.*

5. *The values of the aerodynamic forces per unity of longitude that the model with chord $c = 2$ m would produce in the wind tunnel experiments with an angle of attack of $\alpha = 3°$ and incident current with Mach number M=0.3. Consider ISA conditions at an altitude of $h = 1000$ m.*

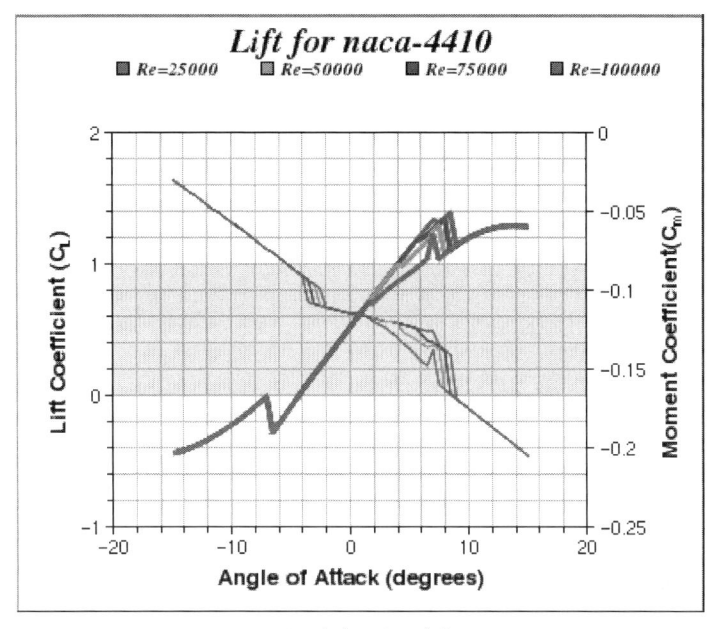

(a) $c_l(\alpha)$ and $c_m(\alpha)$.

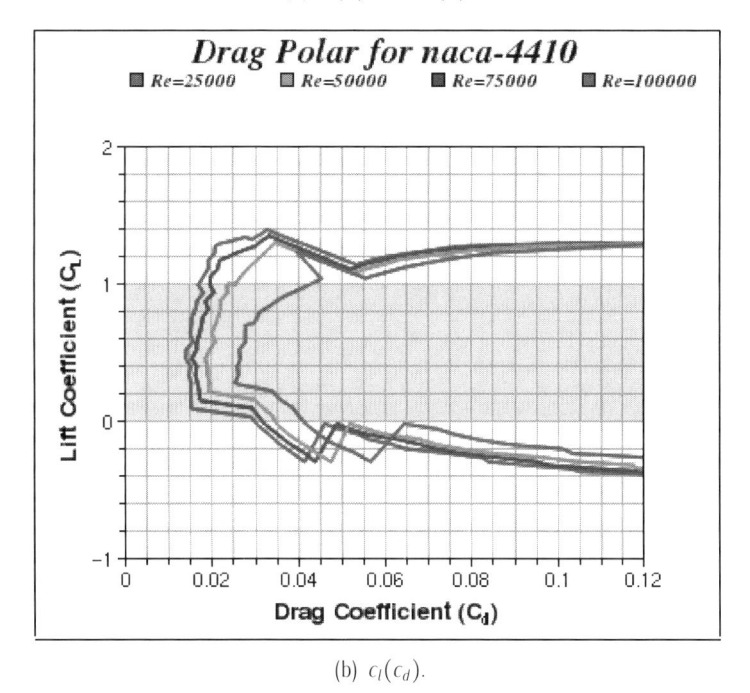

(b) $c_l(c_d)$.

Figure 3.30: Characteristic curves of a NACA 4410 airfoil.

Solution to Problem 3.2:

We want to approximate the experimental data given in Figure 3.30, respectively to a straight line and a parabolic curve. Therefore, to univocally define such curves, we must choose:

- Two pair of points (c_l, α) of the $c_l(\alpha)$ curve in Figure 3.30.a.
- Three pair of points (c_l, c_d) of the $c_l(c_d)$ curve in Figure 3.30.b.

According to Figure 3.30 for Re=100000 we choose (any other combination properly chosen must work):

c_l	α
0.5	0°
1	4°

Table 3.3: Data obtained from Figure 3.30.a.

c_l	c_d
0	0.03
1	0.0175
1.4	0.0325

Table 3.4: Data obtained from Figure 3.30.b.

1. **The expression of the lift curve in the linear range in the form:** $c_l = c_{l0} + c_{l\alpha}\alpha$:
   With the data in Table 3.3:

$$c_{l0} = 0.5; \qquad (3.50)$$
$$c_{l\alpha} = 7.16 \cdot 1/rad. \qquad (3.51)$$

The required curve yields then:

$$c_l = 0.5 + 7.16\alpha \quad [\alpha \text{ in rad}] \qquad (3.52)$$

3.5 PROBLEMS

2. **The expression of the parabolic polar of the airfoil in the form:** $c_d = c_{d0} + bc_l + kc_l^2$:

   With the data in Table 3.4 we have a system of three equations with three unkonws that is to be solved. It yields:

$$c_{d0} = 0.03; \tag{3.53}$$
$$b = -0.048; \tag{3.54}$$
$$k = 0.0357. \tag{3.55}$$

The expression of the parabolic polar yields:

$$c_d = 0.03 - 0.048 c_l + 0.0357 c_l^2. \tag{3.56}$$

3. **The angle of attack and the coefficient of lift corresponding to the minimum coefficient of drag:**

   In order to do so, we seek the minimum of the parabolic curve:

$$\frac{dc_l}{dc_d} = 0 = b + 2 \cdot kc_l. \tag{3.57}$$

Substituting in Equation (3.56):

$$\frac{dc_l}{dc_d} = 0 = -0.048 + 2 \cdot 0.0357 c_l \rightarrow (c_l)_{c_{d_{min}}} = 0.672. \tag{3.58}$$

Substituting $(c_l)_{c_{d_{min}}}$ in Equation (3.52), we obtain:

$$(\alpha)_{c_{d_{min}}} = 0.024 \, rad \, (1.378°).$$

4. **The angle of attack, the coefficient of lift and the coefficient of drag corresponding to the maximum aerodynamic efficiency:**

   The aerodynamic efficiency is defined as:

$$E = \frac{l}{d} = \frac{c_l}{c_d}. \tag{3.59}$$

Substituting the parabolic polar curve in Equation (3.59), we obtain:

$$E = \frac{c_l}{c_{d0} + bc_l + kc_l^2}. \tag{3.60}$$

In order to seek the values corresponding to the maximum aerodynamic efficiency, one must derivate and make it equal to zero, that is:

$$\frac{dE}{dc_l} = 0 = \frac{c_{d0} - kc_l^2}{(c_{d0} + bc_l + kc_l^2)^2} \rightarrow (c_l)_{E_{max}} = \sqrt{\frac{c_{d0}}{k}}. \quad (3.61)$$

Substituting according to the values previously obtained ($c_{d0} = 0.03$, $k = 0.0357$): $(c_l)_{E_{max}} = 0.91$. Substituting in Equation (3.52) and Equation (3.56), we obtain:

- $(\alpha)_{E_{max}} = 0.058$ rad (3.33°);
- $(c_d)_{E_{max}} = 0.01588$.

5. **The values of the aerodynamic forces per unity of longitude that the model with chord $c = 2$ m would produce in the wind tunnel experiments with an angle of attack of $\alpha = 3°$ and incident current with Mach number M=0.3:**

According to ISA:

- $\rho(h = 1000) = 0.907$ Kg/m^3;
- $a(h = 1000) = \sqrt{\gamma_{air} R(T_0 - \lambda h)} = 336.4$ m/s;

where a corresponds to the speed of sound, $\gamma_{air} = 1.4$, $R = 287$ J/KgK, $T_0 = 288.15$ k and $\lambda = 6.5 \cdot 10^{-3}$.

Since the experiment is intended to be at M=0.3:

$$V = M \cdot a = 100.92 \text{ m/s}. \quad (3.62)$$

Since the experiment is intended to be at $\alpha = 3°$, using Equation (3.52) and Equation (3.56):

$$c_l = 0.87; \quad (3.63)$$
$$c_d = 0.01526. \quad (3.64)$$

Finally:

$$l = c_l \frac{1}{2} \rho c V^2 = 8036.77 \text{ N/m}; \quad (3.65)$$

$$d = c_d \frac{1}{2} \rho c V^2 = 140.96 \text{ N/m}. \quad (3.66)$$

3.5 Problems

Problem 3.3: Wings

We want to analyze the aerodynamic performances of a trapezoidal wing with a plant-form as in Figure 3.31 and an efficiency factor of the wing of $e = 0.96$. Moreover, we will employ a NACA 4415 airfoil with the following characteristics:

- $c_l = 0.2 + 5.92\alpha$.
- $c_d = 6.4 \cdot 10^{-3} - 1.2 \cdot 10^{-3} c_l + 3.5 \cdot 10^{-3} c_l^2$.

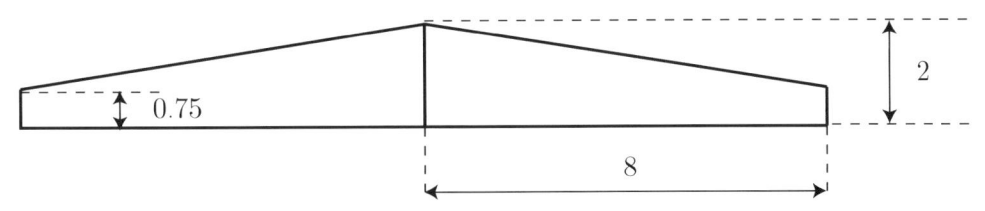

Figure 3.31: Plant-form of the wing (Dimensions in meters)

Calculate:

1. *The following parameters of the wing[4]: chord at the root; chord at the tip; mean chord; wing-span; wet surface; enlargement.*

2. *The lift curve of the wing in the linear range.*

3. *The polar of the wing assuming that it can be calculated as $C_D = C_{D_0} + C_{D_i} C_L^2$.*

4. *Calculate the optimal coefficient of lift, $C_{L_{opt}}$, for the wing. Compare it with the airfoils's one.*

5. *Calculate the optimal coefficient of drag, $C_{D_{opt}}$, for the wing. Compare it with the airfoils's one.*

6. *Maximum aerodynamic efficiency, E_{max}, for the wing. Compare it with the airfoils's one.*

7. *Discuss the differences observed in $C_{L_{opt}}$, $C_{D_{opt}}$ and E_{max} between the wing and the airfoil.*

---
[4]Based on the given data in Figure 3.31.

AERODYNAMICS

Solution to Problem 3.3:

1. **Chord at the root; chord at the tip; mean chord; wing-span; wet surface; enlargement:**

   According to Figure 3.31:

   - The wing-span, b, is $b = 16$ m.

   - The chord at the tip, c_t, is $c_t = 0.75$ m.

   - The chord at the root, c_r, is $c_r = 2$ m.

   We can also calculate the wet surface of the wing calculating twice the area of a trapezoid as follows:

   $$S_w = 2\left(\frac{(c_r + c_t)}{2}\frac{b}{2}\right) = 22 \ m^2.$$

   The mean chord, $\bar{c}$, can be calculated as $\bar{c} = \frac{S_w}{b} = 1.375$ m; and the enlargement, A, as $A = \frac{b}{\bar{c}} = 11.63$.

2. **Wing's lift curve:**

   The lift curve of a wing can be expressed as follows:

   $$C_L = C_{L_0} + C_{L_\alpha}\alpha, \qquad (3.67)$$

   and the slope of the wing's lift curve can be expressed related to the slope of the airfoil's lift curve as:

   $$C_{L_\alpha} = \frac{C_{l_\alpha}}{1 + \frac{C_{l_\alpha}}{\pi A}} e = 4.89 \cdot 1/rad.$$

   In order to calculate the independent term of the wing's lift curve, we must consider the fact that the zero-lift angle of attack of the wing coincides with the zero-lift angle of attack of the airfoil, that is:

   $$\alpha(L = 0) = \alpha(l = 0). \qquad (3.68)$$

   First, notice that the lift curve of an airfoil can be expressed as follows

   $$C_l = C_{l_0} + C_{l_\alpha}\alpha. \qquad (3.69)$$

3.5 PROBLEMS

Therefore, with Equation (3.67) and Equation (3.68) in Equation (3.69), we have that:

$$C_{L_0} = C_{l_0} \frac{C_{L_\alpha}}{C_{l_\alpha}} = 0.165.$$

The required curve yields then:

$$C_L = 0.165 + 4.89\alpha \quad [\alpha \text{ in } rad].$$

3. **The expression of the parabolic polar of the wing:**

   Notice first that the statement of the problem indicates that the polar should be in the following form:

$$C_D = C_{D0} + C_{D_i} C_L^2. \tag{3.70}$$

   For the calculation of the parabolic drag of the wing we can consider the parasite term approximately equal to the parasite term of the airfoil, that is, $C_{D_0} = C_{d_0}$.

   The induced coefficient of drag can be calculated as follows:

$$C_{D_i} = \frac{1}{\pi A e} = 0.028.$$

   The expression of the parabolic polar yields then:

$$C_D = 0.0064 + 0.028 C_L^2. \tag{3.71}$$

4. **The optimal coefficient of lift, $C_{L_{opt}}$, for the wing. Compare it with the airfoils's one.**

   The optimal coefficient of lift is that making the aerodynamic efficiency maximum. The aerodynamic efficiency is defined as:

$$E = \frac{L}{D} = \frac{C_L}{C_D}. \tag{3.72}$$

   Substituting the parabolic polar curve given in Equation (3.70) in Equation (3.72), we obtain:

$$E = \frac{C_L}{C_{D_0} + C_{D_i} C_L^2}. \tag{3.73}$$

   In order to seek the values corresponding to the maximum aerodynamic efficiency, one

89

Aerodynamics

must derivate and make it equal to zero, that is:

$$\frac{dE}{dC_L} = 0 = \frac{C_{D_0} - C_{D_i}C_L^2}{(C_{D_0} + C_{D_i}C_L^2)^2} \rightarrow (C_L)_{E_{max}} = C_{L_{opt}} = \sqrt{\frac{C_{D_0}}{C_{D_i}}}. \tag{3.74}$$

For the case of an airfoil, the aerodynamic efficiency is defined as:

$$E = \frac{l}{d} = \frac{c_l}{c_d}. \tag{3.75}$$

Substituting the parabolic polar curve given in the statement in the form $c_{d_0} + bc_l + kc_l^2$ in Equation (3.75), we obtain:

$$E = \frac{c_l}{c_{d_0} + bc_l + kc_l^2}. \tag{3.76}$$

In order to seek the values corresponding to the maximum aerodynamic efficiency, one must derivate and make it equal to zero, that is:

$$\frac{dE}{dC_l} = 0 = \frac{c_{d_0} - kc_l^2}{(c_{d0} + bc_l + kc_l^2)^2} \rightarrow (c_l)_{E_{max}} = (c_l)_{opt} = \sqrt{\frac{c_{d_0}}{k}}. \tag{3.77}$$

According to the values previously obtained ($C_{D_0} = 0.0064$ and $C_{D_i} = 0.028$) and the values given in the statement for the airfoil's polar ($c_{d_0} = 0.0064$, $k = 0.0035$), substituting them in Equation (3.74) and Equation (3.77), respectively, we obtain:

- $(C_L)_{opt} = 0.478$;
- $(c_l)_{opt} = 1.35$.

5. **The optimal coefficient of drag, $C_{D_{opt}}$ for the wing. Compare it with the airfoils's one:**
Once the optimal coefficient of lift has been obtained for both airfoil and wing, simply by substituting their values into both parabolic curves given respectively in the statement and in Equation (3.71), we obtain:

$$C_{D_{opt}} = 0.0064 + 0.028 C_{L_{opt}}^2 = 0.01279.$$
$$c_{d_{opt}} = 6.4 \cdot 10^{-3} - 1.2 \cdot 10^{-3} c_{l_{opt}} + 3.5 \cdot 10^{-3} c_{l_{opt}}^2 = 0.01115.$$

6. **Maximum aerodynamic efficiency E_{max} for the wing. Compare it with the airfoils's one:**

The maximum aerodynamic efficiency can be obtained as:

$$E_{max_{wing}} = \frac{C_{L_{opt}}}{C_{D_{opt}}} = 37.$$

$$E_{max_{airfoil}} = \frac{c_{l_{opt}}}{c_{d_{opt}}} = 121.$$

7. **Discuss the differences observed in $C_{L_{opt}}$, $C_{D_{opt}}$ and E_{max} between the wing and the airfoil.**

According to the results it is straightforward to see that meanwhile the optimum coefficient of drag is similar for both airfoil and wing, the optimum coefficient of lift is approximately three times lower that the airfoil's one. Obviously this results in an approximately three time lower efficiency for the wing when compared to the airfoil's one.

What does it mean? A three dimensional aircraft made of 2D airfoils generates much more drag than the 2D airfoil in order to achieve a required lift. Therefore, we can not simply extrapolate the analysis of an airfoil to the wing.

Such loss of efficiency is due to the so-called induced drag by lift. The explanation behind this behavior is due to the difference of pressures between extrados and intrados. In particular, in the region close to the marginal border, there is an air current surrounding the marginal border which passes from the intrados, where the pressure is higher, to extrados, where the pressure is lower, giving rise to two vortexes, one in each border rotating clockwise and counterclockwise. This phenomena produces downstream a whirlwind trail.

The presence of this trail modifies the fluid filed and, in particular, modifies the velocity each wing airfoil "sees". In addition to the freestream velocity u_∞, a vertical induced velocity u_i must be added (See Figure 3.23). The closer to the marginal border, the higher the induced velocity is. Therefore, the effective angle of attack of the airfoil is lower that the geometric angle, which explains both the reduction in the coefficient of lift (with respect to the bi-dimensional coefficient) and the appearance of and induced drag (See Figure 3.24).

Problem 3.4: Airfoils and Wings

1. In a wind tunnel experiment we have measured the distribution of pressures over a symmetric airfoil with angle of attack 6°. The distributions of the coefficient of pressures for intrados (C_{pI}) and extrados (C_{pE}) can be approximated by the following functions:

$$C_{pI}(x) = \begin{cases} 10\frac{x}{c}, & 0 \leq \frac{x}{c} \leq \frac{1}{10}, \\ 2 - 10\frac{x}{c}, & \frac{1}{10} \leq \frac{x}{c} \leq \frac{1}{5}; \\ 0, & \frac{1}{5} \leq \frac{x}{c} \leq 1. \end{cases}$$

$$C_{pE}(x) = \begin{cases} -15\frac{x}{c}, & 0 \leq \frac{x}{c} \leq \frac{1}{5}, \\ \frac{-15}{4}(1 - \frac{x}{c}), & \frac{1}{5} \leq \frac{x}{c} \leq 1. \end{cases}$$

   (a) Draw the curve $-C_p(\frac{x}{c})$.
   (b) Considering $c = 1$ [m], calculate the coefficient of lift of the airfoil.

2. Based on the previous airfoil as transversal section, we want to design a rectangular wing with wing-span b and constant chord $c = 1$ m. The distribution of the coefficient of lift along the wing-span for angle of attack 6° can be approximated by the following function:

$$c_l(y) = c_{l_{airfoil}} \cdot \left(1 - \frac{4}{A} \cdot \left(\frac{y}{b}\right)^2\right), \quad -\frac{b}{2} \leq y \leq \frac{b}{2},$$

being $c_{l_{airfoil}}$ the coefficient of lift of the airfoil previously calculated and A the enlargement of the wing.

   (a) Calculate the coefficient of lift of the wing as a function of the enlargement A.
   (b) Calculate the coefficient of lift of the wing for $A=1$, $A=8$ y $A=\infty$.
   (c) Draw the distribution of the coefficient of lift along the wing-span for $A=1$, $A=8$ y $A=\infty$. Discuss the results.

3.5 PROBLEMS

Solution to Problem 3.4:

1. **Airfoil:**
   a) The curve is as follows:

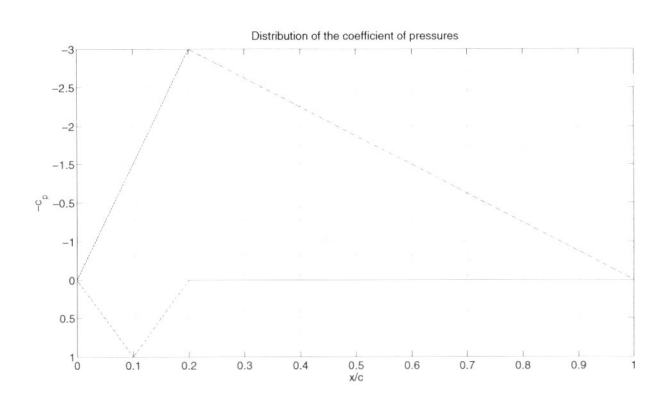

   Figure 3.32: Distribution of the coefficient of pressures (Problem 3.4).

   b) The coefficient of lift of the airfoil for $\alpha = 6°$ can be calculated as follows:

   $$c_l = \frac{1}{c}\int_{x_{le}}^{x_{te}} (c_{pI}(x) - c_{pE}(x))dx. \tag{3.78}$$

   In this case, with $c = 1$ and the given distributions of pressures of intrados and extrados, Equation (3.78) becomes:

   $$c_l = \frac{1}{c}\left[\int_0^{1/10} (10x)dx + \int_{1/10}^{1/5} (2-10x)dx + \int_{1/5}^{1} (0)dx \right.$$
   $$\left. - \int_0^{1/5} (-15x)dx - \int_{1/5}^{1} (-15/4(1-x))dx\right] = 1.6.$$

2. **Wing:**
   a) The coefficient of lift for the wing for $\alpha = 6°$ can be calculated as follows:

   $$C_L = \frac{1}{S_w}\int_{-b/2}^{b/2} c(y)c_l(y)dy. \tag{3.79}$$

Substituting in Equation (3.79) considering c(y)=1:

$$C_L = \frac{1}{b}\int_{-b/2}^{b/2} 1.6\left(1 - \frac{4}{A}(\frac{y}{b})^2\right) dy = 1.6(1 - \frac{1}{3A}).$$

b) The values of C_L for the different enlargements are:

- $A = 1 \rightarrow C_L = 1.06$.
- $A = 8 \rightarrow C_L = 1.53$.
- $A = \infty \rightarrow C_L = 1.6$.

Considering, for instance, a wing-span b=20 m, the curve is as follows:

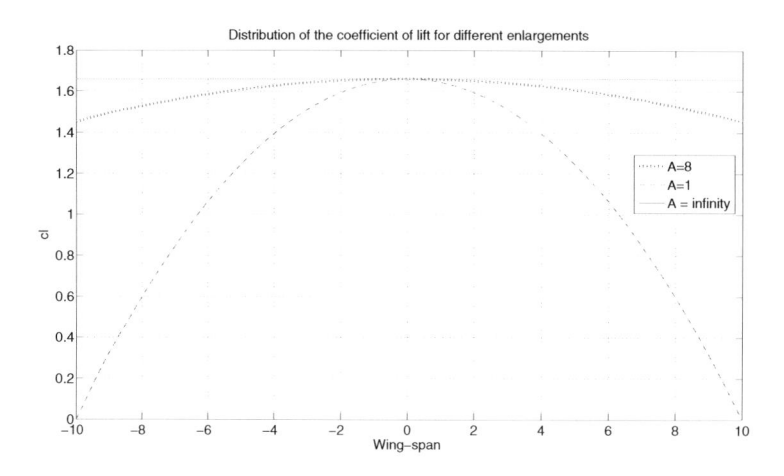

Figure 3.33: Coefficient of lift along the wingspan (Problem 3.4).

c) The discussion has to do with the differences in lift generation between finite and infinity wing.

It can be observed that if the enlargement is infinity the wing behaves as the bi-dimensional airfoil y ($c_l(y)$ constant). On the other hand, if the enlargement is finite, $c_l(y)$ shows a maximum in the root of the wing ($y = 0$) and goes to zero in the tip of the wing ($y/c = A/2$). As the enlargement decreases, the maximum $c_l(y)$ also decreases.

The explanation behind this behavior is due to the difference of pressures between extrados and intrados. In particular, in the region close to the marginal border, there is an air current surrounding the marginal border which passes from the intrados, where the pressure is higher, to extrados, where the pressure is lower, giving rise to two vortexes, one in each border rotating clockwise and counterclockwise. This phenomena produces downstream a whirlwind trail.

The presence of this trail modifies the fluid field and, in particular, modifies the velocity each wing airfoil "sees". In addition to the freestream velocity, u_∞, a vertical induced velocity, u_i, must be added (See Figure 3.23). The closer to the marginal border, the higher the induced velocity is. Therefore, the effective angle of attack of the airfoil is lower that the geometric angle, which explains both the reduction in the coefficient of lift (with respect to the bi-dimensional coefficient) and the fact that this reduction is higher when one gets closer to the marginal border.

Problem 3.5: High-Lift devices

1. We want to analyze the aerodynamic performances of a trapezoidal wing with a plant-form as in Figure 3.34. The wing mounts two triple slotted Fowler flaps. The efficiency factor (Oswald factor) of the wing is $e = 0.96$. The wing is build employing NACA 4415 airfoils with the following characteristics:
   - $c_l = 0.2 + 5.92\alpha$. (α en radianes)
   - $c_d = 6.4 \cdot 10^{-3} - 1.2 \cdot 10^{-3} c_l + 3.5 \cdot 10^{-3} c_l^2$.

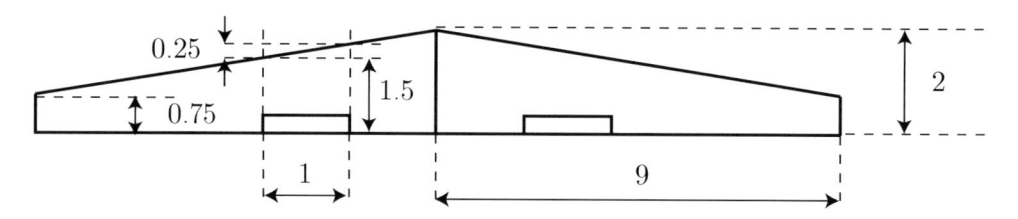

Figure 3.34: Plant-form of the wing (dimensions in meters).

On regard of the effects of the Fowler flaps in the maximum coefficient of lift, it is known that:

- The increase of the maximum coefficient of lift of the airfoil ($\Delta c_{l_{max}}$) can be approximated by the following expression:

$$\Delta c_{l_{max}} = 1.9 \frac{c'}{c}, \quad (3.80)$$

being c the chord in the root and c' the extended chord (consider $c' = 3$ [m]).

- The increase of the maximum coefficient of lift of the wing ($\Delta C_{L_{max}}$) can be related to the increase of the maximum coefficient of lift of the airfoil ($\Delta c_{l_{max}}$) by means of the following expression:

$$\Delta C_{L_{max}} = 0.92 \Delta c_{l_{max}} \frac{S_{fw}}{S_w} \cos \Lambda. \quad (3.81)$$

Based on the data given in Figure 3.34, calculate:

(a) *Chord in the root and tip of the wing. wing-span and enlargement. Wet wing surface (S_w) and surface wet by the flaps (S_{fw}). Aircraft swept (Λ) measured from the leading edge.*

Assuming a clean configuration (no flap deflection), typical of cruise conditions, and knowing also that the stall of the airfoil takes place at an angle of attack of 15°:

b) *calculate the maximum coefficient of lift of the airfoil.*

c) *calculate the expression of the lift curve of the wing in its linear range.*

d) *calculate the maximum coefficient of lift of the wing (assume that the aircraft (wing) stalls also at an angle of attack of 15°)*

Assuming a configuration with flaps fully deflected, typical of a final approach, calculate:

e) *the maximum coefficient of lift of the wing.*

It is known that the mass of the aircraft is 4500 kg. For sea level ISA conditions and force due to gravity equal to 9.81 m/s²:

f) *calculate the stall speeds of the aircraft for both configurations (clean and full).*

g) *compare and discuss the results.*

AERODYNAMICS

Solution to Problem 3.5:

a. **Chord at the root; chord at the tip; wing-span and enlargement; wing wet surface and flap wet surface; swept:**

According to Figure 3.34:

- The wing-span, b, is $b = 18$ m.
- The chord at the tip, c_t, is $c_t = 0.75$ m.
- The chord at the root, c_r, is $c_r = 2$ m.

We can also calculate the wet surface of the wing calculating twice the area of a trapezoid as follows:

$$S_w = 2\left(\frac{(c_r + c_t)}{2}\frac{b}{2}\right) = 24.75 \; m^2.$$

In the same way, the flap wet surface (S_{fw}) can be calculated as follows:

$$S_{fw} = 2\left(1 \cdot 1.5 + \frac{1}{2}1 \cdot 0.25\right) = 3.25 \; m^2.$$

The mean chord, $\bar{c}$, can be calculated as $\bar{c} = \frac{S_w}{b} = 1.375$ m; and the enlargement, A, as $A = \frac{b}{\bar{c}} = 13.09$.

Finally, the swept of the wing (Λ) measured from the leading edge is:

$$\Lambda = \arctan(\frac{0.25}{1}) = 14°.$$

b. $c_{l_{max}}$

According to the expression given in the statement for the airfoil's lift curve: $c_l = 0.2 + 5.92\alpha$, and given that the airfoils stalls at $\alpha = 15°$, the maximum coefficient of lift will be given by the value of the coefficient of lift at the stall angle:

$$c_{l_{max}} = 0.2 + 5.92 \cdot 15\frac{2\pi}{360} = 1.74. \tag{3.82}$$

c. **Wing's lift curve:**

The lift curve of a wing can be expressed as follows:

$$C_L = C_{L_0} + C_{L_\alpha}\alpha, \tag{3.83}$$

98

and the slope of the wing's lift curve can be expressed related to the slope of the airfoil's lift curve as:

$$C_{L_\alpha} = \frac{C_{l_\alpha}}{1 + \frac{C_{l_\alpha}}{\pi A}} e = 4.96 \cdot 1/rad.$$

In order to calculate the independent term of the wing's lift curve, we must consider the fact that the zero-lift angle of attack of the wing coincides with the zero-lift angle of attack of the airfoil, that is:

$$\alpha(L=0) = \alpha(l=0). \tag{3.84}$$

First, notice that the lift curve of an airfoil can be expressed as follows

$$C_l = C_{l_0} + C_{l_\alpha}\alpha. \tag{3.85}$$

Therefore, with Equation (3.83) and Equation (3.84) in Equation (3.85), we have that:

$$C_{L_0} = C_{l_0}\frac{C_{L_\alpha}}{C_{l_\alpha}} = 0.1678. \tag{3.86}$$

The required curve yields then:

$$C_l = 0.1678 + 4.96\alpha \quad [\alpha \text{ in } rad].$$

d. $C_{L_{max}}$ in clean configuration:

Given the expression in Equation (3.86), and given that the aircraft (wing) stalls at $\alpha = 15°$, the maximum coefficient of lift will be given by:

$$C_{L_{max}} = 0.1678 + 4.96 \cdot 15\frac{2\pi}{360} = 1.466. \tag{3.87}$$

e. $C_{L_{max}}$ in full configuration (with all flaps deflected):

In order to obtain the maximum coefficient of lift for full configuration ($C_{L_{max}}$) we have that:

$$C_{L_{max_f}} = C_{L_{max}} + \Delta C_{L_{max}}. \tag{3.88}$$

As it was given in the Equation (3.81), $\Delta C_{L_{max}}$ can be expressed as:

$$\Delta C_{L_{max}} = 0.92\Delta c_{l_{max}}\frac{S_{fw}}{S_w}\cos\Lambda,$$

where Λ, S_{fw}, and S_w are already known and $c_{l_{max}}$ was given in Equation (3.80). Therefore,

$\Delta C_{L_{max}} = 0.334$.

$C_{L_{max_f}}$ yields then 1.8.

f. V_{stall}:

Knowing that $L = C_L \cdot \frac{1}{2}\rho S_w V^2$, that the flight can be considered to be equilibrated, i.e., $L = m \cdot g$, and that the stall speed takes place when the coefficient of lift is maximum, we have that:

$$V_{stall} = \sqrt{\frac{m \cdot g}{\frac{1}{2}\rho S_w C_{L_{max}}}} = 44.56 \; m/s.$$

For the case of full configuration, we use $C_{L_{max_f}}$ and consider the wing-and-flap wet surface (notice that we are not including the chord extension in the wing wet surface). The stall velocity yields:

$$V_{stall_f} = \sqrt{\frac{m \cdot g}{\frac{1}{2}\rho (S_w + S_{fw}) C_{L_{max_f}}}} = 37.8 \; m/s.$$

g. *Discussion:*

High-lift devices are designed to increase the maximum coefficient of lift. By their deployment, they increase the aerodynamic chord and the camber of the airfoil, modifying thus the geometry of the airfoil so that the stall speed during specific phases of flight such as landing or take-off is reduced significantly, allowing to flight slower than in cruise.

Deploying high-lift devices also increases the drag coefficient of the aircraft. Therefore, for any given weight and airspeed, deflected flaps increase the drag force. Flaps increase the drag coefficient of an aircraft because of higher induced drag caused by the distorted span-wise lift distribution on the wing with flaps extended. Some devices increase the planform area of the wing and, for any given speed, this also increases the parasitic drag component of total drag.

By decreasing operating speed and increasing drag, high-lift device shorten takeoff and landing distances as well as improve climb rate. Therefore, these devices are fundamental during take-off (reduce the velocity at which the aircraft's lifts equals aircraft's weight), during the initial phase of climb (increases the rate of climb so that obstacles can be avoided) and landing (decrease the impact velocity and help braking the aircraft).

REFERENCES

[1] ANDERSON, J. (2012). *Introduction to flight, seventh edition.* McGraw-Hill.

[2] ANDERSON, J. D. (2001). *Fundamentals of aerodynamics*, volume 2. McGraw-Hill New York.

[3] FRANCHINI, S. and GARCÍA, O. (2008). *Introducción a la ingeniería aeroespacial.* Escuela Universitaria de Ingeniería Técnica Aeronáutica, Universidad Politécnica de Madrid.

[4] FRANCHINI, S., LÓPEZ, O., ANTOÍN, J., BEZDENEJNYKH, N., and CUERVA, A. (2011). *Apuntes de Tecnología Aeroespacial.* Escuela de Ingeniería Aeronáutica y del Espacio. Universidad Politécnica de Madrid.

[5] PRANDTL, L. (1904). Über Flüssigkeitsbewegung bei sehr kleiner Reibung. In *Proceedings of 3rd International Mathematics Congress, Heidelberg.*

4
AIRCRAFT STRUCTURES

Contents

4.1	Generalities		104
4.2	Materials		108
	4.2.1	Properties	108
	4.2.2	Materials in aircraft	109
4.3	Loads		113
	4.3.1	Fuselage loads	113
	4.3.2	Wing and tail loads	114
	4.3.3	Landing gear loads	114
	4.3.4	Other loads	114
4.4	Structural components of an aircraft		114
	4.4.1	Structural elements and functions of the fuselage	115
	4.4.2	Structural elements and functions of the wing	117
	4.4.3	Tail	118
	4.4.4	Landing gear	118
	References		119

A Structure holds things together, carries loads, and provides integrity. Structural engineering is the application of statics and solid mechanics to devise structures with sufficient strength, stiffness, and useful life to fulfill a mission without failure with a minimum amount of weight. Aerospace engineers pay particular attention to designing light structures due to the strong dependence of weight on operational costs.

The aim of this chapter is to give an overview of aircraft structures. The chapter starts with some generalities in Section 4.1. Then material properties are analyzed in Section 4.2 focusing on aircraft materials. The Loads that appear in an aircraft structure will be described in Section 4.3. Finally, Section 4.4 will be devoted to describe the fundamental structural components of an aircraft. Thorough references on the matter are, for instance, MEGSON [3] and CANTOR et al. [1].

↑ F   ↓ F

Traction   Compression

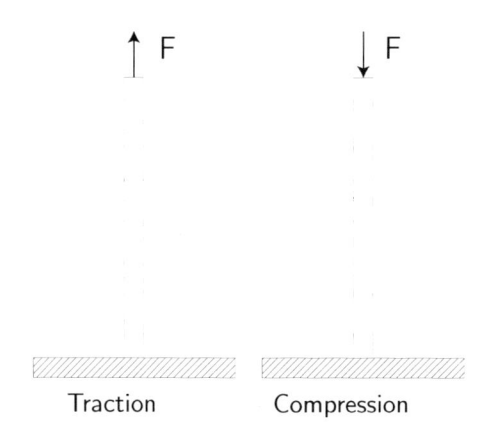

Figure 4.1: Normal stress. Adapted from FRANCHINI et al. [2].

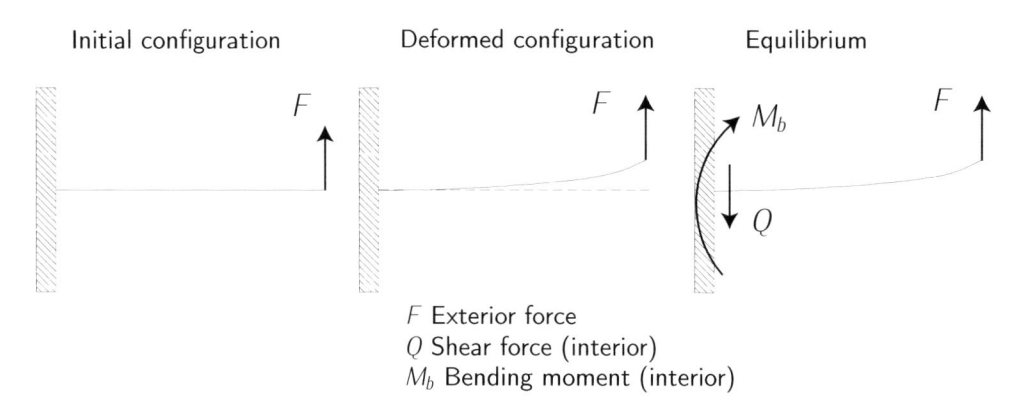

Initial configuration   Deformed configuration   Equilibrium

F Exterior force
Q Shear force (interior)
M_b Bending moment (interior)

Figure 4.2: Bending. Adapted from FRANCHINI et al. [2].

4.1 GENERALITIES

It is the task of the designer to consider all possible loads. The combination of materials and design of the structure must be such that it can support loads without failure. In order to estimate such loads one can take measurements during the flight, take measurements of a scale-model in a wind tunnel, do aerodynamic calculations, and/or perform test-flights with a prototype. Aircraft structures must be able to withstand all flight conditions and be able to operate under all payload conditions.

A force applied lengthwise to a piece of structure will cause normal stress, being either tension (also refereed to as traction) or compression stress. See Figure 4.1. With tensile loads, all that matters is the area which is under stress. With compressive loads, also the shape is important, since buckling may occur. Stress is defined as load per area, being $\sigma = F/A$.

4.1 GENERALITIES

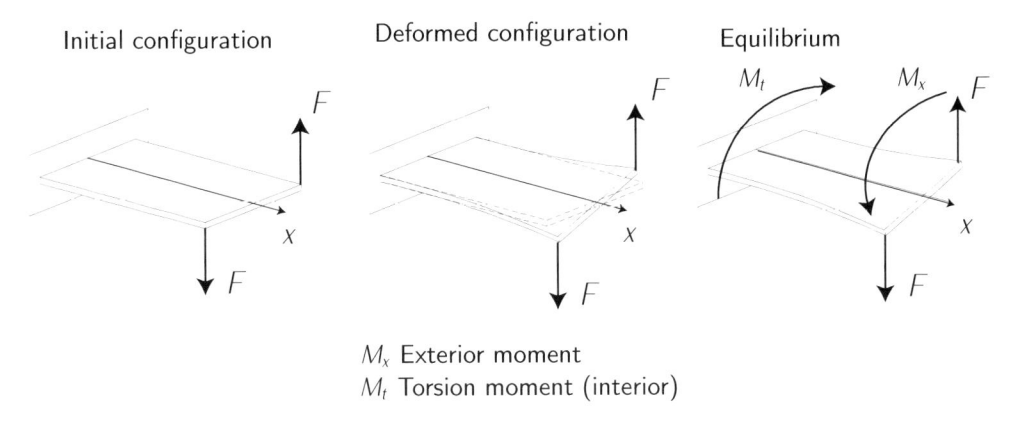

Initial configuration   Deformed configuration   Equilibrium

M_x Exterior moment
M_t Torsion moment (interior)

Figure 4.3: Torsion. Adapted from FRANCHINI et al. [2].

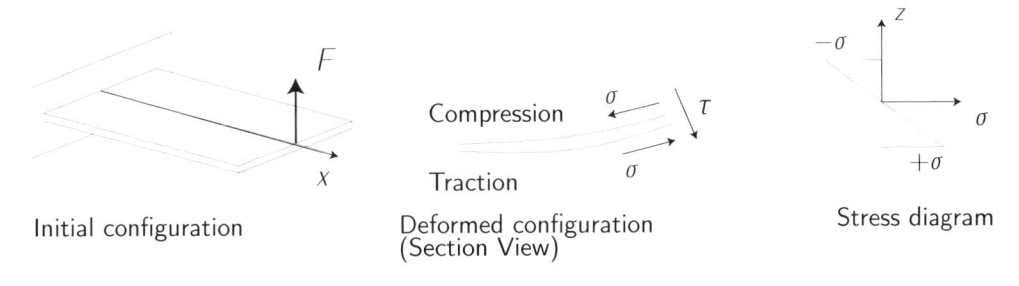

Initial configuration   Deformed configuration (Section View)   Stress diagram

Figure 4.4: Shear stress due to bending. Adapted from FRANCHINI et al. [2].

If a force is applied at right angles (say perpendicular to the lengthwise of a beam), it will apply shear stress and a bending moment. See Figure 4.2. If a force is offset from the line of a beam, it will also cause torsion. See Figure 4.3. Both bending and torsion causes shear stresses. Shear is a form of loading which tries to tear the material, causing the atoms or molecules to slide over one another. See Figure 4.4 and Figure 4.5. Overall, a prototypical structure suffers from both normal (σ) and shear (τ) stresses. See Figure 4.6 in which an illustrative example of the stresses over a plate is shown.

Structures subject to normal or shear stresses may also be deformed. See Figure 4.7 and Figure 4.8.

Strain, $\epsilon = \frac{\Delta l}{l_i} = \frac{l - l_i}{l_i}$ is the proportional deflection within a material as a result of an applied stress. It is impossible to be subjected to stress without experiencing strain. For elastic deformation, which is present below the elastic limit, Hooke's law applies: $\sigma = E\epsilon$, where E is refereed to as the modulus of Young, and it is a property of the material. The stresses within a structure must be kept below a defined permitted level, depending of the requirements of the structure (in general, stresses must no exceed the elastic limit, σ_y). See Figure 4.9.

105

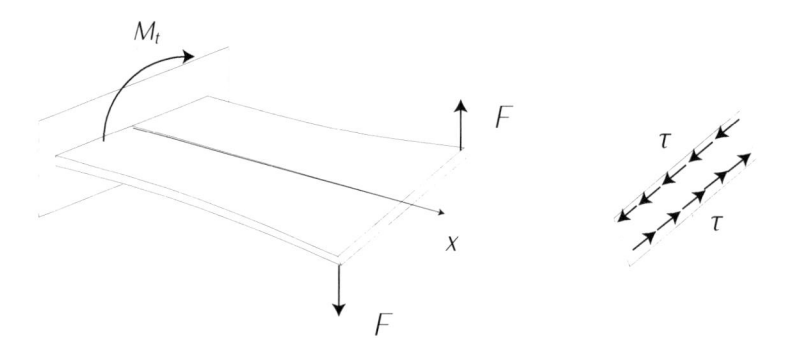

Figure 4.5: Shear stress due to torsion. Adapted from Franchini et al. [2].

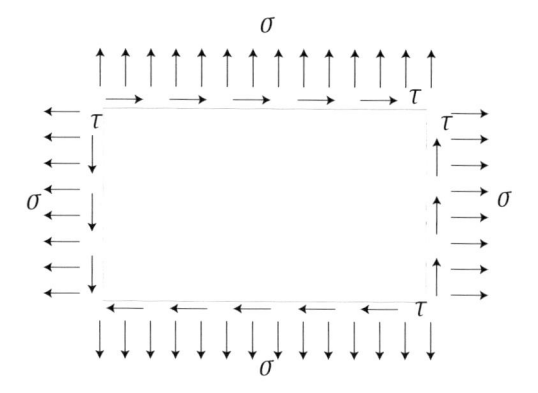

Figure 4.6: Stresses in a plate. Adapted from Franchini et al. [2].

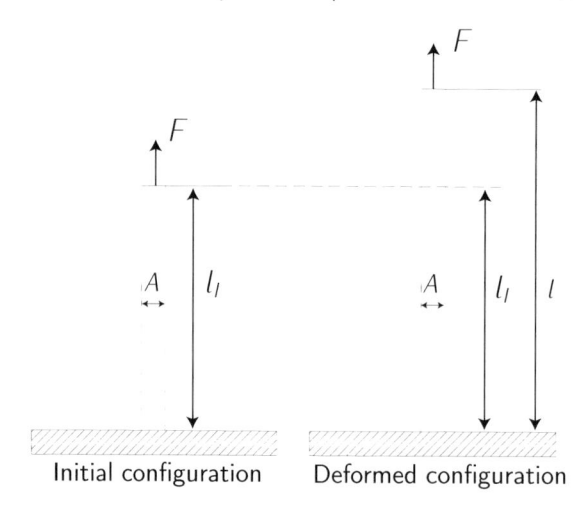

Figure 4.7: Normal deformation. Adapted from Franchini et al. [2].

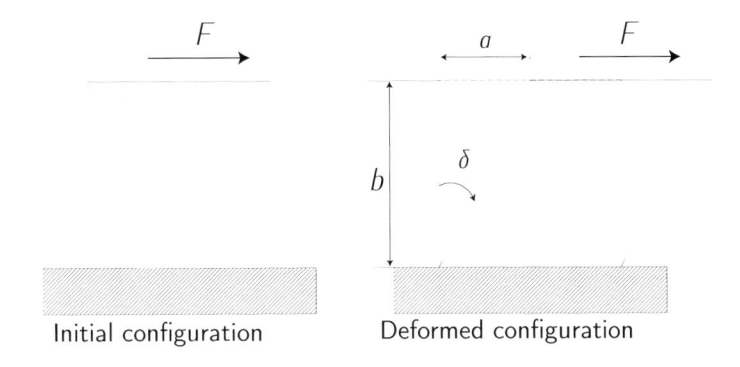

Figure 4.8: Tangential deformation. Adapted from FRANCHINI et al. [2].

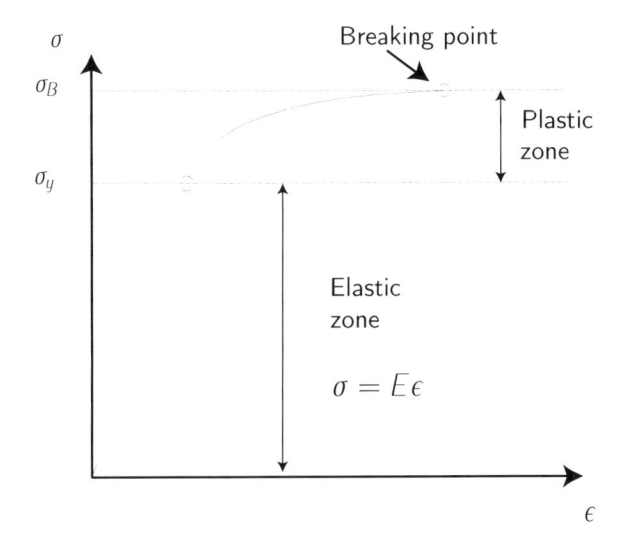

σ_B Breaking stress

σ_y Elastic limit

Figure 4.9: Behavior of an isotropic material. Adapted from FRANCHINI et al. [2].

4.2 Materials

As a preliminary to the analysis of loads and basic aircraft structural elements presented in subsequent sections, we shall discuss some of the properties of materials and the materials themselves that are used in aircraft construction. Readers are referred to MEGSON [3].

4.2.1 Properties

Several factors influence the selection of the structural materials for an aircraft. The most important one is the combination of strength and lightness. Other properties with different importance (sometimes critical) are stiffness, toughness, resistance to corrosion and fatigue, ease of fabrication, availability and consistency of supply, and cost (also very important). A brief description of some of the most important properties is given in the sequel:

Ductility: Ductility refers to a solid material's ability to deform under tensile stress, withstanding large strains before fracture occurs. These large strains are accompanied by a visible change in cross-sectional dimensions and therefore give warning of impending failure.

Strength: The strength of a material is its ability to withstand an applied stress without failure. The applied stress may be tensile, compressive, or shear. Strength of materials is a subject which deals with loads, deformations, and the forces acting on a material. Looking at Figure 4.9, it is associated to σ_B (breaking stress); the greater σ_B is the more strengthless is the material.

Toughness: Toughness is the ability of a material to absorb energy and plastically deform without fracturing. Toughness requires a balance of strength and ductility. Strength indicates how much force the material can support, while toughness indicates how much energy a material can absorb before fracturing. Looking at Figure 4.9, it is associated to the difference between σ_B and σ_y; the greater this difference is the more capacity the material ha to absorb impact energy by plastic deformation.

Brittleness: A brittle material exhibits little deformation before fracture, the strain normally being below 5%. Brittle materials therefore may fail suddenly without visible warning. Brittleness and toughness are antonyms.

Elasticity: A material is said to be elastic if deformations disappear completely on removal of the load. Looking at Figure 4.9, this property is associated to σ_y (elastic limit); the greater σ_y is the more elastic the material. Notice that, within the elastic zone, stress and strain are linearly related with the Young Modulus (E), i.e, $\sigma = E \cdot \epsilon$.

Stiffness: Stiffness is the resistance of an elastic body to deformation by an applied force. Looking at Figure 4.9, this property is associated to σ_y (elastic limit); the lower σ_y is the more stiff the material. Elasticity and Stiffness are antonyms.

Plasticity: A material is perfectly plastic if no strain disappears after the removal of load. Ductile materials are elastoplastic and behave in an elastic manner until the elastic limit is reached after which they behave plastically. When the stress is relieved the elastic component of the strain is recovered but the plastic strain remains as a permanent set.

Fatigue: Mechanical fatigue occurs due to the application of a very large number of relatively small cyclic forces (always below the breaking stress σ_B) which results in material failure. For instance, every single flight of an aircraft can be considered as a cycle. In this manner, the aircraft can regularly withstand the nominal loads (always below the breaking stress σ_B), but after a large amount of cycles some parts of the structure might fail due to mechanical fatigue. For these reasons, aircraft may be tested for three times its life-cycle. In order to be able to withstand such testing, many aircraft components may be made stronger than is strictly necessary to meet the static strength requirements. The parts that might suffer from mechanical fatigue are termed fatigue-critical.

Corrosion: Corrosion is the gradual destruction of materials (usually metals) by chemical reaction with its environment. Roughly speaking, it has to do with the oxidation of the material and thus the loss of some of its properties. Corrosion resistance is an important factor to consider during material selection. Methods to prevent corrosion include: painting, which however incorporates an important amount of weight; anodizing, in which the aircraft is treated with a stable protective oxide layer; cladding, which basically consists of adding a layer of pure aluminum to the surface material (essentially, to attach a less noble material to a more noble material); and finally cadmium plating, which consists of covering the surface material with a more noble material (assuming the structure is made of a less noble material). These ideas are based on having two different materials with very different properties in terms of oxidation, so that if one suffers corrosion, the other does not.

4.2.2 Materials in aircraft

The main groups of materials used in aircraft construction nowadays are steel, aluminum alloys, titanium alloys, and fibre-reinforced composites.

Titanium alloys

Titanium alloys possess high specific properties, have a good fatigue strength/tensile strength ratio with a high fatigue limit, and some retain considerable strength at temperatures up to 400–500°C. Generally, there is also a good resistance to corrosion and corrosion fatigue although properties are adversely affected by exposure to temperature and stress in a salty environment. The latter poses particular problems in the engines of carrier operated aircraft. Further disadvantages are a relatively high density so that weight penalties are imposed if the alloy is extensively used, coupled with high costs (of the material itself and due to its fabrication), approximately seven times those of aluminum and steel. Therefore, due its very particular characteristics (good fatigue strength/tensile strength at very high temperatures), titanium alloys are typically used in the most demanding elements of jet engines, e.g., the turbine blades.

Steels

Steels result of alloying Iron (Fe) with Carbon (C). Steels were the materials of the primary and secondary structural elements in the 30s. However, they were substituted by aluminum alloys as it will be described later on. Its high specific density prevents its widespread use in aircraft construction, but it has retained some value as a material for castings of small components demanding high tensile strengths, high stiffness, and high resistance to damage. Such components include landing gear pivot brackets, wing-root attachments, and fasteners.

Aluminum alloys

If one thinks in pure aluminum, the first thought is that it has virtually no structural application. It has a relatively low strength and it is extremely flexible. Nevertheless, when alloyed with other metals its mechanical properties are improved significantly, preserving its low specific weight (a key factor for the aviation industry). The typical alloying elements are copper, magnesium, manganese, silicon, zinc, and lithium. Aluminum alloys substituted steel as primary and secondary structural elements of the aircraft after World War II and thereafter. Four groups of aluminum alloy have been used in the aircraft industry for many years and still play a major role in aircraft construction: Al-Cu (2000 series); Al-Mg (5000 series); Al-Mg-Si (6000 series); Al-Zn-Mg (7000 series)[1]. The latest aluminum alloys to find general use in the aerospace industry are the aluminum–lithium (Al-Li, 8000 series) alloys.

Alloys from each of the above groups have been used extensively for airframes, skins, and other stressed components. Fundamentally, because all of them have a very low

---

[1] The following aluminum alloys are commonly used in aircraft and other aerospace structures: 7075 aluminum; 6061 aluminum; 6063 aluminum; 2024 aluminum; 5052 aluminum.

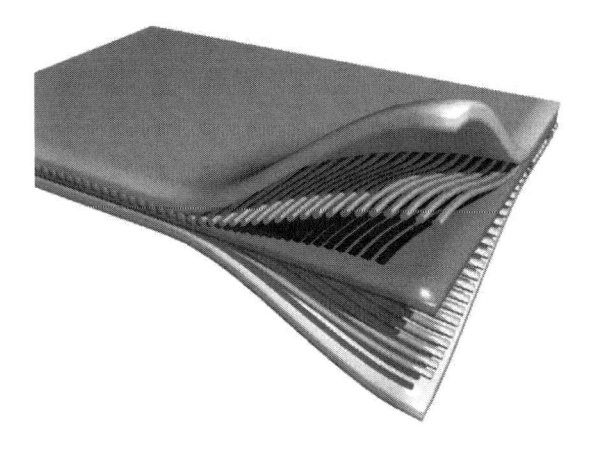

Figure 4.10: Sketch of a fibre-reinforced composite materials. © PerOX / Wikimedia Commons / Public Domain.

specific weight. Regarding the mechanical properties of the different alloys, the choice has been influenced by factors such as strength (proof and ultimate stress), ductility, easy of manufacture (e.g. in extrusion and forging), resistance to corrosion and suitability for protective treatments (e.g., anodizing), fatigue strength, freedom from liability to sudden cracking due to internal stresses, and resistance to fast crack propagation under load.

Unfortunately, as one particular property of aluminum alloys is improved, other desirable properties are sacrificed. Since the alloying mechanisms/process are complicated (basically micro-structural/chemical processes), finding the best trade-off is a challenging engineering problem. In the last 10 years, aluminum alloys are being systematically substituted by fibre-reinforced composite materials, first in the secondary structures, and very recently also in the primary structural elements (as it is the case of A350 or B787 Dreamliner).

Fibre-reinforced composite materials

Composite materials are materials made from two or more constituent materials with significantly different physical or chemical properties, that when combined produce a material with characteristics different from the individual components. In particular, the aircraft manufacturing industry uses the so-called fibre-reinforced composite materials, which consist of strong fibers such as glass or carbon set in a matrix of plastic or epoxy resin, which is mechanically and chemically protective.

A sheet of fibre-reinforced material is anisotropic, i.e. its properties depend on the direction of the fibers working at traction-compression. Therefore, in structural form two or more sheets are sandwiched together to form a lay-up so that the fibre directions match those of the major loads. This lay-up is embedded into a matrix of plastic or epoxy resin that fits things together and provides structural integrity to support both bending and shear stresses.

In the early stages of the development of fibre-reinforced composite materials, glass fibers were used in a matrix of epoxy resin. This glass-reinforced plastic (GRP) was used for helicopter blades but with limited use in components of fixed wing aircraft due to its low stiffness. In the 1960s, new fibre-reinforcements were introduced; Kevlar, for example, is an aramid material with the same strength as glass but is stiffer. Kevlar composites are tough but poor in compression and difficult to manufacture, so they were used in secondary structures. Another composite, using boron fibre, was the first to possess sufficient strength and stiffness for primary structures. These composites have now been replaced by carbon-fibre-reinforced plastics (CFRP), which have similar properties to boron composites but are very much inexpensive.

Typically, CFRP has a Young modulus of the order of three times that of GRP, one and a half times that of a Kevlar composite and twice that of aluminum alloy. Its strength is three times that of aluminum alloy, approximately the same as that of GRP, and slightly less than that of Kevlar composites. Nevertheless, CFRP does suffer from some disadvantages. It is a brittle material and therefore does not yield plastically in regions of high stress concentration. Its strength is reduced by impact damage which may not be visible and the epoxy resin matrices can absorb moisture over a long period which reduces its matrix-dependent properties, such as its compressive strength; this effect increases with increase of temperature. On the contrary, the stiffness of CFRP is much less affected than its strength by the absorption of moisture and it is less likely to fatigue damage than metals.

Replacing 40% of an aluminum alloy structure by CFRP results, roughly, in a 12% saving in total structural weight. Indeed, nowadays the use of composites has been extended up to 50% of the total weight of the aircraft, covering most of the secondary structures of the aircraft and also some primary structures. For instance, in the case of the Airbus A350XWB, the empennage and the wing are manufactured essentially based on CRPF. Also, some parts of the nose and the fuselage are manufactured on CRPF. The A350XWB material breakdown is as follows (in percentage of its structural weight) according to Airbus:

- 52% fiber-reinforced composites.

- 20% aluminum alloys.

- 14% titanium.

- 7% steel.

- 7% miscellaneous.

4.3 LOADS

The structure of a typical commercial aircraft is required to support two distinct classes of loads: the first, termed ground loads, include all loads encountered by the aircraft during movement or transportation on the ground such as taxiing, landing, or towing; while the second, air loads, comprise loads imposed on the structure during flight operations[2].

The two above mentioned classes of loads may be further divided into surface forces which act upon the surface of the structure, e.g., aerodynamic forces and hydrostatic pressure, and body forces which act over the volume of the structure and are produced by gravitational and inertial effects, e.g., force due to gravity. Calculation of the distribution of aerodynamic pressure over the various surfaces of an aircraft's wing was presented in Chapter 3.

Basically, all air loads are the different resultants of the corresponding pressure distributions over the surfaces of the skin produced during air operations. Generally, these resultants cause direct loads, bending, shear, and torsion in all parts of the structure.

4.3.1 FUSELAGE LOADS

The fuselage will experience a wide range of loads from a variety of sources.

The weight of both structure of the fuselage and payload will cause the fuselage to bend downwards from its support at the wing, putting the top in tension and the bottom in compression. In maneuvering flight, the loads on the fuselage will usually be greater than for steady flight. Also landing loads may be significant. The structure must be designed to withstand all loads cases in all circumstances, in particular in critical situations.

Most of the fuselage of typical commercial aircraft is usually pressurized (this also applies for other types of aircraft). The pressure inside the cabin corresponds, during the cruise phase, to that at an altitude of 2000–2500 [m] (when climbing/descending below/above that altitude, it is usually changed slowly to adapt it to terrain pressure). Internal pressure will generate large bending loads in fuselage frames. The structure in these areas must be reinforced to withstand these loads. Also, for safety, the designer must consider what would happen if the pressurization is lost. The damage due to depressurization depends on the rate of pressure loss. For very high rates, far higher loads would occur than during normal operation.

Doors and hatches are a major challenge when designing an aircraft. Depending on their design, doors will or will not carry some of the load of the fuselage structure. Windows, since they are very small, do not create a severe problem. On the floor of the fuselage also very high localized loads can occur, especially from small-heeled shoes. Therefore floors need a strong upper surface to withstand high local stresses.

---

[2]In Chapter 7 we will examine in detail the calculation of ground and air loads for a variety of cases.

4.3.2 Wing and tail loads

The lift produced by the wing creates a shear force and a bending moment, both of which are at their highest values at the root of the wing. Indeed, the root of the wing is one (if not the most) structurally demanding elements of the aircraft. The structure at this point needs to be very strong (high strenght) to resist the loads and moments, but also quite stiff to reduce wing bending. Thus, the wing is quite thick at the root.

Another important load supported by the wing is, in the case of wing-mounted engines, that of the power plant. Moreover, the jet fuel is typically located inside the wing. Therefore, an appropriate location of the power plant weight together with a correct distribution of the jet fuel (note that it is being consumed during the flight) contribute to compensate the lift forces during the flight, reducing the shear force and bending moment at the wing root. Fuel load close to the tips reduces this moment. Therefore the order in which the tanks are emptied is from the root to the tip. Nevertheless, when the aircraft is on the ground the lift is always lower than weight (when the aircraft is stopped, there is no lift), and all three forces, i.e., its structural weight, fuel, and power plant, can not be compensated by upwards lift. Therefore, the wing must also be design to withstand these loads which requires a design compromise.

The tailplane, rudder, and ailerons also create lift, causing a torsion in the fuselage. Since the fuselage is cylindrical, it can withstand torsion very effectively.

4.3.3 Landing gear loads

The main force caused by the landing gear is an upward shock during landing. Thus, shock-absorbers are present, absorbing the landing energy and thus reducing the force done on the structure. The extra work generated during a hard landing results in a very large increase in the force on the structure.

4.3.4 Other loads

Other loads include engine thrust on the wings or fuselage which acts in the plane of symmetry but may, in the case of engine failure, cause severe fuselage bending moments; concentrated shock loads during a catapult launch for fighters; and hydrodynamic pressure on the fuselages or floats of seaplanes.

4.4 Structural components of an aircraft

Aircraft are generally built up from the basic components of wings, fuselages, tail units and control surfaces, landing gear, and power plant.

4.4 STRUCTURAL COMPONENTS OF AN AIRCRAFT

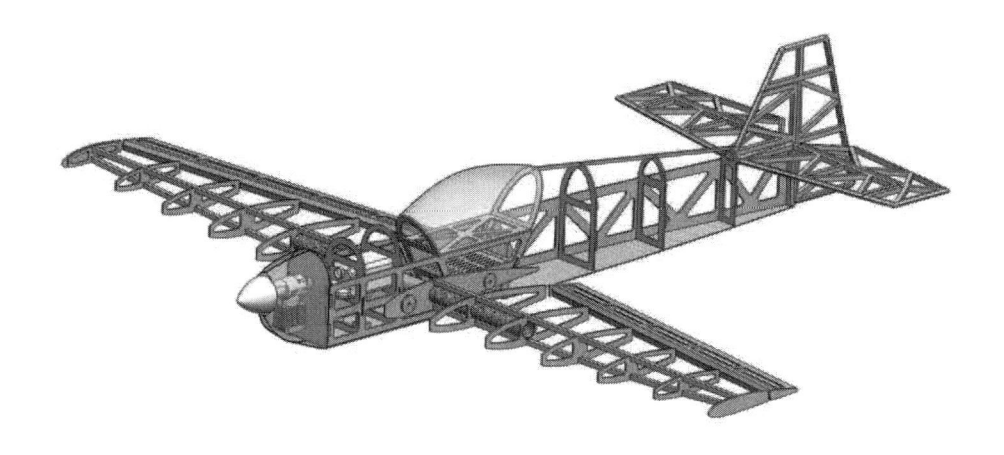

Figure 4.11: Aircraft monocoque skeleton.

The structure of an airplane is the set of those elements whose mission is to transmit and resist the applied loads; to provide an aerodynamic shape and to protect passengers, payload, systems, etc. from the environmental conditions encountered during the flight. These requirements, in most aircraft, result in thin shell structures where the outer surface or skin of the shell is usually supported by longitudinal stiffening elements and transverse frames to enable it to resist bending, compressive, and torsional loads without buckling. Such structures are known as semi-monocoque, while thin shells which rely entirely on their skins for their capacity to resist loads are referred to as monocoque.

4.4.1 STRUCTURAL ELEMENTS AND FUNCTIONS OF THE FUSELAGE

The fuselage should carry the payload, and is the main body to which all parts are connected. It must be able to resist bending moments (caused by weight and lift from the tail), torsional loads (caused by fin and rudder), and cabin pressurization. The structural strength and stiffness of the fuselage must be high enough to withstand these loads. At the same time, the structural weight must be kept to a minimum.

In transport aircraft, the majority of the fuselage is cylindrical or near-cylindrical, with tapered nose and tail sections. The semi-monocoque construction, which is virtually standard in all modern aircraft, consists of a stressed skin with added stringers to prevent buckling, attached to hoop-shaped frames. See Figure 4.12.

The fuselage has also elements perpendicular to the skin that support it and help keep its shape. These supports are called frames if they are open or ring-shaped, or bulkheads if they are closed.

Disturbances in the perfect cylindrical shell, such as doors and windows, are called cutouts. They are usually unsuitable to carry many of the loads that are present on the

Figure 4.12: Semimonocoque Airbus A340 rear fuselage, seen from inside. © Sovxx / Wikimedia Commons / CC-BY-SA-3.0.

surrounding structure. The direct load paths are interrupted and as a result the structure around the cut-out must be reinforced to maintain the required strength.

In aircraft with pressurized fuselages, the fuselage volume both above and below the floor is pressurized, so no pressurization loads exist on the floor. If the fuselage is suddenly de-pressurized, the floor will be loaded because of the pressure difference. The load will persist until the pressure in the plane has equalized, usually via floor-level side wall vents. Sometimes different parts of the fuselage have different radii. This is termed a double-bubble fuselage. Pressurization can lead to tension or compression of the floor-supports, depending on the design.

Frames give the fuselage its cross-sectional shape and prevent it from buckling when it is subjected to bending loads. Stringers give a large increase in the stiffness of the skin under torsion and bending loads, with minimal increase in weight. Frames and stringers make up the basic skeleton of the fuselage. Pressure bulkheads close the pressure cabin at both ends of the fuselage, and thus carry the loads imposed by pressurization. They may take the form of flat discs or curved bowls. Fatigue-critical areas are at the fuselage upper part and at the joints of the fuselage frames to the wing spars.

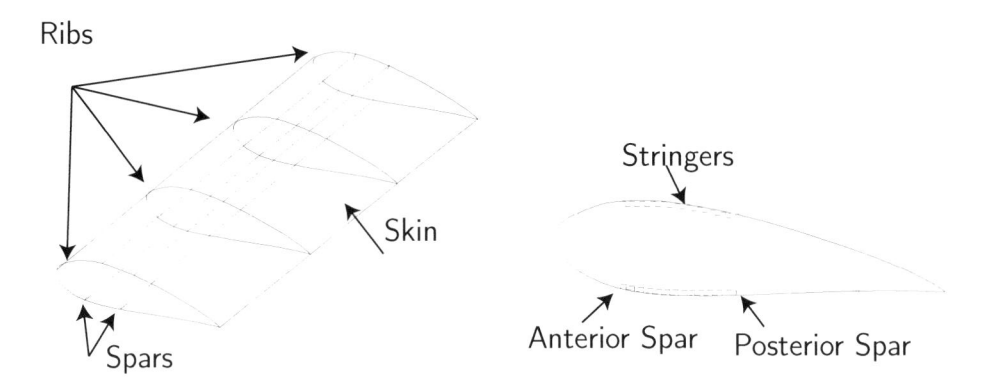

Figure 4.13: Structural wing sketch.

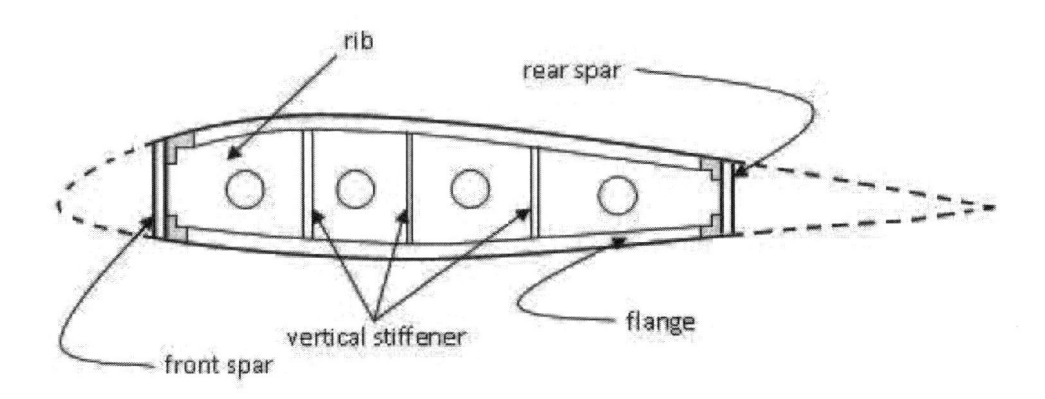

Figure 4.14: Structural wing torsion box. © User Eas4200c.f08.aero6.inman / Wikimedia Commons / Public Domain.

4.4.2 Structural elements and functions of the wing

Providing lift is the main function of the wing of an aircraft. A wing consists of two essential parts. The internal wing structure, consisting of spars, ribs, and stringers, and the external wing, which is the skin.

Ribs give the shape to the wing section, support the skin (prevent buckling), and act to prevent the fuel flowing around as the aircraft maneuvers. Its primary structural function is to withstand bending moments (the moment resultant of aerodynamic forces) and shear stresses (due to the vertical and horizontal resultant of forces). They serve as attachment points for the control surfaces, flaps, landing gear, and engines. They also separate the individual fuel tanks within the wing.

The wing stringers (also referred to as stiffeners) are thin strips of material (a beam) to which the skin of the wing is fastened. They run spanwise and are attached between the ribs. Their job is to stiffen the skin so that it does not buckle when subjected to compression loads caused by wing bending and twisting, and by loads from the aerodynamic effects of lift and control-surface movements.

The ribs also need to be supported, which is done by the spars. These are simple beams that usually have a cross-section similar to an I-beam. The spars are the most heavily loaded parts of an aircraft. They carry much more force at its root, than at the tip. Since wings will bend upwards, spars usually carry shear forces and bending moments.

Aerodynamic forces not only bend the wing, they also twist it. To prevent this, a second spar is introduced. Torsion now induces bending of the two spars. Modern commercial aircraft often use two-spar wings where the spars are joined by a strengthened section of skin, forming the so-called torsion-box structure. The skin in the torsion-box structure serves both as a spar-cap (to resist bending), as part of the torsion box (to resist torsion) and to transmit aerodynamic forces.

4.4.3 Tail

For the structural components of the stabilizers of the tail, fundamentally all exposed for the wing holds.

4.4.4 Landing gear

The landing gear (also referred to as undercarriage) of an aircraft supports the aircraft on the ground, provide smooth taxiing, and absorb shocks of taxiing and landings. It has no function during flight, so it must be as small and light as possible, and preferably easily retractable.

Due to the weight of the front (containing cabin and equipment) and rear parts (where the empennage is located) of the aircraft, large bending moments occur on the centre section of the fuselage. Therefore, to withstand these bending moments, a strong beam is located. This reduces the space in which the landing gear can be retracted.

When an aircraft lands, a large force is generated on the landing gear as it touches the ground. To prevent damage to the structure, this shock must be absorbed and dissipated as heat by the landing gear. If the energy is not dissipated, the spring system might just make the aircraft bounce up again.

After touchdown, the aircraft needs to brake. Disc brakes are primarily used. The braking of an aircraft can be supplemented by other forms of braking, such as air brakes, causing a large increase in drag, or reverse thrust, thrusting air forward.

References

[1] Cantor, B., Assender, H., and Grant, P. (2010). *Aerospace materials*. CRC Press.

[2] Franchini, S., López, O., Antoín, J., Bezdenejnykh, N., and Cuerva, A. (2011). *Apuntes de Tecnología Aeroespacial*. Escuela de Ingeniería Aeronáutica y del Espacio. Universidad Politécnica de Madrid.

[3] Megson, T. (2007). *Aircraft structures for engineering students*. A Butterworth–Heinemann Title.

AIRCRAFT INSTRUMENTS AND SYSTEMS

Contents

5.1		Aircraft instruments	122
	5.1.1	Sources of data	123
	5.1.2	Instruments requirements	126
	5.1.3	Instruments to be installed in an aircraft	126
	5.1.4	Instruments layout	130
	5.1.5	Aircrafts' cockpits	131
5.2		Aircraft systems	135
	5.2.1	Electrical system	135
	5.2.2	Fuel system	137
	5.2.3	Hydraulic system	139
	5.2.4	Flight control systems: Fly-By-Wire	140
	5.2.5	Air conditioning & pressurisation system	142
	5.2.6	Other systems	143
	References		145

In this chapter the goal is to give a brief overview of the different instruments, systems, and subsystems that one can find in a typical aircraft. First, in Section 5.1, the focus will be on the instruments. Notice that modern aircraft are becoming more and more sophisticated and classical instruments are being substituted by electronic displays. Aircraft systems will be briefly analyzed in Section 5.2. Again, many elements of classical mechanical (pneumatic, hydraulic) systems are being substituted by electronics. Therefore, in modern terminology, the discipline that encompasses instruments (as electronic displays) and electronic systems is referred to as avionics. Nonetheless, still many small aircraft use instruments and also some important aircraft systems are not based on electronics, e.g., fuel system, hydraulic system. An introductory reference is FRANCHINI et al. [1], by which this chapter is inspired. Thorough references on aircraft systems are MOIR and SEABRIDGE [4], KOSSIAKOFF et al. [2], TOOLEY and WYATT [5], and LANGTON et al. [3].

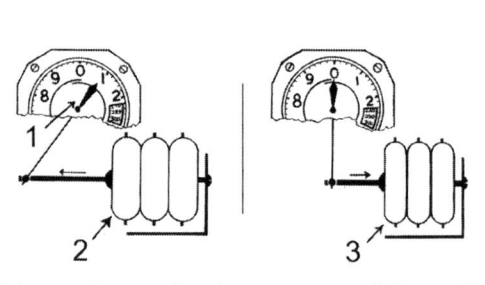

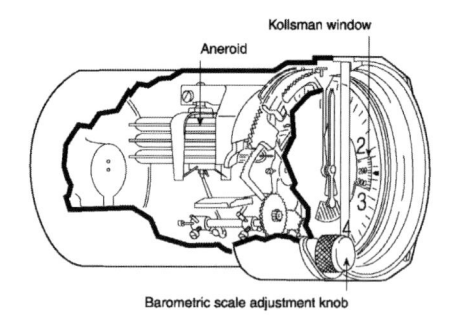

(a) Line art diagram of an altimeter in action.1) Pointer; 2) Aneroid cell expanded; 3) Aneroid cell contracted. Author: Pearson Scott Foresman. / Wikimedia Commons / Public Domain.

(b) Schematic barometric altimeter. Author: User:Dhaluza / Wikimedia Commons / Public Domain.

Figure 5.1: Barometric altimeter.

5.1 Aircraft instruments

Flight instruments are specifically referred to as those instruments located in the cockpit of an aircraft that provide the pilot with the information about the flight situation of the aircraft, such as position, speed, and attitude. The flight instruments are of particular use in conditions of poor visibility, such as in clouds, when such information is not available from visual reference outside the aircraft. The term is sometimes used loosely as a synonym for cockpit instruments as a whole, in which context it can include engine instruments, navigational instruments, and communication equipment.

Historically, the first instruments needed on board were the magnetic compass and a clock in order to calculate directions of flight and times of flight. To calculate the remaining fuel in the tanks, a glass pipe showing the level of fuel was presented on the cockpit. Before World War I, cockpits begin to present altimeters, anemometers, tachometers, etc. In the period between wars (1919-1939), the era of the pioneers, more and more sophisticated instruments were demanded to fulfill longer and longer trips: the directional gyro (heading indicator) and the artificial horizon (attitude indicator) appeared, and the panel of instruments started to have a standard layout.

Nowadays, in the era of electronics and information technologies, the cockpits present the information in on-board computers, using digital indicators and computerized elements of measure. Since instruments play a major role in controlling the aircraft and performing safe operations in compliance with air navigation requirements, it is necessary to present data in a clean and standard layout, so that the pilot can interpret them rapidly and clearly. The design of on board instruments requires knowing the physical variables one wants to measure, and the concepts and principles within each instrument.

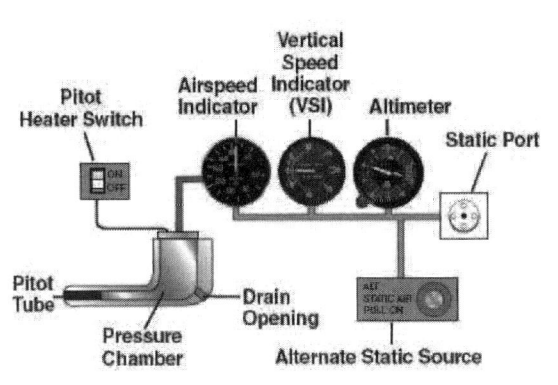

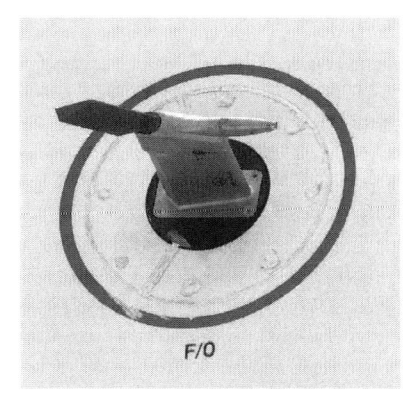

(a) Diagram of en:pitot-static system, including static port and pitot tube as well as the pitot static instruments. Author User:Giggy. / Wikimedia Commons / Public Domain.

(b) A380 Pitot tube. © David Monniaux / Wikimedia Commons / CC-BY-SA-3.0.

Figure 5.2: Pitot tube.

5.1.1 Sources of data

Different sources of information are needed for the navigation of an aircraft in the air.

Certain data come by measuring physical magnitudes of the air surrounding the aircraft, such as the pressure (barometric altimeter) or the velocity of air (pitot tube). Other data are obtained by measuring the accelerations of the aircraft using accelerometers. Also the angular changes (changes in attitude) and changes in the angular velocity can be measured using gyroscopes. The course of the aircraft is calculated through the measure of the direction of the magnetic field of the Earth.

A barometric altimeter is an instrument used to calculate the altitude based on pressure measurements. As already studied in Chapter 2, ISA relates pressure and altitude. An aneroid barometer measures the atmospheric pressure from a static port outside the aircraft. As already mentioned in Chapter 2, the aneroid altimeter can be calibrated in three manners (QNE, QNH, QFE) to show the pressure directly as an altitude above a reference (101225 [Pa] level, sea level, the airport, respectively). Figure 5.1 illustrates how a barometric altimeter works and how it looks like. Details on how the altimeter indicator works will be given later on when analyzing the altimeter.

A pitot tube is an instrument used to measure fluid flow velocity. A basic pitot tube consists of a tube pointing directly into the fluid flow, in which the fluid enters (at aircraft's airspeed). The fluid is brought to rest (stagnation). This pressure is the stagnation pressure of the fluid, which can be measured by an aneroid. The measured stagnation pressure cannot itself be used to determine the fluid velocity (airspeed in aviation). Using Bernoulli's equation (see Equation (3.7)), the velocity of the incoming flow (thus the airspeed of the aircraft, since the pitot tube is attached to the aircraft) can be calculated. Figure 5.2

(a) Diagram of en:pitot-static system, including static port and pitot tube as well as the pitot satic instruments. Author User:LucasVB. / Wikimedia Commons / Public Domain.

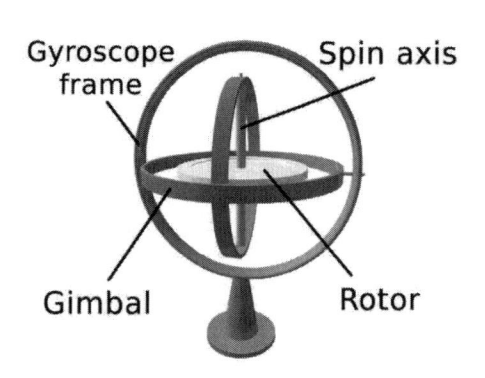

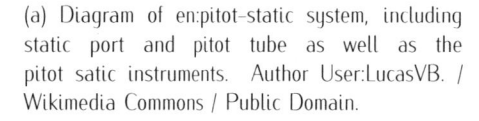

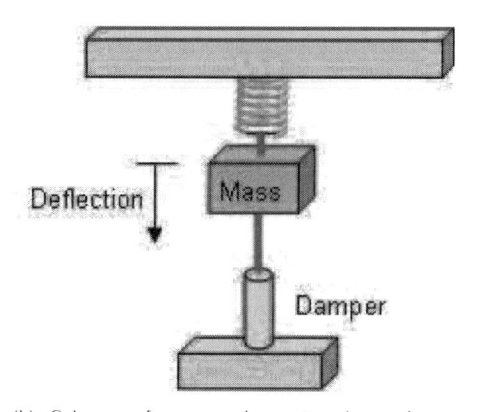

(b) Scheme of an accelerometer (mass-damper system). Author Rex07 / Wikimedia Commons / Public Domain.

Figure 5.3: Gyroscope and accelerometes.

illustrates how a pitot tube works and how it looks like. Details on how the airspeed indicator works will be given later on.

A gyroscope is a mechanical (also exist electronic) device based on the conservation of the kinetic momentum, i.e., a spinning cylinder with high inertia rotating at high angular velocity, so that the kinetic momentum is very high and it is not affected by external actions. Thus, the longitudinal axis of the cylinder points always in the same direction. Figure 5.3.a illustrates it. An accelerometer is a device that calculates accelerations based on displacement measurements. It is typically composed by a mass-damper system attached to a spring as illustrated in Figure 5.3.b. When the accelerometer experiences an acceleration, the mass is displaced. The displacement is then measured to give the acceleration (applying basic physics and the Second Newton Law). A typical accelerometer works in a single direction, so that a set of three is needed to cover the three directions of the space. The duple gyroscopes and accelerometer conforms the basis of an Inertial Measurement Unit (IMU), an element used for inertial navigation (to be studied in Chapter 10), i.e., three accelerometers measure the acceleration in the three directions and three gyroscopes measure the angular acceleration in the three axis; with an initial value of position and attitude and via integration, current position, velocity, attitude, and angular velocity can be calculated[1]. Figure 5.4 illustrates it.

The aircraft can also send electromagnetic waves to the exterior to know, for instance, the altitude with respect to the ground (radio-altimeter), or the presence of clouds in

---

[1]Notice that these calculations are complicated, since the values need to be projected in the adequate reference frames, and also the gravity, which is always accounted by the accelerometer in the vertical direction, needs to by considered. This is not covered in this course and will be studied in more advanced courses of navigation.

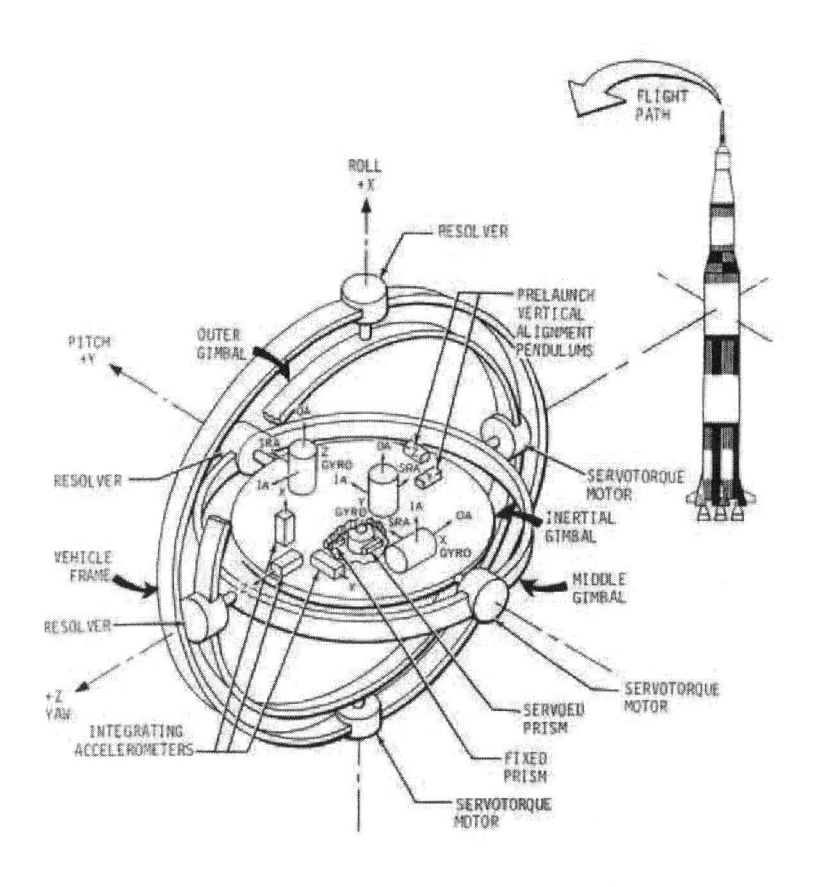

Figure 5.4: Diagram of ST-124 gimbals with accelerometers and gyroscopes (conforming the basic elements of a Inertial Measurement Unit). Author NASA/MSFC / Wikimedia Commons / Public Domain.

the intended trajectory (meteorologic radar). It can also receive electromagnetic waves from specific aeronautical radio-infrastructures, both for en-route navigation (VOR[2], NDB, etc.), and for approach and landing phases (ILS, MLS, etc). Also, the new systems of satellite navigation (GPS, GLONASS, and the future GALILEO) will be key in the future for more precise and reliable navigation. Aircraft have on-board instruments (the so-called navigation instruments) to receive, process, and present this information to the pilot.

---

[2]VHF Omnidirectional Radio range (VOR), Non-Directional Beacon (NDB); Instrumental Landing System (ILS), Microwaves Landing System (MLS); Global Position System (GPS), Global Navigation Satellite System (GLONASS). These systems will be studied in Chapter 10.

5.1.2 Instruments requirements

ICAO establishes the criteria (some are rules, other recommendations) to design, manufacture, and install the instruments. Some of these recommendations are:

- All instruments should be located in a way that can be read clearly and easily by the pilot (or the corresponding member of the crew).
- The illumination should be enough to be able to read without disturbance nor reflection at dark.
- The flight instruments, navigation instruments, and engines instruments to be used by the pilot must be located in front of his/her view.
- All flight instruments must be grouped together in the instrument panel.
- All engine instruments should be conveniently grouped to be readable by the appropriate member of the crew.
- The multiengine aircraft must have identical instruments for each engine, and be located in a way that avoids any possible confusion.
- The instruments should be installed so that are subject to minimal vibrations.

5.1.3 Instruments to be installed in an aircraft

The instruments to be installed in an aircraft are, on the one hand, flight and navigation instruments, and, on the other, instruments of the power plant[3].

Flight and navigation instruments

ICAO establishes that the minimum required flight and navigation instruments are:

Airspeed indicator: The airspeed indicator presents the aircraft's speed (usually in knots) relative to the surrounding air. It works by measuring the pressure (static and dynamic) in the aircraft's pitot tube. The indicated airspeed must be corrected for air density (using barometric and temperature data) in order to obtain the true airspeed, and for wind conditions in order to obtain the ground speed.

Attitude indicator (artificial horizon): The attitude indicator (also known as an artificial horizon) presents the aircraft's attitude relative to the horizon. This instrument provides information to the pilot on, for instance, whether the wings are leveled or whether the aircraft nose is pointing above or below the horizon.

---

[3]the content of this subsection has been partially based on Wikipedia flight instruments.

5.1 Aircraft instruments

(a) Airspeed indicator.

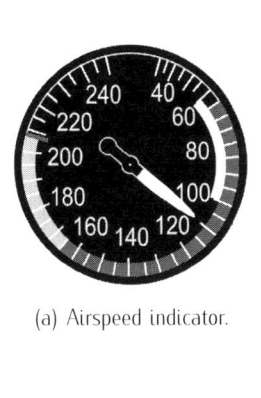

(b) Attitude indicator.

(c) Altimeter.

(d) Heading indicator.

Figure 5.5: Airspeed indicator: © Mysid / Wikimedia Commons / CC-BY-SA-3.0; Attitude indicator: © El Grafo / Wikimedia Commons / GNU-3.0; Altimeter: © Bsayusd / Wikimedia Commons / Public Domain; Heading indicator: © Oona Rłisłnen / Wikimedia Commons / CC-BY-SA-3.0.

Altimeter: The altimeter presents the altitude of the aircraft (in feet) above a certain reference (typically sea-level, destination airport, or 101325 isobar according to the three different barometric settings studied in Chapter 2) by measuring the difference between the pressure in aneroid capsules inside the barometric altimeter and the atmospheric pressure obtained through the static ports. The variations in volume of the aneroid capsule, which contains a gas, due to pressure differences are traduced into altitude by a transducer. If the aircraft ascends, the capsule expands as the static pressure drops causing the altimeter to indicate a higher altitude. The opposite occurs when descending.

Heading indicator (directional gyro): The heading indicator (also known as the directional gyro) displays the aircraft's heading with respect to the magnetic north. The principle of operation is based on a gyroscope.

Magnetic compass: The compass shows the aircraft's heading relative to magnetic north. It a very reliable instrument in steady level flight, but it does not work well when turning, climbing, descending, or accelerating due to the inclination of the Earth's magnetic field. The heading indicator is used instead (based on gyroscopes, more reliable instruments).

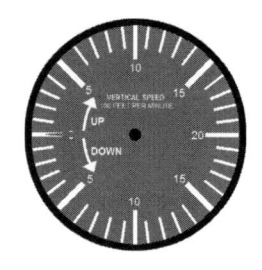

Aircraft instruments and systems

(a) Turn and slip.   (b) Variometer.   (c) Magnetic compass.

Figure 5.6: Turn and slip: Author User:Dhaluza / Wikimedia Commons / Public Domain; Variometer: © User:The High Fin Sperm Whale / Wikimedia Commons / CC-BY-SA-3.0; Magnetic Compass: © User:Chopper / Wikimedia Commons / CC-BY-SA-3.0.

Turn indicator (turn and slip): The turn indicator (also known as turn and slip) displays direction of turn and rate of turn. The direction of turn displays the rate that the aircraft's heading is changing. The internally mounted inclinometer (some short of balance indicator or *ball*) displays *quality* of turn, i.e. whether the turn is correctly coordinated, as opposed to an uncoordinated turn, wherein the aircraft would be in either a slip or a skid.

Vertical speed indicator (variometer): The vertical speed indicator (also referred to as variometer) displays the rate of climb or descent typically in feet per minute. This is done by sensing the change in air pressure.

Additionally, an indicator of exterior air temperature and a clock are also required.

Additional panel instruments: Obviously, most aircraft have more than the minimum required instruments. Additional panel instruments that may not be found in smaller aircraft are:

The **Course Deviation Indicator (CDI):** is an instrument used in aircraft navigation to determine an aircraft's lateral position in relation to a track, which can be provided, for instance, by a VOR or an ILS. This instrument can also be integrated with the heading indicator in a horizontal situation indicator.

A **Radio Magnetic Indicator (RMI):** is generally coupled to an Automatic Direction Finder (ADF), which provides bearing for a tuned NDB. While simple ADF displays may have only one needle, a typical RMI has two, coupling two different ADF receivers, allowing the pilot to determine the position by bearing interception.

ILS Instrumental Landing System: This system is nowadays fundamental for the phases of final approach and landing in instrumental conditions. The on-board ILS instrumental system indicates a path angle and an alignment with the axis of the runway, i.e., it assists pilots in vertical and lateral navigation.

(a) An aircraft Course Deviation Indicator (CDI). (b) An aircraft Radio Magnetic Indicator (RMI). ©
© User:Wessmann.clp / Wikimedia Commons / User:Wessmann.clp / Wikimedia Commons / GNU-3.0.
CC-BY-SA-3.0.

Figure 5.7: Navigation instruments: Course Deviation Indicator (CDI) and Radio Magnetic Indicator (RMI). Notice that the lecture of these instruments is not covered in this course. Nevertheless, in Chapter 10 an interpretation is provided.

Power plant instruments

ICAO also establishes a minimum required set of instruments for the power plant. We will just mention a few, not going into details:

- Tachometer for measuring the velocity of turn of the crankshaft (or the compressor in a jet).

- Indicator of the temperature of air entering the carburetor (just for piston aircraft).

- Indicator of the temperature of oil at the entrance and exit.

- Indicator of the temperature at the entrance of the turbine and the exit gases (just for jet aircraft).

- Indicator of fuel pressure and oil pressure.

- Indicator of tank level.

- Indicator of thrust (jets) and motor-torque (propellers).

Aircraft instruments and systems

Figure 5.8: Six basic instruments in a light twin-engine airplane arranged in a *basic-T*. From top left: airspeed indicator, attitude indicator, altimeter, turn coordinator, heading indicator, and vertical speed indicator. © User:Meggar / Wikimedia Commons / CC-BY-SA-3.0.

5.1.4 Instruments layout

Flight and navigation instruments layout

Most aircraft are equipped with a standard set of flight instruments which provide the pilot with information about the aircraft's attitude, airspeed, and altitude.

Most aircraft built since the 50s have four of the flight instruments located in a standardized pattern called the T-arrangement, which has become throughout the years a standard. The attitude indicator is at the top center, airspeed indicator to the left, altimeter to the right, and heading indicator below the attitude indicator. The other two, turn indicator and vertical speed indicator, are usually found below the airspeed indicator and altimeter, respectively, but for these two there is no common standard. The magnetic compass will be above the instrument panel. In newer aircraft with electronic displays substituting conventional instruments, the layout of the displays conform to the basic T-arrangement. The basic T-arrangement can be observed in Figure 5.5 and Figure 5.8.

Power plant instruments layout

This instruments layout is less standardized and we will not go into detail.

(a) Aircraft cockpit. Author User:Arpingstone / Wikimedia Commons / Public Domain.

(b) Airbus A380 cockpit. © User:Ssolbergj / Wikimedia Commons / CC-BY-SA-2.0.

Figure 5.9: Aircraft cockpit.

5.1.5 AIRCRAFTS' COCKPITS

The content of this section is inspired by WIKIPEDIA [6].

A cockpit or flight deck is the area, usually in the nose of an aircraft, from which the cabin crew (pilot and co-pilots) commands the aircraft. Except for some small aircraft, modern cockpits are physically separated from the cabin. The cockpit contains the flight instruments on an instrument panel, and the controls which enable the pilot to fly the aircraft, i.e., the control yoke (also known as a control column) that actuates on the elevator and ailerons[4], the pedals that actuates on the rudder, and the throttle level position to adjust thrust.

The layout of cockpits in modern airliners has become largely unified across the industry. The majority of the systems-related actuators (typically some short of switch), are usually located in the ceiling on an overhead panel. These are for instance, actuators for the electric system, fuel system, hydraulic system, and pressurization system. Radio communication systems are generally placed on a panel between the pilot's seats known as the pedestal. The instrument panel or instrument display is located in front of the pilots, so that all displays are visible. In modern electronic cockpits, the block displays usually regarded as essential are Mode Control Panel (MCP), Primary Flight Display (PFD), Navigation Display (ND), Engine Indicator and Crew Alerting System (EICAS), Flight Management System (FMS), and back-up instruments. Thus, these five elements (together with the back-ups) compose the instrument panel (containing all flight and navigation instruments as electronic displays) in a modern airliner. Notice that mechanical instruments have been substituted by electronic displays, and this is why this discipline is now referred to as avionics systems (the electronics on board the aircraft).

---

[4]An alternative to the yoke in most modern aircraft is the centre stick or side-stick (colloquially known as joystick).

Aircraft instruments and systems

(a) B-747 Mode Control Panel. Author User:Snowdog / Wikimedia Commons / Public Domain.

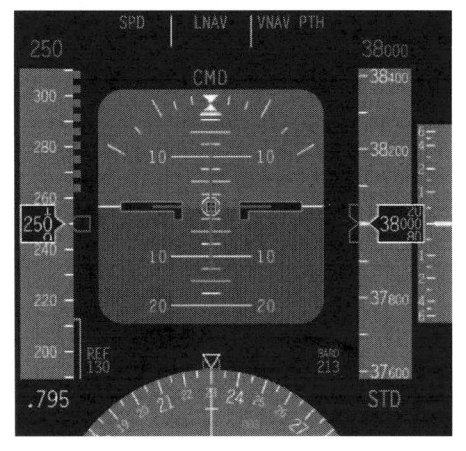

(b) Example of a typical PFD on an aircraft with glass cockpit. © User:Denelson83 / Wikimedia Commons / CC-BY-SA-3.0.

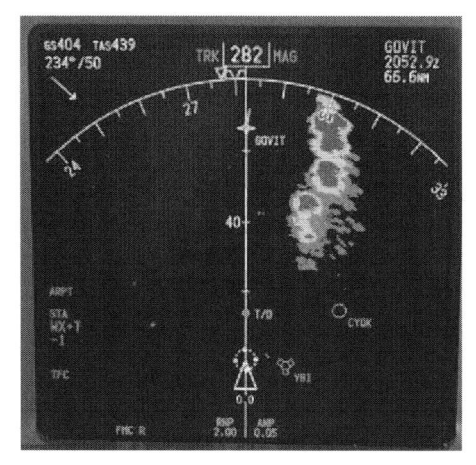

(c) B-737 Navigation Display with weather radar. © User:Shawn / Wikimedia Commons / CC-BY-SA-2.0.

Figure 5.10: Aircraft glass cockpit displays: MCP, PFD, and ND.

Mode Control Panel (MCP): A MCP is an instrument panel that permits cabin crew to control the autopilot and related systems. It is a long narrow panel located centrally in front of the pilot, just above the PFD and rest of displays. The panel covers a long but narrow area usually referred to as the *glareshield panel* as illustrated in Figure 5.10.a. The MPC contains the elements (mechanical or digital) that allow the cabin crew to select the autopilot mode, i.e., to specify the autopilot to hold a specific altitude, to change altitude at a specific rate, to maintain a specific heading, to turn to a new heading, to follow a route of waypoints, etc., and to engage or disengage the auto-throttle. Thus, it permits activating different levels of automation in flight (from fully automated to fully manual). Notice that MCP is a Boeing designation (that has been informally adopted as a generic name); the same unit with the same functionalities on an Airbus aircraft is referred to as the FCU (Flight Control Unit).

5.1 Aircraft instruments

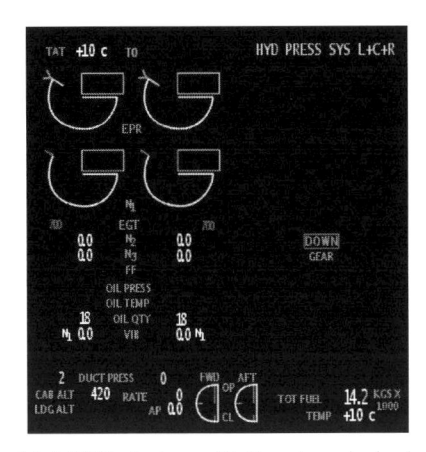

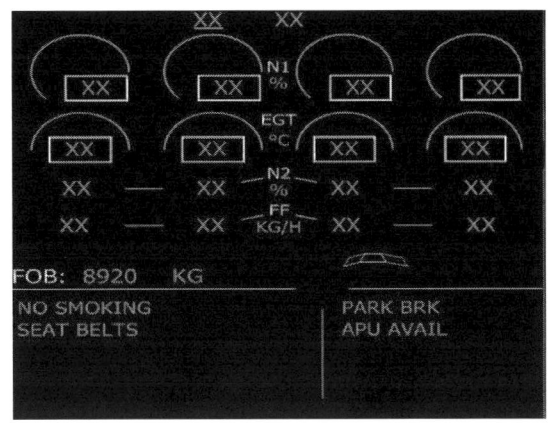

(a) EICAS display. © User:Anynobody / Wikimedia Commons / CC-BY-SA-3.0.

(b) Airbus A340-300 ECAM Display. Author: User:Trainler / Wikimedia Commons / Public Domain.

Figure 5.11: EICAS/ECAM cockpit displays.

Primary Flight Display (PFD): The PFD is a modern, electronic based aircraft instrument dedicated to flight information. It combines the older instruments arrangement (T-arrangement or T-arrangement plus turn and slip and variometer) into one compact display, simplifying pilot tasks. It is located in a prominent position, typically centered in the cockpit for direct view. It includes in most cases a digitized presentation of the attitude indicator (artificial horizon), air speed indicator, altitude indicator, and the vertical speed indicator (variometer). Also, it might include some form of heading indicator (directional gyro) and ILS/VOR deviation indicators (CDI). Figure 5.10.b illustrates it.

Navigation Display (ND): The ND is an electronic based aircraft instrument showing the route, information on the next waypoint, current wind speed and wind direction. It can also show meteorological data such as incoming storms, navaids[5] located on earth. This electronic display is sometimes referred to as MFD (multi-function display). Figure 5.10.c illustrates how it looks like.

Engine Indication and Crew Alerting System (EICAS) (used by Boeing) or Electronic Centralized Aircraft Monitor (ECAM) (by Airbus): The EICAS/ECAM displays information about the aircraft's systems, including its fuel, electrical, and propulsion systems (engines). It allows the cabin crew to monitor the following information: values for the different engines, fuel temperature, fuel flow, the electrical system, cockpit or cabin temperature and pressure, control surfaces and so on. The pilot may select display of

---

[5]navaids refers to navigational aids and will be studied in Chapter 10. It includes VORs, DMEs, ILS, NDB, etc.

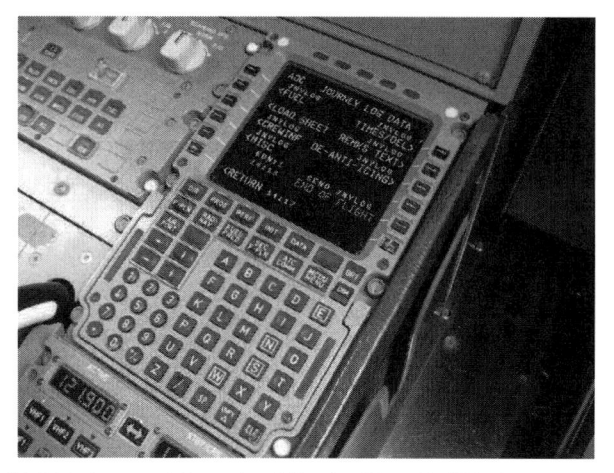

(a) FMS Control Display Unit of a B-737-300. © User:PresLoiLoi / Wikimedia Commons / CC-BY-SA-3.0.

(b) Multifunctional Control and Display Unit (MCDU) of an Airbus A320. Author: Christoph Paulus / Wikimedia Commons / Public Domain.

Figure 5.12: Flight Management System (FMS) Control Display Unit.

information by means of button press. The EICAS/ECAM display improves situational awareness by allowing the cabin crew to view complex information in a graphical format and also by alerting the crew to unusual or hazardous situations. For instance, for the EICAS display, if an engine begins to lose oil pressure, an alert sounds, the display switches to the page with the oil system information and outline the low oil pressure data with a red box.

Flight Management System (FMS): The FMS is a specialized computer system that automates a wide variety of in-flight tasks, reducing the workload on the flight crew. Its primary function is in-flight management of the flight plan[6], which is uploaded before departure and updated via data-link communications. Another function of the FMS is to guide the aircraft along the flight plan. This is done by measuring the current state (position, velocity, heading angle, etc.) of the aircraft, comparing them with the desired one, and finally setting a guidance law. From the cockpit, the FMS is normally controlled through a Control Display Unit (CDU) which incorporates a small screen and keyboard or touchscreen. The FMS sends the flight plan for display to the Navigation Display (ND) and other electronic displays in order them to present the following flight plan information: waypoints, altitudes, speeds, bearings, navaids, etc.

---

[6]the flight plan will be studied in Chapter 10.

5.2 Aircraft systems

In order an aircraft to fulfill its mission, e.g., to transport passengers from one city to another in a safe, comfortable manner, many systems and subsystems are needed. These systems and subsystems must be fully integrated since most of them are interdependent. In this section we present some of the main systems that can be found in an aircraft, e.g., electrical system, fuel system, hydraulic, flight control system, etc. More detailed information can be consulted, for instance, in MOIR and SEABRIDGE [4] and LANGTON et al. [3].

5.2.1 Electrical system

The electrical system is of great importance, since multitude of elements run with electric energy, among which we can cite: Indicator instruments, navigation and communication equipments, electro-actuators, electro-pneumatic mechanisms, illumination, passenger comfort (meals, entertainment, etc).

The electrical system is formed by the unities and basic components which generate, store, and distribute the electric energy to all systems that need it. Generally, in aircraft the primary source is Alternating Current (AC), and the secondary source in Direct Current (DC). The typical values of the AC are 115 V and 400 Hz, while the typical value of DC is 28 V. Due to safety reasons, the principal elements of the systems must be redundant (back-up systems), at least be double. Therefore, we can distinguish:

- Power generation elements (AC generation).
- Primary power distribution and protection (AC distribution).
- Power conversion and energy storage (AC to DC and storage).
- Secondary power distribution and protection elements (DC distribution).

There are different power generation sources for aircraft. They can be either for nominal conditions, for redundancy, or to handle emergency situations. These power sources include:

- Engine driven AC generators.
- Auxiliary Power Units (APU).
- External power, also referred to as Ground Power Unit (GPU).
- Ram Air Turbines (RAT).

The engine driven AC generators are the primary source of electrical energy. Each of the engines on an aircraft drives an AC generator. The produced power using the rotation of the turbine in nominal flight is used to supply the entire aircraft with electrical energy.

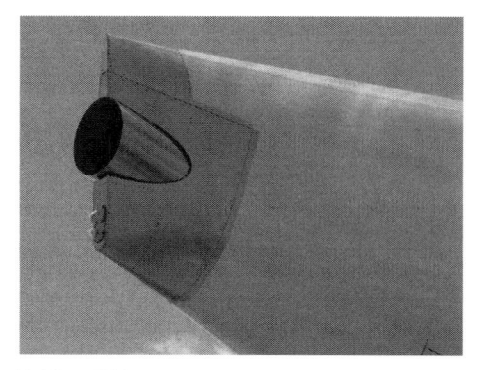

(a) Ground electrical power delivered to a Boeing 787. © Olivier Cleynen / Wikimedia Commons / CC-BY-SA-3.0.

(b) The APU exhaust at the tail end of an Airbus A380. © David Monniaux / Wikimedia Commons / CC-BY-SA-3.0.

Figure 5.13: Aircraft electrical generation sources: ground unit and APU.

When the aircraft is on the ground, the main generators do not work, but still electrical energy is mandatory for handling operation, maintenance actions, or engine starting. Therefore, it is necessary to extract the energy from other sources. Typically, the aircraft might use an external source such a GPU, or the so-called Auxiliary Power Unit (APU). The APU is a turbine engine situated in the rear part of the aircraft body which produces electrical energy. The APU is typically used on the ground as primary source of electrical energy, while on air is a back-up power source.

Some aircraft are equipped with Ram Air Turbines (RAT). The RAT is an air-driven turbine, normally stowed in the aircraft ventral or nose section. The RAT is used as an emergency back-up element, which is deployed when the conventional power generation elements are unavailable, i.e., in case of failure of the main generator or the APU when on air or ground, respectively.

Engine driven generators, GPU, APU, and RAT produce AC current. This AC current is distributed throughout the system to feed the elements of the aircraft that require electrical input, e.g., lighting, heating, communication systems, etc. However, it is also necessary to store energy in case of emergency. The energy must be stored in batteries working in DC. Therefore, the energy must be converted (AC to DC), for which one needs transformation units. The most frequently used method of power conversion in modern aircraft electrical system is the Transformer Rectifier Unit (TRU), which converts a three-phase 115V AC current into 28V DC current. Then the DC current stored in batteries can be distributed by means of a secondary DC distribution system, used also to feed certain elements of the aircraft that require electrical input. Notice that both the generation and the distribution need protection elements (for the case of AC current: under/over-voltage protection, under/over frequency protection, differential current protection, current phase protection) and control elements (in order to regulate voltage).

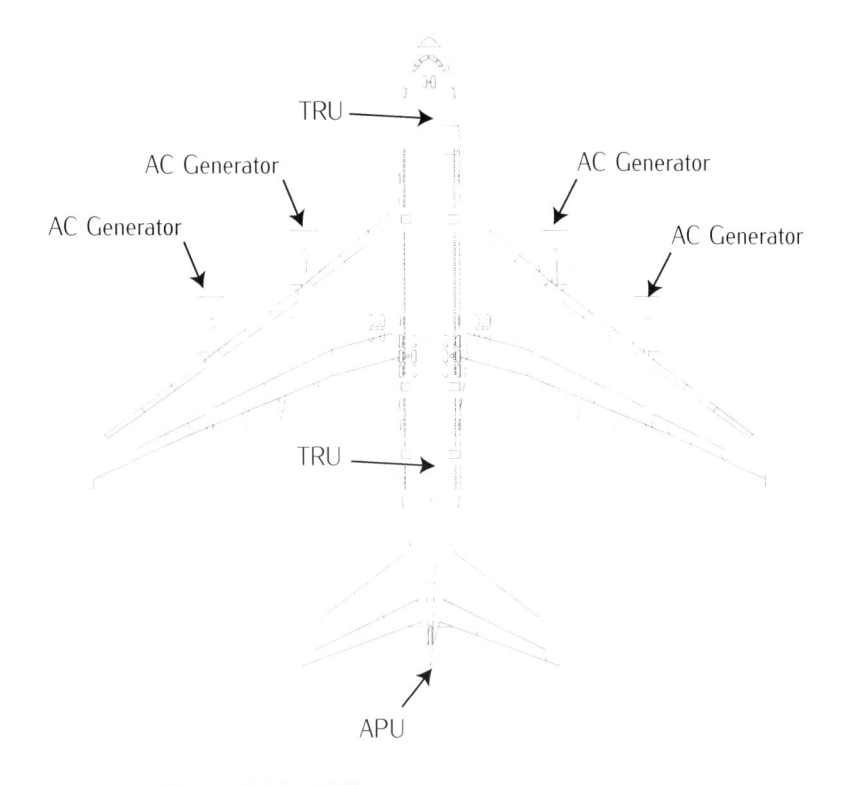

Figure 5.14: A380 power system components.

5.2.2 Fuel system

The main purpose of an aircraft fuel system is to provide a reliable supply of fuel to the power plant. Given that an aircraft with no fuel (or with no properly supplied fuel) can not fly (unless gliding), this system is key to ensure safe operations. The commonly used fuel is high octane index gasoline for piston aircraft, and some type of kerosene for jet aircraft. Even though fuel systems differ greatly due to the type of fuel and the type of mission, one can distinguish the following needs: refuel and defuel; storage; fuel pressurization; fuel transfer; engine feed; etc. Thus, the system is fundamentally composed by:

- tanks;
- fuel hydrants;
- feeding pumps;
- pipes and conducts;
- valves and filters;
- sensors, indicators, and control elements.

Aircraft instruments and systems

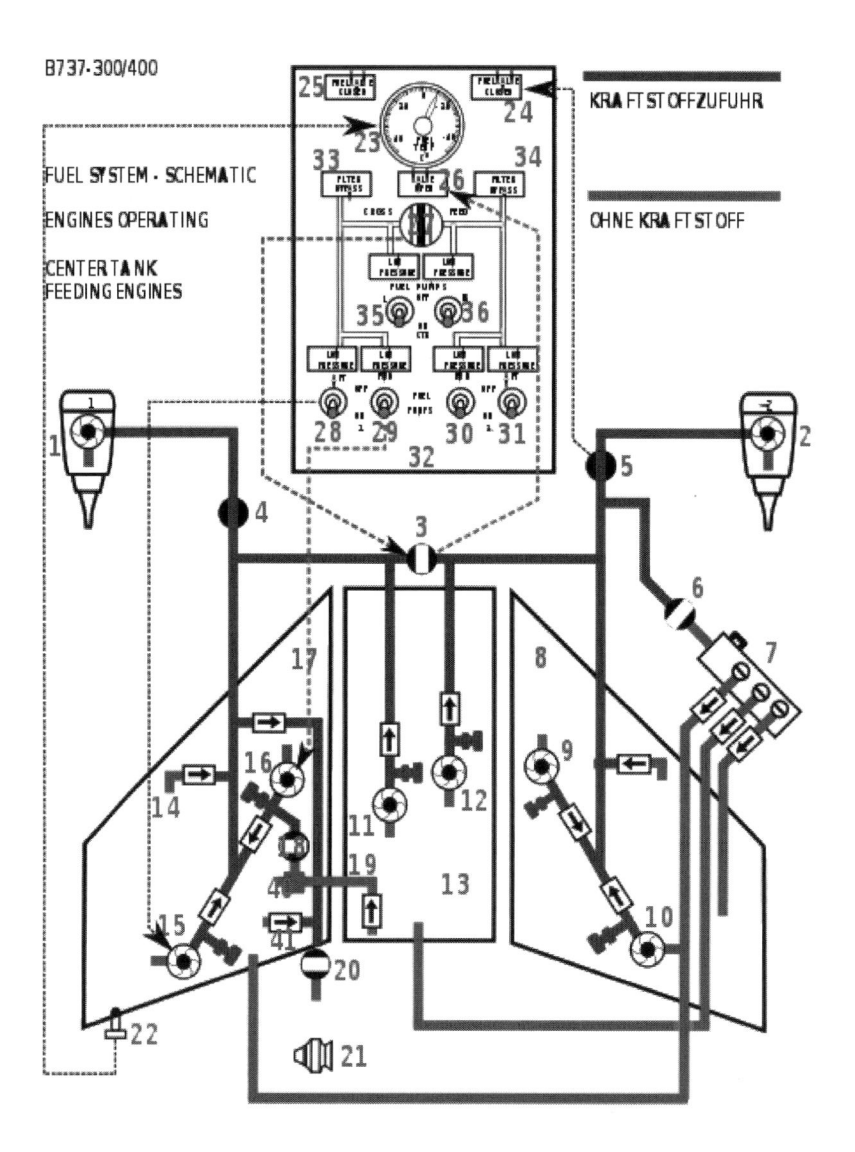

Figure 5.15: Diagrammatic representation of the Boeing 737-300 fuel system. © User:RosarioVanTulpe / Wikimedia Commons / Public Domain.

Tanks are used to storage fuel. Three main types can be distinguished: independent tanks; integrated tanks; interchangeable tanks. The independent tanks (concept similar to car tanks) are nowadays obsolete, just present in regional aircraft. The most extended in commercial aviation are integrated tanks, meaning that the tank is also part of the structure of (typically) the wing. The integral tanks are painted internally with a anti-corrosion substance and sealing all union and holes. The interchangeable tanks are those installed for determined missions.

The filling up and emptying process is centralized in a unique point, the fuel hydrant, which supplies fuel to all tanks thanks to feeding pumps which pump fuel throughout the pipes and conducts conforming the distribution network of the system. To be more precise, there are two fundamental types of pumps: the fuel transfer pumps, which perform the task of transferring fuel between the aircraft tanks, and the fuel booster pumps (also referred to as engine feed pumps), which are used to boost (preventing from flameouts and other inconveniences) the fuel flow form the fuel system to the engine.

The system is completed with valves, filters, sensors, indicators, and control elements. Valves can be simply transfer valves or non-return valves (to preserve the logic direction of fuel flow) or vent valves (to eliminate air during refueling). Filters are used to remove contaminants in the system. Last, different sensors are located within the system to measure different performance parameters (fuel quantity, fuel properties, fuel level, etc). The measurements are displayed in several indicators, some of them shown directly to the pilot, some others analyzed in a control unit. Both pilot and control unit (the later automatically) might actuate on the system to modify some of the performances. Notice that the subsystems that encompass the indicators, displays, and control unit might be also seen as part of an electronic or avionics system.

Figure 5.15 shows a diagrammatic representation of the Boeing 737-300 fuel system[7].

5.2.3 Hydraulic system

Hydraulic systems have been used since the early 30s and still nowadays play an important role in modern airliners. The basic function of the aircraft hydraulic system is to provide the required power to hydraulic consumers, such for instance: primary flight controls (ailerons, rudder, and elevator); secondary flight controls (flaps, slats, and spoilers); other systems, such landing gear system (extension and retraction, braking, steering, etc.), or door opening, etc.

The main advantages of hydraulic systems are:

- relative low weight in comparison with the required force to apply;
- simplicity in the installation;
- low maintenance;
- high efficiency with low losses, just due to liquid friction.

The main components of a hydraulic system are:

---

[7] 1 Engine Driven Fuel Pump - Left Engine; 2 Engine Driven Fuel Pump - Right Engine; 3 Crossfeed Valve; 4 Left Engine Fuel Shutoff Valve; 5 Right Engine Fuel Shutoff Valve; 6 Manual Defuling Valve; 7 Fueling Station; 8 Tank No. 2 (Right); 9 Forward Fuel Pump (Tank No. 2); 10 Aft Fuel Pump (Tank No. 2); 11 Left Fuel Pump (Center Tank); 12 Right Fuel Pump (Center Tank); 13 Center Tank; 14 Bypass Valve; 15 Aft Fuel Pump (Tank No. 1); 16 Forward Fuel Pump (Tank No. 1); 17 Tank No. 1 (Left); 18 Fuel Scavenge Shutoff Valve; 20 APU Fuel Shutoff Valve; 21 APU; 22 Fuel Temperature Sensor; 23-36 Indicators.

Aircraft instruments and systems

(a) Rear landing gear hydraulics.
© User:BrokenSphere /
Wikimedia Commons / CC-BY-SA-3.0.

(b) Hydraulic aileron actuator. © Hannes Grobe /
Wikimedia Commons / CC-BY-SA-3.0.

Figure 5.16: Aircraft hydraulic system: aileron actuator and landing gear actuator.

- a source of energy (any of the sources of the electrical system, i.e., engine driven alternator, APU, RAT);
- a reservoir or tanks to store the hydraulic fluid;
- a filter to maintain clean the hydraulic fluid;
- a mean of storing energy such as an accumulator (high density fluid tank);
- pipeline manifold (pipe or chamber branching into several openings);
- pumps (engine driven or electric), pipes, and valves;
- a mechanism for hydraulic oil cooling;
- pressure and temperature sensors;
- actuators (actuate mechanically on the device).

5.2.4 Flight control systems: Fly-By-Wire (Wikipedia [7])

Fly-by-wire Wikipedia [7] is a system that replaces the conventional manual flight controls of an aircraft with an electronic interface. The movements of flight controls are converted to electronic signals transmitted by wires (hence the fly-by-wire term), and flight control computers determine how to move the actuators at each control surface to provide the adequate response. The fly-by-wire system also allows automatic signals sent by the aircraft's computers to perform functions without the pilot's input, as in systems that automatically help stabilize the aircraft.

(a) Diagram showing the linkage between the control column and wheel and the various control surfaces of the Pfitzner Flyer (conventional control system). © User:Flight Magazine. / Wikimedia Commons / CC0 1.0.

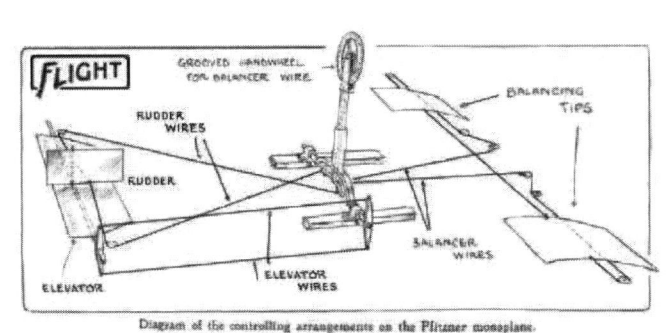

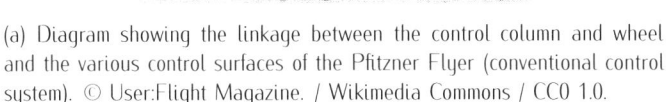

(b) Fly-by-wire joystick in exposition. © User:russavia / Wikimedia Commons / CC-BY-SA-2.0.

Figure 5.17: Flight control system: conventional and flight by wire.

Mechanical and hydro-mechanical flight control systems are relatively heavy and require careful routing of flight control cables through the aircraft by systems of pulleys, cranks, tension cables, and hydraulic pipes. Both systems often require redundant backups to deal with failures, which again increases weight. Furthermore, both have limited ability to compensate for changing aerodynamic conditions. Dangerous characteristics such as stalling, spinning, and pilot-induced oscillation, which depend mainly on the stability and structure of the aircraft concerned rather than the control system itself, can still occur with these systems.

The term fly-by-wire implies a purely electrically-signaled control system. However, it is used in the general sense of computer-configured controls, where a computer system is interposed between the operator and the final control actuators or surfaces. This modifies the manual inputs of the pilot in accordance with control parameters.

Command

Fly-by wire systems are quite complex; however their operation can be explained in relatively simple terms. When a pilot moves the control column (also referred to as sidestick or joystick), a signal is sent to a computer through multiple wires or channels (a *triplex* is when there are three channels). The computer receives the signals, which are then sent to the control surface actuator, resulting in surface motion. Potentiometers in the actuator send a signal back to the computer reporting the position of the actuator. When the actuator reaches the desired position, the two signals (incoming and outgoing) cancel each other out and the actuator stops moving.

Automatic Stability Systems

Fly-by-wire control systems allow aircraft computers to perform tasks without pilot input. Automatic stability systems operate in this way. Gyroscopes fitted with sensors are mounted in an aircraft to sense movement changes in the pitch, roll, and yaw axes. Any movement results in signals to the computer, which automatically moves control actuators to stabilize the aircraft to nominal conditions.

Digital Fly-By-Wire

A digital fly-by-wire flight control system is similar to its analog counterpart. However, the signal processing is done by digital computers and the pilot literally can "fly-via-computer". This also increases the flexibility of the flight control system, since the digital computers can receive input from any aircraft sensor, e.g., altimeters and pitot tube. This also increases the electronic stability, because the system is less dependent on the values of critical electrical components in an analog controller. The computers sense position and force inputs from pilot controls and aircraft sensors. They solve differential equations to determine the appropriate command signals that move the flight controls to execute the intentions of the pilot. The Airbus Industries Airbus A320 became the first airliner to fly with an all-digital fly-by-wire control system.

Main advantages

Summing up, the main advantages of fly-by-wire systems are:

- decrease in weight, which results in fuel savings;
- reduction in maintenance time (instead of adjusting the system, pieces are simply changed by new ones, so that maintenance is made more agile);
- better response to air gusts, which results in more comfort for passengers;
- automatic control of maneuvers (the systems avoid the pilot executing maneuvers with exceed of force in the controls).

5.2.5 AIR CONDITIONING & PRESSURISATION SYSTEM

The cabin air conditioning seeks keeping the temperature and humidity of the air in the cabin within certain range of values, avoidance ice and steam formation, the air currents and bad smells. The flight at high altitudes also force to pressurize the cabin, so that passengers can breath sufficient oxygen (remember that human being rarely can reach 8.000 m mountains in the Himalayan, only after proper natural conditioning not to suffer from altitude disease). That is why, cabins have an apparent atmosphere bellow 2500 m. Both systems can be built independently.

5.2.6 Other systems

There are many other systems in the aircraft, some are just in case of emergency. For the sake of brevity, we just mention some of them providing a brief description. Notice also that this taxonomy is not standard and thus the reader might encounter it in a different way in other textbooks.

Pneumatic system: Pursues the same function as hydraulic systems, actuating also in control surfaces, landing gears, doors, etc. The only difference is that the fluid is air.

Oxygen system: Emergency system in case the cabin is depressurized.

Ice and rain protection system: In certain atmospheric conditions, ice can be formed rapidly with influence in aerodynamic surfaces. There exist preventive systems which heat determined zones, and also corrective systems that meld ice once formed.

Fire protection system: Detection and extinction system in case of fire.

Information and communication system: Provides information and permits both internal communication with the passengers (musical wire), and external communication (radio, radar, etc). External communication includes VHF and HF communication equipment. Also, flight-deck audio systems might be included. Information system refers also to data-link communications, including all kind of data broadcasted from the ATM units, but also all aircraft performance data (speed, pressure, altitude, etc.) that can be recorded using Flight Data Recorder (FDR), Automatic Dependent Surveillance Broadcast (ADSB), and track-radar.

Air navigation system: It includes all equipment needed for safe navigation. All navigation instruments and electronic displays already described in Section 5.1 might be seen as part of this system. The Traffic Alert and Collision Avoidance System (TCAS) can be also included in this system.

Avionics system: This term is somehow confuse, since avionics refers to the electronics on board the aircraft. As it has been exposed throughout the chapter, electronics is becoming more and more important in modern aircraft. Practically every single system in an aircraft has electronic elements (digital signals, displays, controllers, etc.) to some extent. Therefore, is not clear whether avionics should be a system by itself, but is becoming more and more popular to use the term avionics systems to embrace all electronics on board the aircraft (and sometimes also the earth-based equipment that interrogates the aircraft). Characteristic elements of avionics are microelectronic devices (microcontrollers), data buses, fibre optic buses, etc. It is also important to pay attention to system design and integration since the discipline is transversal to all elements in the aircraft.

REFERENCES

[1] FRANCHINI, S., LÓPEZ, O., ANTOÍN, J., BEZDENEJNYKH, N., and CUERVA, A. (2011). *Apuntes de Tecnología Aeroespacial*. Escuela de Ingeniería Aeronáutica y del Espacio. Universidad Politécnica de Madrid.

[2] KOSSIAKOFF, A., SWEET, W. N., SEYMOUR, S., and BIEMER, S. M. (2011). *Systems engineering principles and practice*, volume 83. Wiley.com.

[3] LANGTON, R., CLARK, C., HEWITT, M., and RICHARDS, L. (2009). *Aircraft fuel systems*. Wiley Online Library.

[4] MOIR, I. and SEABRIDGE, A. (2008). *Aircraft systems: mechanical, electrical and avionics subsystems integration*, volume 21. Wiley. com.

[5] TOOLEY, M. and WYATT, D. (2007). Aircraft communication and navigation systems (principles, maintenance and operation).

[6] WIKIPEDIA (2013a). *Cockpit*. http://en.wikipedia.org/wiki/Cockpit. Last accesed 25 feb. 2013.

[7] WIKIPEDIA (2013b). *Fly-by-wire*. http://en.wikipedia.org/wiki/Fly-by-wire. Last accesed 25 feb. 2013.

AIRCRAFT PROPULSION

Contents

6.1	The propeller		148
	6.1.1	Propeller propulsion equations	148
6.2	The jet engine		150
	6.2.1	Some aspects about thermodynamics	151
	6.2.2	Inlet	153
	6.2.3	Compressor	154
	6.2.4	Combustion chamber	156
	6.2.5	Turbine	158
	6.2.6	Nozzles	160
6.3	Types of jet engines		162
	6.3.1	Turbojets	162
	6.3.2	Turbofans	164
	6.3.3	Turboprops	165
	6.3.4	After-burning turbojet	166
References			167

Aircraft require to thrust themselves to accelerate and thus counteract drag forces. In this chapter, we look at the way aircraft engines work. All aircraft propulsion systems are based on the principle of reaction of airflow through a power plant system. The two means for accelerating the airflow surrounding the aircraft that are presented in this chapter are through propellers and jet expansion, which give rise to the so-called propeller engines and jet engines to be studied in Section 6.1 and Section 6.2, respectively. In Section 6.3 the different types of jet engines will be studied. A third type of propulsion systems are the rocket engines, but they are used in spacecrafts and lay beyond the scope of this course. An introductory reference on the topic is NEWMAN [3, Chapter 6]. Thorough references are, for instance, MATTINGLY et al. [2] and JENKINSON et al. [1].

AIRCRAFT PROPULSION

The design of an aircraft engine must satisfy diverse needs. The first one is to provide sufficient thrust to counteract the aerodynamic drag of the aircraft, but also to exceed it in order to accelerate. Moreover, it must provide enough thrust to fulfill with the operational requirements in all circumstances (climbs, turns, etc.). Moreover, commercial aircraft focus also on high engine efficiency and low fuel consumption rates. On the contrary, fighter aircraft might require an important excess of thrust to perform sharp, aggressive maneuvers in combat.

6.1 THE PROPELLER

Typically, general aviation aircraft are powered by propellers and internal combustion piston engines (similar to those used in the automobile industry). The basic working principles are as follows: the air in the surroundings enters the engine, it is mixed with fuel and burned, thereby releasing a tremendous amount of energy in the mix (air and fuel) that is employed in increasing its energy (heat and molecular movement). This mix at high speed is exhausted to move a piston that is attached to a crankshaft, which in turn acts rotating a propeller.

The process of combustion in the engine provides very little thrust. Rather, the thrust is produced by the propeller due to aerodynamics. Propellers have various (two, three, or four) blades with an airfoil shape. The propeller acts as a rotating wing, creating a lift force due to its motion in the air. The aerodynamics of blades, i.e., the aerodynamics of helicopters, are slightly different than those studied in Chapter 3, and lay beyond the scope of this course. Nevertheless, the same principles apply: the engine rotates the propeller, causing a significant change in pressure across the propeller blades, and finally producing a net balance of forwards *lift* force.

6.1.1 PROPELLER PROPULSION EQUATIONS

A schematic of a propeller propulsion system is shown in Figure 6.1. As the reader would notice, this illustration has strong similarities with the continuity equation illustrated in Figure 3.3. The thrust force is generated due to the change in velocity as the air moves across the propeller between the inlet (0) and outlet (e). As studied in Chapter 3, the mass flow into the propulsion system (considered as a stream tube) is a constant.

The fundamentals of propelled aircraft flights are based on Newton's equations of motion and the conservation of energy and momentum:

Attending at conservation of momentum principle, the force or thrust is equal to the mass flow times the difference between the exit and inlet velocities, expressed as:

$$F = \dot{m} \cdot (u_e - u_0), \tag{6.1}$$

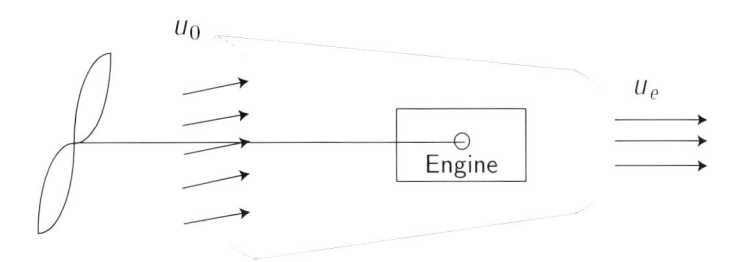

Figure 6.1: Propeller schematic.

where u_0 is the inlet velocity, u_e is the exit velocity, and $\dot{m}$ is the the mass flow. The exit velocity is higher than the inlet velocity because the air is accelerated within the propeller.

Attending at the conservation of energy principle for ideal systems, the output power of the propeller is equal to the kinetic energy flow across the propeller. This is expressed as follows:

$$P = \dot{m} \cdot \left(\frac{u_e^2}{2} - \frac{u_0^2}{2}\right) = \frac{\dot{m}}{2} \cdot (u_e - u_0) \cdot (u_e + u_0), \qquad (6.2)$$

where P denotes propeller power.

As real systems do not behave ideally, the propeller efficiency can be defined as:

$$\eta_{prop} = \frac{F \cdot u_0}{P}. \qquad (6.3)$$

where $F \cdot u_0$ is the useful work, and P refers to the input power, i.e., the power that goes into the engine. In other words, the efficiency η is a ratio between the real output power generated to move the aircraft and the input power demanded by the engine to generate it. In an ideal system $F \cdot u_0 = P$. In real systems $F \cdot u_0 < P$ due to, for instance, mechanical losses in transmissions, etc.

Operating with Equation (6.1) and Equation (6.2) and substituting in Equation (6.3) it yields:

$$\eta_{prop} = \frac{2 \cdot u_0}{u_e + u_0}. \qquad (6.4)$$

In order to obtain a high efficiency ($\eta_{prop} \sim 1$), one wants to have u_e as close as possible to u_0. However, looking at Equation (6.1), at very close values for input and input velocities, one would need a much larger mass flow to achieve a desired thrust. Therefore, there are certain limits on how efficient an aircraft engine can be and one would always need to find a compromise. Rewriting Equation (6.1) as:

$$\frac{F}{\dot{m}u_0} = \frac{u_e}{u_0} - 1, \qquad (6.5)$$

leads to a relation for propulsive efficiency. Notice that, if $u_e = u_0$, there is no thrust. For higher values of thrust, the efficiency drops dramatically.

Besides the propeller efficiency, other effects contribute to decrease the efficiency of the system. This is the case of the thermal effects in the engine. The thermal efficiency can be defined as:

$$\eta_t = \frac{P}{\dot{m}_f \cdot Q}, \qquad (6.6)$$

where P is power, $\dot{m}f$ is the mass flow of fuel, and Q is the characteristic heating value of the fuel.

Finally, the overall efficiency can be defined combining both as follows:

$$\eta_{overall} = \eta_t \cdot \eta_{prop} = \frac{F \cdot u_0}{\dot{m}_f \cdot Q}. \qquad (6.7)$$

6.2 THE JET ENGINE

Even though there are various types of jet engines (also referred to as gas turbine engines) as it is to be studied in Section 6.3, all of them share the same core elements, i.e., inlet, compressor, burner, turbine, and nozzle. Figure 6.2 illustrates schematically a jet engine with its core elements and the canonical engine station numbers, which are typically used to notate the airflow characteristics (T, p, ρ, etc.) through the different components. In this Figure, the station 0 represent the freestream air flow; 1 represents the entrance of the inlet; 2 and 3 represent the entrance and exit of the compressor, respectively; 4 and 5 represent the entrance and exit of the turbine, respectively; 6 and 7 represent the entrance and exit of the after-burner[1] (in case there is one, which is not generally the case), respectively; and finally 8 represents the exit of nozzle.

Roughly speaking, the inlet brings freestream air into the engine; the compressor increases its pressure; in the burner fuel is injected and combined with high-pressure air, and finally burned; the resulting high-temperature exhaust gas goes into the power turbine generating mechanical work to move the compressor and producing thrust when passed through a nozzle (due to action-reaction Newton's principle). Details of these engine core components are given in the sequel.

---

[1] Notice that in the figure there is not after-burner, but however the station numbers 6 and 7 have been added for the sake of generalizing.

6.2 THE JET ENGINE

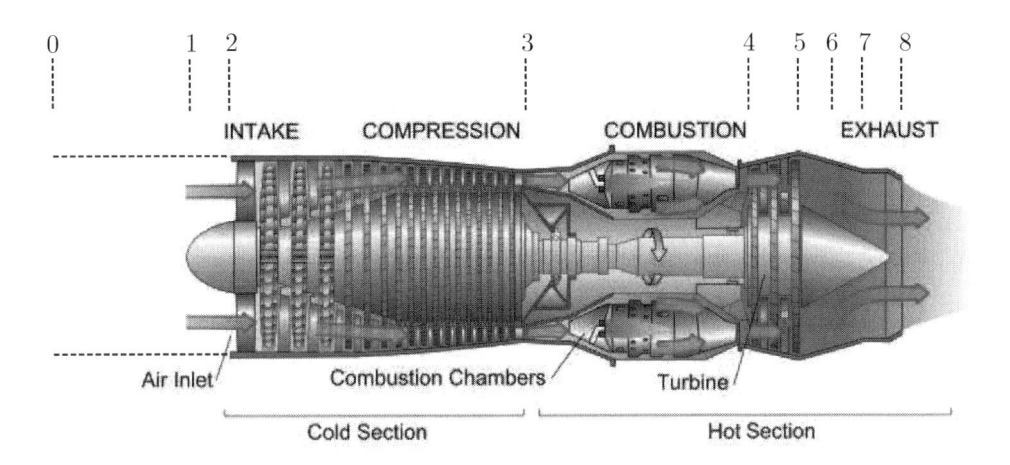

Figure 6.2: Jet engine: Core elements and station numbers. Adapted from: © Jeff Dahl / Wikimedia Commons / CC-BY-SA-3.0.

6.2.1 SOME ASPECTS ABOUT THERMODYNAMICS

Before analyzing the characteristics and equations of the elements of the jet engine, let us briefly explain some basic concepts regarding thermodynamics (useful to understand what follows in the section). For more insight, the reader is referred to any undergraduate text book on thermodynamics.

The first law of the thermodynamics can be stated as follows:

$$\Delta E = Q + W, \tag{6.8}$$

where E denotes de energy of the system, Q denotes de heat, and W denotes the work. In other words, an increase (decrease) in the energy of the system results in heat and work.

The energy of the system can be expressed as:

$$E = U + \frac{mV^2}{2} + mgz, \tag{6.9}$$

where U denotes the internal energy, the term $\frac{mV^2}{2}$ denotes the kinetic energy, and the term mgz denotes the potential energy (with z being the altitude). In the case of a jet engine, z can be considered nearly constant, and the potential term thus neglected. Also, it is typical to use stagnation values for pressure and temperature of the gas, i.e., the values that the gas would have considering $V = 0$ as already presented in Chapter 3. Under this assumption, the kinetic terms can be also neglected. Therefore, Equation 6.8

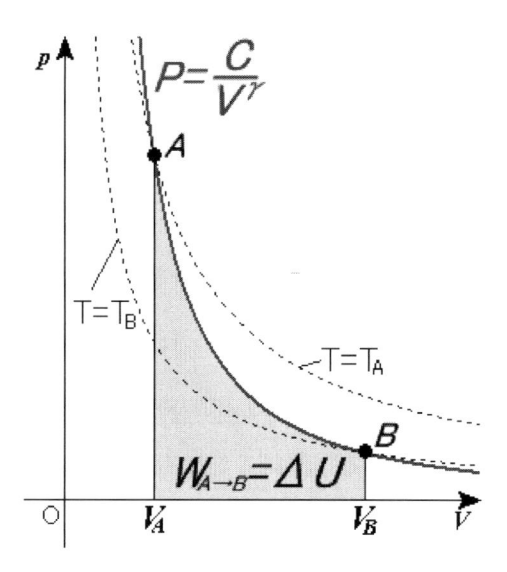

Figure 6.3: Sketch of an adiabatic process. © Yuta Aoki / Wikimedia Commons / CC-BY-SA-3.0.

can be expressed as:

$$\Delta U = Q + W. \tag{6.10}$$

Now, the work can be divided into two terms: mechanical work (W_{mech}) and work needed to expand/contract the gas ($\Delta(PV)$), i.e.,

$$W = W_{mech} + \Delta(PV). \tag{6.11}$$

Also, the enthalpy (h) of the system can be defined as: $h = U + PV$. In sum, the energy equation (Equation (6.8)) can be expressed as follows:

$$\Delta h = Q + W_{mech}. \tag{6.12}$$

An increase of enthalpy can be expressed as follows:

$$\Delta h = c \cdot \Delta T; \tag{6.13}$$

where c is the specific heat of the gas and T is the temperature.

Moreover, we state now how the stagnation values are related to the real values:

$$h_t = h + \frac{V^2}{2}; \tag{6.14}$$

$$p_t = p + \frac{1}{2} \cdot \rho \cdot V^2. \tag{6.15}$$

Notice that, if the process is adiabatic then $Q = 0$, and thus the increase (decrease) in enthalpy is all turned into mechanical work. Adiabatic processes will be assumed for the stages at the compressor and turbine. Moreover, for an adiabatic process there are some relations between pressure and temperature for an ideal gas, i.e.,

$$P \cdot V^\gamma = constant \rightarrow \frac{p_a}{p_b} = \left(\frac{T_a}{T_b}\right)^{\frac{\gamma}{\gamma-1}}, \tag{6.16}$$

being a and b state conditions within the adiabatic process and γ the ratio of specific heats[2].

6.2.2 Inlet

The free-stream air enters the jet engine at the inlet (also referred to as intake). There exist a variety of shapes and sizes dependent on the speed regime of the aircraft. For subsonic regimes, the inlet design in typically simple and short (e.g., for most commercial and cargo aircraft). The surface front is called the inlet lip, which is typically thick in subsonic aircraft. See Figure 6.4.a.

On the contrary, supersonic aircraft inlets have a relatively sharp lip as illustrated in Figure 6.4.b. This sharpened lip minimizes performance losses from shock waves due to supersonic regimes. In this case, the inlet must slow the flow down to subsonic speeds before the air reaches the compressor.

An inlet must operate efficiently under all flight conditions, either at very low or very high speeds. At low speeds the freestream air must be pulled into the engine by the compressor. At high speeds, it must allow the aircraft to properly maneuver without disrupting flow to the compressor.

Given that the inlet does no thermodynamic work, the total temperature through the inlet is maintained constant, i.e.:

$$\frac{T_{2t}}{T_{1t}} = \frac{T_{2t}}{T_0} = 1. \tag{6.17}$$

The total pressure through the inlet changes due to aerodynamic flow effects. The ratio of change is typically characterized by the inlet pressure recovery (IPR), which measures how much of the freestream flow conditions are recovered and can be expressed as follows:

$$IPR = \frac{p_{2t}}{p_{0t}} = \frac{p_{2t}}{p_0}; \quad M < 1.$$

---

[2] This ratio is also referred to as heat capacity ratio or adiabatic index.

Aircraft propulsion

(a) The inlet of a CF6-80C2B2 turbofan engine from an All Nippon Airways aircraft in Tokyo International Airport. © Noriko SHINAGAWA / Wikimedia Commons / CC-BY-SA-2.0.

(b) English Electric Lightning (XN776), a British supersonic jet fighter aircraft of the Cold War era, in the National Museum of Flight in East Fortune, East Lothian, Scotland. © Ad Meskens / Wikimedia Commons / CC-BY-SA-3.0.

Figure 6.4: Types of inlets

As pointed out before, the shape of the inlet, the speed of the aircraft, the airflow characteristics that the engine demands, and aircraft maneuvers are key factors to obtain a high pressure recovery, which is also related to the efficiency of the inlet expressed as:

$$\eta_i = \frac{p_{2t}}{p_{1t}} = \frac{p_{2t}}{p_0}.$$

6.2.3 Compressor

In the compressor, the pressure of the incoming air is increased by mechanical work. There are two fundamental types of compressors: axial and centrifugal. See Figure 6.5 as illustration of these two types.

In axial compressors the flow goes parallel to the rotation axis, i.e., parallel to the axial direction. In a centrifugal compressor the airflow goes perpendicular to the axis of rotation. The very first jet engines used centrifugal compressors, and they are still used on small turbojets. Modern turbojets and turbofans typically employ axial compressors. An axial compressor is composed by a duple rotor–stator (if the compressor is multistage,

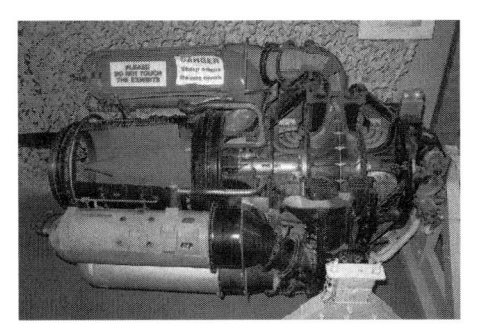

(a) A Rolls-Royce Welland jet engine cut away showing a centrifugal compressor. © user:geni / Wikimedia Commons / CC-BY-SA-3.0.

(b) A General Electric J85-GE-17A turbojet engine showing a multistage axial compressor. ©Sanjay Acharya / Wikimedia Commons / CC-BY-SA-3.0.

Figure 6.5: Types of jet compressors

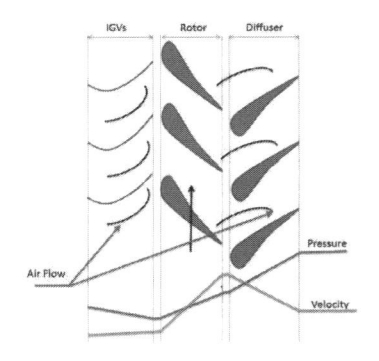

(a) Elements of an axial compressor: rotor and stator (difussor). © Sachin roongta / Wikimedia Commons / CC-BY-SA-3.0.

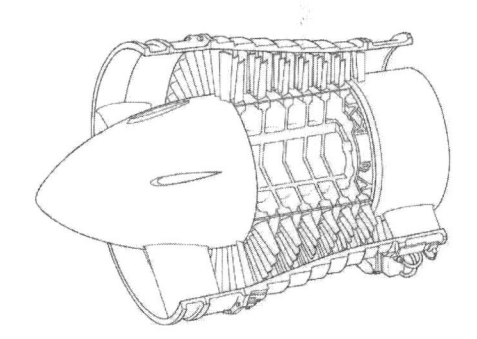

(b) Sketch of an axial compressor. Author User:Flanker / Wikimedia Commons / Public Domain.

Figure 6.6: Axial compressor.

then there will one duple per stage). In short, the rotor increases the absolute velocity of the fluid and the stator converts this into pressure increase as Figure 6.6 illustrates.

A typical, single-stage, centrifugal compressor increases the airflow pressure by a factor of 4. A similar single stage axial compressor will produce a pressure increase of between 15% and 60%, i.e., pressure ratios of 1.15-1.6 (small when compared to the centrifugal one). The fundamental advantage of axial compressor is that several stages can be easily linked together, giving rise to a multistage axial compressor, which can supply air with a pressure ratio of 40. It is much more difficult to produce an efficient multistage centrifugal compressor and therefore most high-compression jet engines incorporate multistage axial compressors. If only a moderate amount of compression is required, the best choice would be a centrifugal compressor.

Aircraft propulsion

Let us now focus on the equations that govern the evolution of the airflow over the compressor.

The pressure increase is quantified in terms of the so-called compressor pressure ratio (CPR), which is the ratio between exiting and entering air pressure. Using the station numbers of Figure 6.2, the CPR can be expressed as the stagnation pressure at stage 3 (p_{3t}) divided by the stagnation pressure at stage 2 (p_{2t}):

$$CPR = \frac{p_{3t}}{p_{2t}}.$$

The process can be considered adiabatic. Thus, according to the thermodynamic relation between pressure and temperature given in Equation (6.16), CPR can be also expressed as follows:

$$CPR = \frac{p_{3t}}{p_{2t}} = \left(\frac{T_{3t}}{T_{2t}}\right)^{\frac{\gamma}{(\gamma-1)}},$$

where γ is the ratio of specific heats ($\gamma \approx 1.4$ for air).

Referring the reader to Section 6.2.1 and doing some algebraic operations, the mechanical work consumed by the compressor can be expressed:

$$W_{comp} = \frac{cT_{2t}}{\eta_c}(CPR^{\frac{(\gamma-1)}{\gamma}} - 1), \qquad (6.18)$$

where c is the specific heat of the gas and η_c is the compressor efficiency. The efficiency factor is included to account for the real performance as opposed to the ideal one. Notice that the needed mechanical work is provided by the power turbine, which is connected to the compressor by a central shaft.

6.2.4 Combustion chamber

The combustion chamber (also referred to as burner or combustor) is where combustion occurs. Fuel is mixed with the high-pressure air coming out of the compressor, and combustion occurs. The resulting high-temperature exhaust gas is used to turn the power turbine, producing the mechanical work to move the compressor and eventually producing thrust after passing through the nozzle.

The burner is located between the compressor and the power turbine. The burner is arranged as some short of annulus so that the central engine shaft connecting turbine and compressor can be allocated in the hole. The three main types of combustors are annular; can; and hybrid can-annular.

Can combustors are self-contained cylindrical combustion chambers. Each *can* has its own fuel injector. Each *can* get an air source from individual opening. Like the can type

6.2 THE JET ENGINE

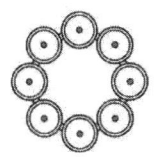

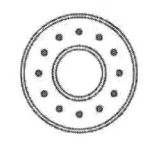

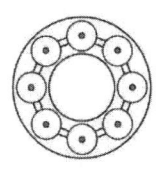

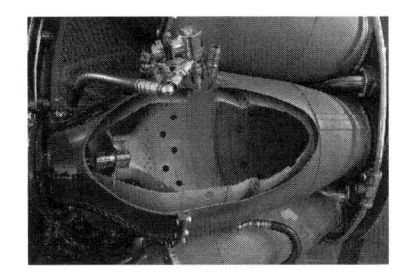

(a) Types of combustor: can (left); annular (center); can-annular (right). © User:Tosaka / Wikimedia Commons / CC-BY-SA-3.0.

(b) A sectioned combustor installed on a Rolls-Royce Nene turbojet engine. © Olivier Cleynen / Wikimedia Commons / CC-BY-SA-3.0.

Figure 6.7: Combustion chamber or combustor.

combustor, can-annular combustors have discrete combustion zones contained in separate liners with their own fuel injectors. Unlike the can combustor, all the combustion zones share a common air casing. Annular combustors do not use separate combustion zones and simply have a continuous liner and casing in a ring (the annulus).

Many modern burners incorporate annular designs, whereas the can design is older, but offers the flexibility of modular cans. The advantages of the can-annular burner design are that the individual cans are more easily designed and tested, and the casing is annular. All three designs are found in modern gas turbines.

The details of mixing and burning the fuel are very complicated and therefore the equations that govern the combustion process will not be studied in this course. For the purposes of this course, the combustion chamber can be considered as the place where the air temperature is increased with a slight decrease in pressure. The pressure in the combustor can be cosidered nearly constant during burning. Using the station numbers from Figure 6.2, the combustor pressure ratio (CPR) is equal to the stagnation pressure at stage 4 (p_{4t}) divided by the stagnation pressure at stage 3 (p_{3t}), i.e.:

$$BPR = \frac{p_{4t}}{p_{3t}} \sim 1.$$

The thermodynamics in the combustion chamber are different from those of the compressor and turbine because in the combustion chamber heat is released during the combustion process. In the compressor and turbine, the processes are adiabatic (there is no heat involved): pressure and temperature are related, and the temperature change is determined by the energy equation.

In the case of the combustion chamber, the process is not adiabatic anymore. Fuel is added in the chamber. The added mass of the fuel can be accounted by using a ratio f of fuel flow to air mass flow, which can be quantified as:

$$f = \frac{\dot{m}_f}{\dot{m}} = \frac{\frac{T_{4t}}{T_{3T}} - 1}{\frac{\eta_b Q}{c T_{3t}} - \frac{T_{4t}}{T_{3T}}}, \qquad (6.19)$$

where $\dot{m}_f$ denotes the mass flow of fuel, Q is the heating constant (which depends on the fuel type), c represents the average specific heat, T_{t3} is the stagnation temperature at the combustor entrance, T_{4t} is the stagnation temperature at the combustor exit, and η_b is the combustor efficiency. This ratio is very important for determining overall aircraft performance because it provides a measure of the amount of fuel needed to burn a determined amount of air flow (at the conditions of pressure and temperature downstream the compressor) and subsequently generate the corresponding thrust.

6.2.5 Turbine

The turbine is located downstream the combustor and transforms the energy from the hot flow into mechanical work to move the compressor (remember that turbine and compressor are linked by a shaft). The turbine is composed of two rows of small blades, one that rotates at very high speeds (the rotor) and the other that remains stationary (the stator). The blades experience flow temperatures of around 1400°K and must, therefore, be either made of special metals (typically titanium alloys) that can withstand the heat.

Depending on the engine type, there may be multiple turbine stages present in the engine. Turbofan and turboprop engines usually employ a separate turbine and shaft to power the fan and gearbox, respectively, and are referred to as two-spool engines. Three-spool configurations exist for some high-performance engines where an additional turbine and shaft power separate parts of the compressor.

The derivation of equations that govern the evolution of the air flow over the turbine are similar to those already exposed for the compressor. As the flow passes through the turbine, pressure and temperature decrease. The decrease in pressure through the turbine is quantified with the so-called turbine pressure ratio (TPR), i.e., the ratio of the exiting to the entering air pressure in the turbine. Using the station numbers of Figure 6.2, the TPR is equal to the stagnation pressure at point 5 (p_{5t}) divided by the stagnation pressure at point 4 (p_{4t}), i.e.,

$$TPR = \frac{p_{5t}}{p_{4t}}.$$

Given that the process can be considered adiabatic, pressure and temperature are related

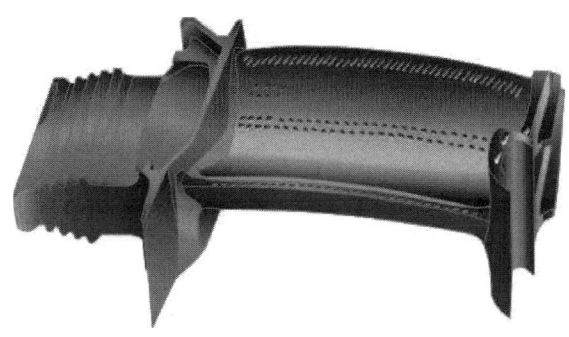

(a) The turbine rotor of a GE J79 turbojet engine. © User:Stahlkocher / Wikimedia Commons / CC-BY-SA-3.0.

(b) A schematic of a high-pressure turbine blade as used in aircraft jet engines. © User:Tomeasy / Wikimedia Commons / CC-BY-SA-3.0.

Figure 6.8: Turbine and schematic blade.

as in Equation (6.16) so that:

$$TPR = \frac{p_{5t}}{p_{4t}} = \left(\frac{T_{5t}}{T_{4t}}\right)^{\frac{\gamma}{\gamma-1}}. \qquad (6.20)$$

Again, teferring the reader to Section 6.2.1 and doing some algebraic operations, the turbine mechanical work W_{turb} can be expressed as follows:

$$W_{turb} = (\eta_t c T_{4t})(1 - TPR^{\frac{(\gamma-1)}{\gamma}}), \qquad (6.21)$$

where c is the specific heat of the gas and η_t is the turbine efficiency.

Compressor and turbine stages work attached one to the other. This relationship can be expressed by setting the work done by the compressor equal to the work done by the turbine, i.e., $W_{turb} = W_{comp}$. Hence, the conservation of energy is ensured. Equating Equation (6.18) and Equation (6.21) yields:

$$\frac{cT_{2t}}{\eta_c}(CPR^{\frac{(\gamma-1)}{\gamma}} - 1) = (\eta_t c T_{4t})(1 - TPR^{\frac{(\gamma-1)}{\gamma}}). \qquad (6.22)$$

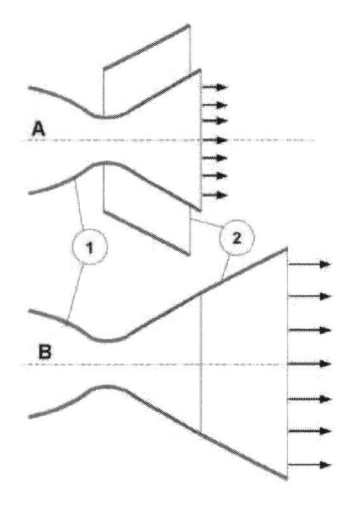

(a) Scheme of a variable extension nozzle. User:IOK / Wikimedia Commons / Public Domain.

(b) variable extension nozzle. © User:Feelfree / Wikimedia Commons / CC-BY-SA-3.0.

Figure 6.9: Variable extension nozzle.

6.2.6 Nozzles

The final stage of the jet engine is the nozzle. The nozzle has three functions, namely: a) to generate thrust; b) to conduct the exhaust gases back to the freestream conditions; and c) to establish the mass flow rate through the engine by setting the exhaust area. The nozzle lays downstream the turbine[3].

There are different shapes and sizes depending on the type of aircraft performance. Simple turbojets and turboprops typically have fixed-geometry convergent nozzles. Turbofan engines sometimes employ a coannular nozzle where the core flow exits the center nozzle while the fan flow exits the annular nozzle. After-burning turbojets and some turbofans often incorporate variable-geometry convergent-divergent nozzles (also referred to as de Laval nozzles), where the flow is first compress to flow through the convergent throat, and then is expanded (typically to supersonic velocities) through the divergent section.

Let us now move on analyzing in brief the equations governing the evolution of the flow in the nozzle. The nozzle exerts no work on the flow, and thus both the stagnation temperature and the stagnation pressure can be considered constant. Recalling the station

---

[3]Notice that in this description of the core elements of a jet engine the after-burner has been omitted. It there is one (fundamentally, for supersonic aircraft), it would located downstream the turbine and upstream the nozzle.

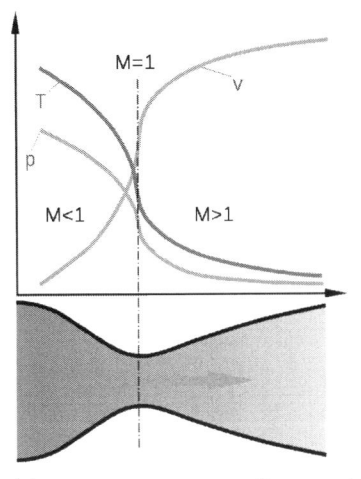

(a) Convergent-divergent nozzle diagram. User:IOK / Wikimedia Commons / Public Domain.

(b) Convergent-divergent rocket nozzle. © User:Jaypee / Wikimedia Commons / CC-BY-SA-3.0.

Figure 6.10: Convergent–divergent nozzle. In the left-hand side, the figure shows approximate flow velocity (v), together with the effect on temperature (T) and pressure (p).

numbers from Figure 6.2, we write:

$$\frac{p_{8t}}{p_{5t}} = \left(\frac{T_{8t}}{T_{5t}}\right)^{\frac{\gamma}{(\gamma-1)}} = 1,$$

where 5 corresponds to the turbine exit and 8 to the nozzle throat.

The stagnation pressure at the exit of the nozzle is equal to the freestream static pressure, unless the exiting flow is expanded to supersonic conditions (a convergent–divergent nozzle). The nozzle pressure ratio (NPR) is defined as:

$$NPR = \frac{p_{8t}}{p_8} = \frac{p_0}{p_8}, \tag{6.23}$$

where p_{8t} is the stagnation nozzle pressure or the free-stream static pressure. In order to determine the total pressure at the nozzle throat p_8, a term referred to as overall engine pressure ratio (EPR) is used. The EPR is defined to be the total pressure ratio across the engine, and can be expressed as follows:

$$EPR = \frac{p_{8t}}{p_{2t}} = \frac{p_{3t}}{p_{2t}} \frac{p_{4t}}{p_{3t}} \frac{p_{5t}}{p_{4t}} \frac{p_{8t}}{p_{5t}}, \tag{6.24}$$

where the compressor, combustor, turbine, and nozzle stages are all represented.

Similarly, the Engine Temperature Ration (ETR) can be expressed as:

$$ETR = \frac{T_{8t}}{T_{2t}} = \frac{T_{3t}}{T_{2t}} \frac{T_{4t}}{T_{3t}} \frac{T_{5t}}{T_{4t}} \frac{T_{8t}}{T_{5t}}, \tag{6.25}$$

from which the nozzle stagnation temperature (T_{8t}) can be calculated.

Considering Equation (6.14), isolating the exit velocity and doing some algebra, it yields:

$$u_e = u_8 = \sqrt{2c\eta_n T_{8t}\left[1 - (\frac{1}{NPR})^{\frac{\gamma-1}{\gamma}}\right]}, \tag{6.26}$$

where η_n is the nozzle efficiency, which is normally very close to 1.

The nozzle performance equations work just as well for rocket engines except that rocket nozzles always expand the flow to some supersonic exit velocity.

Summing up, all the necessary relations between jet engine components have been stated in order to obtain the thrust developed by the jet engine. Notice that, as already pointed out in Equation (6.1), the thrust would be:

$$Thrust = \dot{m} \cdot (u_e - u_0). \tag{6.27}$$

6.3 Types of jet engines

Some of the most important types of jet engines will be now discussed. Specifically, turbojets, turbofans, turboprops, and after-burning turbojets. As a first touch, Figure 6.11 illustrate a sketch of the relative suitability of some of these types of jets. It can be observed that turboprop are more efficient in low subsonic regimes; turbofans are more efficient in high subsonic regimes; and turbojets (also after-burning turbojets) are more efficient for supersonic regimes. If one looks at higher mach numbers (M > 3-4), ramjet, scramjets or rockets will be needed. However these last types are beyond the scope of this course.

6.3.1 Turbojets

A turbojet is basically what has been already exposed in Section 6.2. It is composed by an inlet, a compressor, a combustion chamber, a turbine, and a nozzle. The reader is referred back to Figure 6.2 as illustration. As already mentioned, there are two main types of turbojets depending on the type of compressor: axial or centrifugal. Figures 6.12-6.13 show schematic and real jet engines with centrifugal and axial flow, respectively.

6.3 Types of jet engines

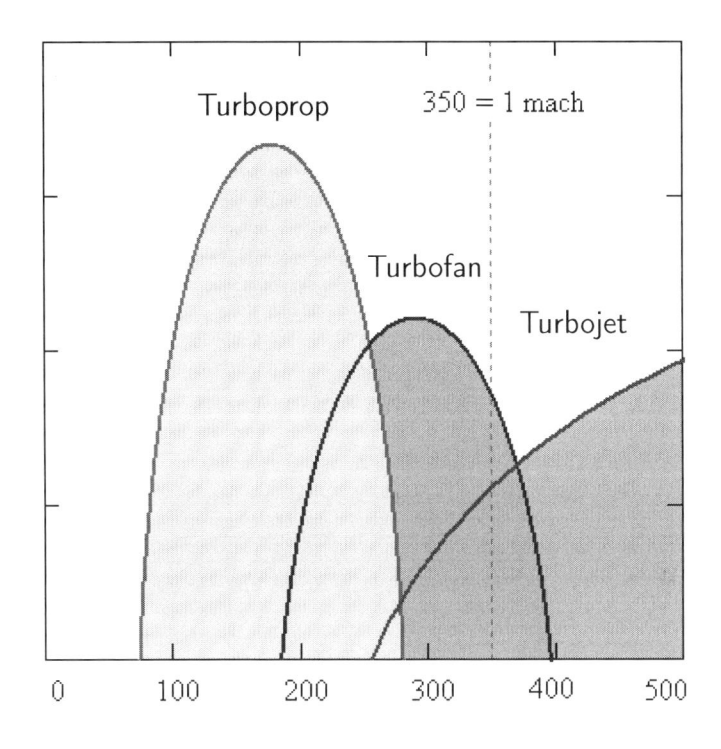

Figure 6.11: Relative suitability of the turboprop, turbofans, and ordinary turbojects for the flight at the 10 km attitude in various speeds. Adapted from Wikimedia Commons / CC-BY-SA-3.0.

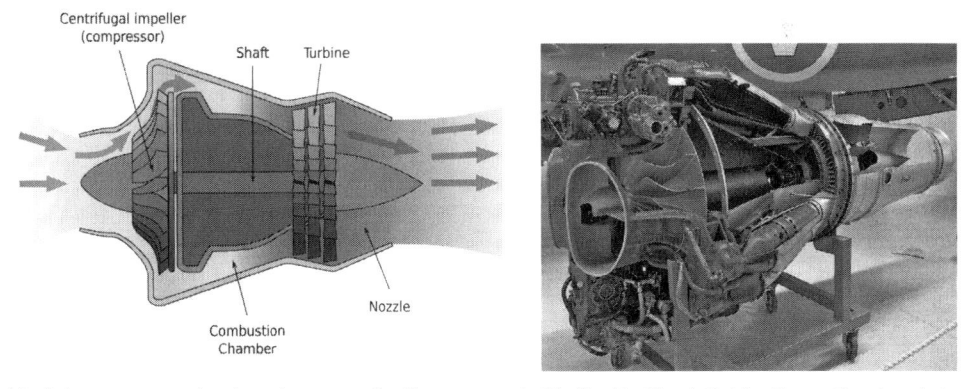

(a) Schematic centrifugal turbojet. Wikimedia Commons / CC-BY-SA-3.0. © Emoscopes / (b) De Havilland Goblin II centrifugal turbojet cut away. © Ian Dunster / Wikimedia Commons / CC-BY-SA-3.0.

Figure 6.12: Turbojet with centrifugal compressor.

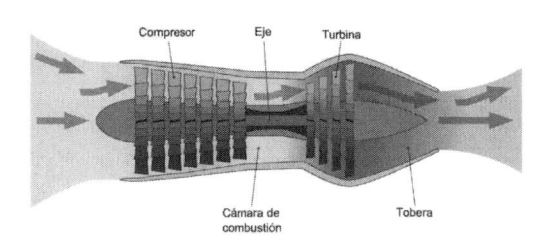

(a) Schematic axial turbojet. © Emoscopes / Wikimedia Commons / CC-BY-SA-3.0.

(b) GE J85 axial turbojet cut away. © Sanjay ach / Wikimedia Commons / CC-BY-SA-3.0.

Figure 6.13: Turbojet with axial compressor.

6.3.2 Turbofans

Most modern commercial aircraft use turbofan engines because of their high thrust and good fuel efficiency at high subsonic regimes. A turbofan engine is similar to a basic jet engine. The only difference is that the core engine is surrounded by a fan in the front and an additional fan turbine at the rear. The fan and fan turbine are connected by an additional shaft. This type of arrangement is called a two-spool engine (one spool for the fan, one spool for the core). Some turbofans might have additional spools for even higher efficiency.

The working principles are very similar to basic jet engines: the incoming air is pulled in by the engine inlet. Some of it passes through the fan and continues on throughout compressor, combustor, turbine, and nozzle, identical to the process in a basic turbojet. The fan causes additional air to flow around (bypass) the engine. This produces greater thrust and reduces specific fuel consumption. Therefore, a turbofan gets some of its thrust from the core jet engine and some from the fan. The ratio between the air mass that flows around the engine and the air mass that goes through the core is called the bypass ratio.

There are two types of turbofans: high bypass and low bypass, as illustrated in Figure 6.14. High bypass turbofans have large fans in front of the engine and are driven by a fan turbine located behind the primary turbine that drives the main compressor. Low bypass turbofans permit a smaller area and thus are more suitable for supersonic regime. A turbofan is very fuel efficient. Indeed, high bypass turbofans are nearly as fuel efficient as turboprops at low speeds. Moreover, because the fan is embedded in the inlet, it operates more efficiently at high subsonic speeds than a propeller. That is why turbofans are found on high-subsonic transportation (typical commercial aircraft) and propellers are used on low-speed transports (regional aircraft).

6.3 Types of jet engines

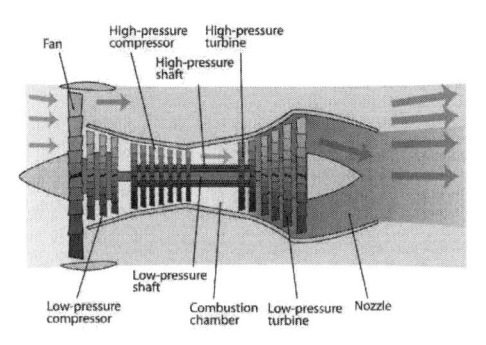

(a) Schematic diagram illustrating the operation of a 2-spool, high-bypass turbofan engine. © K. Aainsqatsi / Wikimedia Commons / CC-BY-SA-3.0.

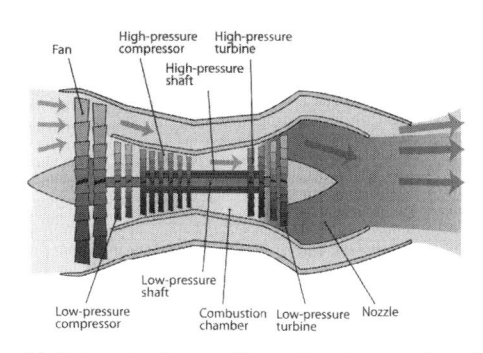

(b) Schematic diagram illustrating the operation of a 2-spool, low-bypass turbofan engine. © K. Aainsqatsi / Wikimedia Commons / CC-BY-SA-3.0.

(c) Turbofan CFM56 (high by-pass). © David Monniaux / Wikimedia Commons / CC-BY-SA-3.0.

(d) Soloviev D-30KU-154 low-bypass turbofan engine. © User:VargaA / Wikimedia Commons / CC-BY-SA-3.0.

Figure 6.14: Turbofan.

6.3.3 Turboprops

Many regional aircraft use turboprop engines. There are two main parts in a turboprop engine: the core engine and the propeller. The core engine is very similar to a basic turbojet except that instead of expanding all the hot exhaust gases through the nozzle to produce thrust, most of this energy is used to turn the turbine. The shaft drives the propeller through gear connections and produces most of the thrust (similarly to a propeller). Figure 6.15 illustrates a turboprop.

The thrust of a turboprop is the sum of the thrust of the propeller and the thrust of the core, which is very small. Propellers become less efficient as the speed of the aircraft increases. Thus, turboprops are only used for low subsonic speed regimes aircraft. A variation of the turboprop engine is the turboshaft engine. In a turboshaft engine, the gearbox is not connected to a propeller but to some other drive device. Many helicopters use turboshaft engines.

Aircraft propulsion

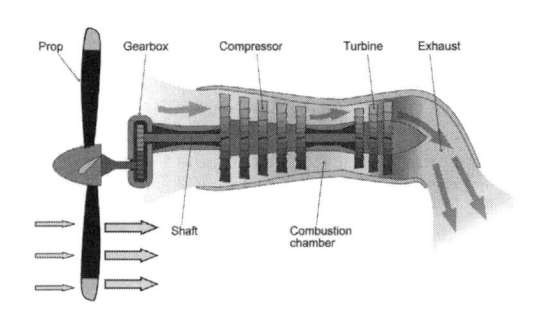

(a) Schematic diagram of the operation of a turboprop engine. © Emoscopes / Wikimedia Commons / CC-BY-SA-3.0.

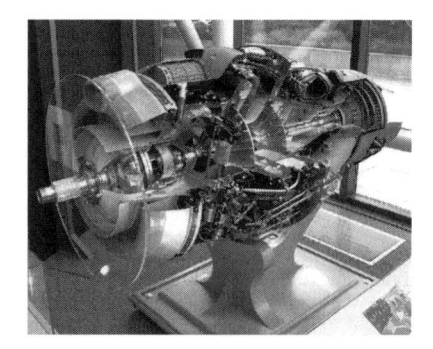

(b) Rolles Royce Dart Turboprop. © Sanjay Acharya / Wikimedia Commons / CC-BY-SA-3.0.

Figure 6.15: Turboprop engines.

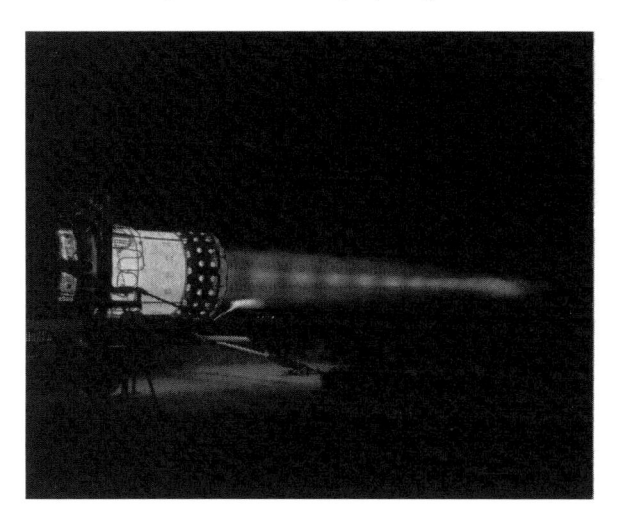

Figure 6.16: A statically mounted Pratt & Whitney J58 engine with full after-burner. Wikimedia Commons / Public Domain.

6.3.4 After-burning turbojet

Modern fighter aircraft typically mount an after-burner. Other alternatives are either a low bypass turbofan or a turbojet. The explanation behind this is that fighters typically need extra thrust to perform sharp maneuvers and fulfill its mission. The after-burner is essentially a long tailpipe into which additional fuel is sprayed directly into the hot exhaust and burned to provide extra thrust. When the after-burner is turned off, the engine performs as a basic turbojet. The exhaust velocity is increased compared to that with after-burner off because higher temperatures are involved.

REFERENCES

[1] JENKINSON, L. R., SIMPKIN, P., RHODES, D., JENKISON, L. R., and ROYCE, R. (1999). *Civil jet aircraft design*, volume 7. Arnold London.

[2] MATTINGLY, J. D., HEISER, W. H., and PRATT, D. T. (2002). *Aircraft engine design*. AIAA.

[3] NEWMAN, D. (2002). *Interactive aerospace engineering and design*. McGraw-Hill.

MECHANICS OF FLIGHT

Contents

7.1	Performances		170
	7.1.1	Reference frames	170
	7.1.2	Hypotheses	170
	7.1.3	Aircraft equations of motion	172
	7.1.4	Performances in a steady linear flight	175
	7.1.5	Performances in steady ascent and descent flight	175
	7.1.6	Performances in gliding	176
	7.1.7	Performances in turn maneuvers	177
	7.1.8	Performances in the runway	179
	7.1.9	Range and endurance	182
	7.1.10	Payload-range diagram	184
7.2	Stability and control		187
	7.2.1	Fundamentals of stability	187
	7.2.2	Fundamentals of control	189
	7.2.3	Longitudinal balancing	191
	7.2.4	Longitudinal stability and control	192
	7.2.5	Lateral-directional stability and control	194
7.3	Problems		196
	References		229

Mechanics of atmospheric flight studies aircraft performances, that is, the movement of the aircraft on the air in response to external forces and torques, and the stability and control of the aircraft's movement, analyzing thus the rotational movement of the aircraft. The study of performances, stability, and control plays a major role in verifying the design requirements. For instance, one must be able to analyze the required power for cruise flight, the required power settings and structural design for climbing with a desired angle at a desired velocity, the range and autonomy of the aircraft, the distances for taking off and landing, the design for making the aircraft stable under disturbances, and so on and so forth. The reader is referred to FRANCHINI and GARCÍA [2] and ANDERSON [1] as introductory references. Appendix A complements the contents of this chapter.

7.1 Performances

7.1.1 Reference frames

Consider the following reference frames[1]:

Definition 7.1 (*Earth Reference Frame*). *An Earth reference frame $F_e(O_e, x_e, y_e, z_e)$ is a rotating topocentric (measured from the surface of the Earth) system. The origin O_e is any point on the surface of Earth defined by its latitude θ_e and longitude λ_e. Axis z_e points to the center of Earth; x_e lays in the horizontal plane and points to a fixed direction (typically north); y_e forms a right-handed thrihedral (typically east).*

Such system it is sometimes referred to as *navigational system* since it is very useful to represent the trajectory of an aircraft from the departure airport.

Definition 7.2 (*Wind Axes Frame*). *A wind axes frame $F_w(O_w, x_w, y_w, z_w)$ is linked to the instantaneous aerodynamic velocity of the aircraft. It is a system of axes centered in any point of the symmetry plane (assuming there is one) of the aircraft, typically the center of gravity. Axis x_w points at each instant to the direction of the aerodynamic velocity of the aircraft $\vec{V}$. Axis z_w lays in to the plane of symmetry, perpendicular to x_w and pointing down according to regular aircraft performance. Axis y_w forms a right-handed thrihedral.*

Orientation angles

There exist several angles used in flight mechanics to orientate the aircraft with respect to a determined reference. The most important ones are:

- Sideslip angle, β, and angle of attack, α: The angles of the aerodynamic velocity, $\vec{V}$, (wind axes reference frame) with respect the body axes reference frame.

- Roll, μ, pitch, γ, and yaw, χ, velocity angles: The angles of the wind axes reference frame with respect of the Earth reference frame. This angles are also referred to as bank angle, flight path angle, and heading angle.

7.1.2 Hypotheses

Consider also the following hypotheses:

Hypothesis 7.1. Flat Earth model: *The Earth can be considered flat, non rotating, and approximate inertial reference frame.*

---

[1]Please, refer to Section 2.4 and/or Appendix A for a more detailed definition of the different reference frames.

7.1 Performances

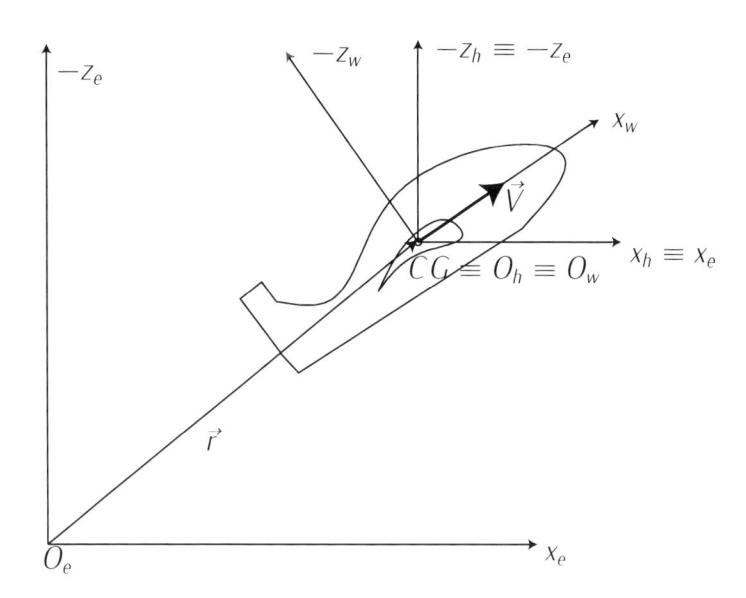

Figure 7.1: Wind axes reference frame.

Hypothesis 7.2. Constant gravity: *The acceleration due to gravity in atmospheric flight of an aircraft can be considered constant ($g = 9.81[m/s^2]$) and perpendicular to the surface of Earth.*

Hypothesis 7.3. Moving Atmosphere: *Wind is taken into account. Vertical component is neglected due its low influence. Only kinematic effects are considered, i.e., dynamic effects of wind are also neglected due its low influence.*

Hypothesis 7.4. 6-DOF model: *The aircraft is considered as a rigid solid with six degrees of freedom, i.e., all dynamic effects associated to elastic deformations, to degrees of freedom of articulated subsystems (flaps, ailerons, etc.), or to the kinetic momentum of rotating subsystems (fans, compressors, etc.), are neglected.*

Hypothesis 7.5. Point mass model: *The translational equations are uncoupled from the rotational equations by assuming that the airplane rotational rates are small and that control surface deflections do not affect forces. This leads to consider a 3 Degree Of Freedom (DOF) dynamic model that describes the point variable-mass motion of the aircraft.*

Hypothesis 7.6. Fixed engines: *We assume the aircraft is a conventional jet airplane with fixed engines.*

Hypothesis 7.7. Variable mass: *The aircraft is modeled as variable mass particle.*

Hypothesis 7.8. Forces acting on an aircraft: *The external actions acting on an aircraft can be decomposed, without loss of generality, into propulsive, aerodynamic, and gravitational.*

Hypothesis 7.9. Symmetric flight: *We assume the aircraft has a plane of symmetry, and that the aircraft flies in symmetric flight, i.e., all forces act on the center of gravity and the thrust and the aerodynamic forces lay on the plane of symmetry.*

Hypothesis 7.10. Small thrust angle of attack: *We assume the thrust angle of attack is small.*

7.1.3 Aircraft equations of motion[3]

3D motion

Under Hypotheses 7.1-7.10, the 3DOF equations governing the translational 3D motion of an airplane are the following:

- 3 dynamic equations relating forces to translational acceleration.
- 3 kinematic equations giving the translational position relative to an Earth reference frame.
- 1 equation defining the variable-mass characteristics of the airplane versus time.

The equation of motion is hence defined by the following Ordinary Differential Equations (ODE) system:

Definition 7.3 (*3DOF equations of 3D motion*).

$$m\dot{V} = T - D - mg\sin\gamma; \tag{7.1a}$$
$$mV\dot{\chi}\cos\gamma = L\sin\mu; \tag{7.1b}$$
$$mV\dot{\gamma} = L\cos\mu - mg\cos\gamma; \tag{7.1c}$$
$$\dot{x}_e = V\cos\gamma\cos\chi + W_x; \tag{7.1d}$$
$$\dot{y}_e = V\cos\gamma\sin\chi + W_y; \tag{7.1e}$$
$$\dot{h}_e = V\sin\gamma; \tag{7.1f}$$
$$\dot{m} = -T\eta. \tag{7.1g}$$

Where in the above:

- the three dynamics equations are expressed in an aircraft based reference frame, the wind axes system $F_w(O, x_w, y_w, z_w)$, usually x_w coincident with the velocity vector.

---
[3] The reader is encouraged to read Appendix A for a better understanding.

7.1 Performances

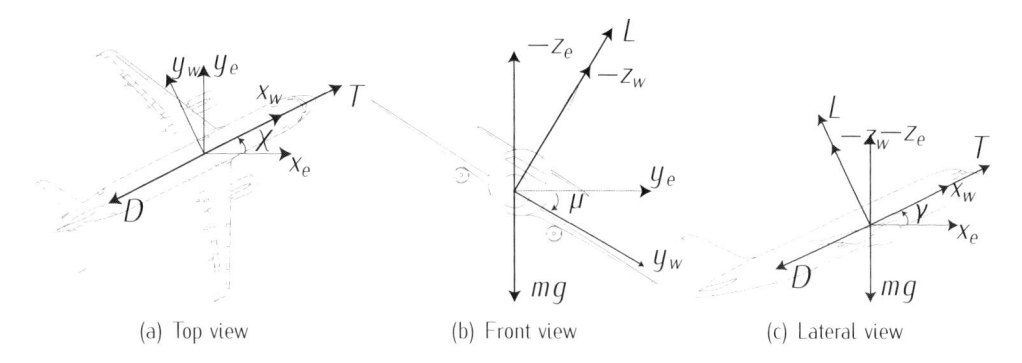

| (a) Top view | (b) Front view | (c) Lateral view |

Figure 7.2: Aircraft forces.

- the three kinematic equations are expressed in a ground based reference frame, the Earth reference frame $F_e(O_e, x_e, y_e, z_e)$ and are usually referred to as down range (or longitude), cross range (or latitude), and altitude, respectively.
- x_e, y_e and h_e denote the components of the position of the center of gravity of the aircraft, the radio vector $\vec{r}$, expressed in an Earth reference frame $F_e(O_e, x_e, y_e, z_e)$.
- W_x, and W_y denote the components of the wind, $\vec{W} = (W_x, W_y, 0)$, expressed in an Earth reference frame $F_e(O_e, x_e, y_e, z_e)$.
- $\mu, \chi,$ and γ are the bank angle, the heading angle, and the flight-path angle, respectively.
- m is the mass of the aircraft and η is the specific fuel consumption.
- g is the acceleration due to gravity.
- V is the true air speed of the aircraft.
- T is the engines' thrust, the force generated by the aircraft's engines. It depends on the altitude h, Mach number M, and throttle π by an assumedly known relationship $T = T(h, M, \pi)$.
- lift, $L = C_L S \hat{q}$, and drag, $D = C_D S \hat{q}$ are the components of the aerodynamic force, where C_L is the dimensionless coefficient of lift and C_D is the dimensionless coefficient of drag, $\hat{q} = \frac{1}{2}\rho V^2$ is referred to as dynamic pressure, ρ is the air density, and S is the wet wing surface. C_L is, in general, a function of the angle of attack, Mach and Reynolds numbers: $C_L = C_L(\alpha, M, Re)$. C_D is, in general, a function of the coefficient of lift: $C_D = C_D(C_L(\alpha, M, Re))$.

Additional assumptions are:

Hypothesis 7.11. Parabolic drag polar *A parabolic drag polar is assumed,* $C_D = C_{D_0} + C_{D_i} C_L^2$.

173

Hypothesis 7.12. Standard atmosphere model *A standard atmosphere is defined with* $\Delta_{ISA} = 0$.

Vertical motion

Considerer the additional hypothesis for a symmetric flight in the vertical plane:

Hypothesis 7.13. Vertical motion

- χ *can be considered constant.*
- *The aircraft performs a leveled wing flight, i.e., $\mu = 0$.*
- *There are no actions out of the vertical plane, i.e, $W_y = 0$.*

Definition 7.4 (3DOF equations of vertical motion). *The 3DOF equations governing the translational vertical motion of an airplane is given by the following ODE system:*

$$m\dot{V} = T - D - mg\sin\gamma, \tag{7.2a}$$
$$mV\dot{\gamma} = L - mg\cos\gamma, \tag{7.2b}$$
$$\dot{x}_e = V\cos\gamma\cos\chi + W_x, \tag{7.2c}$$
$$\dot{h}_e = V\sin\gamma, \tag{7.2d}$$
$$\dot{m} = -T\eta. \tag{7.2e}$$

Horizontal motion

Considerer the additional hypothesis for a symmetric flight in the horizontal plane:

Hypothesis 7.14. Horizontal motion *We consider flight in the horizontal plane, i.e., $\dot{h}_e = 0$ and $\gamma = 0$.*

Definition 7.5 (3DOF equations of horizontal motion). *The 3DOF equations governing the translational horizontal motion of an airplane is given by the following ODE system:*

$$m\dot{V} = T - D, \tag{7.3a}$$
$$mV\dot{\chi} = L\sin\mu, \tag{7.3b}$$
$$0 = L\cos\mu - mg, \tag{7.3c}$$
$$\dot{x}_e = V\cos\chi + W_x, \tag{7.3d}$$
$$\dot{y}_e = V\sin\chi + W_y, \tag{7.3e}$$
$$\dot{m} = -T\eta. \tag{7.3f}$$

7.1.4 Performances in a steady (stationary) linear-horizontal flight

Considerer the additional hypotheses :

- Consider a symmetric flight in the horizontal plane.
- χ can be considered constant.
- The aircraft performs a leveled wing flight, i.e., $\mu = 0$.
- There is no wind.
- The mass and the velocity of the aircraft are constant.

The 3DOF equations governing the motion of the airplane are[4]:

$$T = D, \tag{7.4a}$$
$$L = mg, \ (\text{which implies } n = 1), \tag{7.4b}$$
$$\dot{x}_e = V, \tag{7.4c}$$

Recall the following expressions already exposed in Chapter 3:

- $L = \frac{1}{2}\rho S V^2 C_L(\alpha); \ C_L = C_{L_0} + C_{L_\alpha}\alpha,$
- $D = \frac{1}{2}\rho S V^2 C_D(\alpha); \ C_D = C_{D_0} + k C_L^2,$
- $E = \frac{L}{D} = \frac{C_L}{C_D} = \frac{C_L}{C_{D_0} + k C_L^2}$, with $E_{max} = \frac{1}{2\sqrt{C_{D_0} k}}$ [5].

Considering these expressions, System of equations (7.4) can be expressed as:

$$T = \frac{1}{2}\rho S V^2 C_{D_0} + \frac{2k(mg)^2}{\rho S V^2}, \tag{7.5a}$$
$$mg = \frac{1}{2}\rho S V^2 (C_{L_0} + C_{L_\alpha}\alpha), \tag{7.5b}$$
$$\dot{x}_e = V. \tag{7.5c}$$

Expression (7.5b) says that in order to increase velocity it is necessary to reduce the angle of attack and vice-versa. Expression (7.5a) gives the two velocities at which an aircraft can fly for a given thrust.

7.1.5 Performances in steady (stationary) ascent and descent flight

Considerer the additional hypotheses:

---
[4] $n = \frac{L}{mg}$ is referred to as load factor
[5] remember that E_{max} refers to the maximum efficiency.

Mechanics of Flight

- Consider a symmetric flight in the vertical plane.
- χ can be considered constant.
- The aircraft performs a leveled wing flight, i.e., $\mu = 0$.
- There is no wind.
- The mass, the velocity, and the flight path angle of the aircraft are constant.

The 3DOF equations governing the motion of the airplane are:

$$T = D + mg \sin \gamma, \qquad (7.6a)$$
$$L = mg \cos \gamma, \qquad (7.6b)$$
$$\dot{x}_e = V \cos \gamma \cos \chi, \qquad (7.6c)$$
$$\dot{h}_e = V \sin \gamma, \qquad (7.6d)$$

Typically, commercial and general aviation aircraft have a relation $T/(mg)$ so that flight path angles are small ($\gamma \ll 1$). Therefore, Expression (7.6a) can be expressed as

$$\gamma \cong \frac{T - D}{mg}, \qquad (7.7)$$

and Expression (7.6b) can be expressed as

$$L \cong mg, \rightarrow n \cong 1. \qquad (7.8)$$

Therefore the flight path angle can be controlled by means of the power plant thrust.

Another important characteristic in ascent (descent) flight is the Rate Of Climb (ROC), which is given by Expression (7.6d) as:

$$V_{ROC} = \frac{dh_e}{dt} = V \sin \gamma. \qquad (7.9)$$

7.1.6 Performances in gliding

In all generality, a glider is an aircraft with no thrust. In stationary linear motion in vertical plane, the equations are as follows:

$$D = mg \sin \gamma, \qquad (7.10a)$$
$$L = mg \cos \gamma, \qquad (7.10b)$$

and dividing:

$$\tan \gamma_d = \frac{D}{L} = \frac{C_D}{C_L} = \frac{1}{E(\alpha)}, \qquad (7.11)$$

7.1 Performances

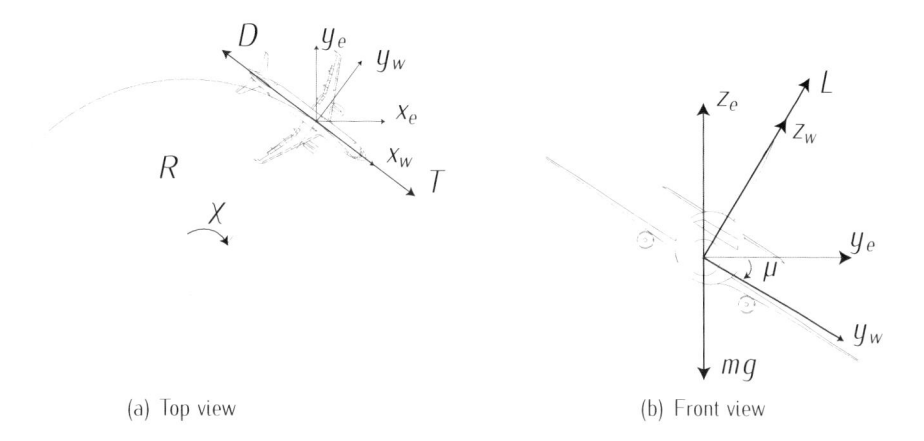

(a) Top view   (b) Front view

Figure 7.3: Aircraft forces in a horizontal loop.

where γ_d is the descent path angle ($\gamma_d = -\gamma$). As in stationary linear-horizontal flight, in order to increase the velocity of a glider it is necessary to reduce the angle of attack. Moreover, the minimum gliding path angle will be obtained flying with the maximum aerodynamic efficiency. The descent velocity of a glider (V_d) can be defined as the loss of altitude with time, that is:

$$V_d = V \sin \gamma_d \cong V \gamma_d. \quad (7.12)$$

7.1.7 Performances in turn maneuvers

Horizontal stationary turn

Considerer the additional hypotheses:

- Consider a symmetric flight in the horizontal plane.
- There is no wind.
- The mass and the velocity of the aircraft are constant.

The 3DOF equations governing the motion of the airplane are:

$$T = D, \quad (7.13a)$$
$$mV\dot{\chi} = L \sin \mu, \quad (7.13b)$$
$$L \cos \mu = mg, \quad (7.13c)$$
$$\dot{x}_e = V \cos \chi, \quad (7.13d)$$
$$\dot{y}_e = V \sin \chi. \quad (7.13e)$$

Mechanics of flight

In a uniform (stationary) circular movement, it is well known that the tangential velocity is equal to the angular velocity ($\dot\chi$) multiplied by the radius of turn (R):

$$V = \dot\chi R. \qquad (7.14)$$

Therefore, System (7.13) can be rewritten as:

$$T = \frac{1}{2}\rho S C_{D_0} + \frac{2kn^2(mg)^2}{\rho V^2 S}, \qquad (7.15a)$$

$$n \sin \mu = \frac{V^2}{gR}, \qquad (7.15b)$$

$$n = \frac{1}{\cos \mu} \to n > 1, \qquad (7.15c)$$

$$\dot x_e = V \cos \chi, \qquad (7.15d)$$

$$\dot y_e = V \sin \chi, \qquad (7.15e)$$

where $n = \frac{L}{mg}$ is the load factor. Notice that the load factor and the bank angle are inversely proportional, that is, if one increases the other reduces and vice versa, until the bank angle reaches 90°, where the load factor is infinity.

The stall speed in horizontal turn is defined as:

$$V_S = \sqrt{\frac{2mg}{\rho S C_{L_{max}}} \frac{1}{\cos \mu}}. \qquad (7.16)$$

Ideal looping

The ideal looping is a circumference of radius R into a vertical plane performed at constant velocity. Considerer then the following additional hypotheses:

- Consider a symmetric flight in the vertical plane.
- χ can be considered constant.
- The aircraft performs a leveled wing flight, i.e., $\mu = 0$.
- There is no wind.
- The mass and the velocity of the aircraft are constant.

The 3DOF equations governing the motion of the airplane are:

$$T = D + mg \sin \gamma, \qquad (7.17a)$$

$$L = mg \cos \gamma + mV\dot\gamma, \qquad (7.17b)$$

$$\dot x_e = V \cos \gamma, \qquad (7.17c)$$

$$\dot h_e = V \sin \gamma, \qquad (7.17d)$$

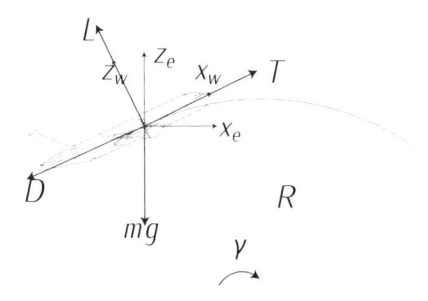

Figure 7.4: Aircraft forces in a vertical loop.

In a uniform (stationary) circular movement, it is well known that the tangential velocity is equal to the angular velocity ($\dot{\gamma}$ in this case) multiplied by the radius of turn (R):

$$V = \dot{\gamma}R. \tag{7.18}$$

The load factor and the coefficient of lift in this case are:

$$n = \cos\gamma + \frac{V^2}{gR}, \tag{7.19a}$$

$$C_L = \frac{2mg}{\rho V^2 S}(\cos\gamma + \frac{V^2}{gR}). \tag{7.19b}$$

Notice that the load factor varies in a sinusoidal way along the loop, reaching a maximum value at the superior point ($n_{max} = 1 + \frac{V^2}{gR}$) and a minimum value at the inferior point ($n_{min} = \frac{V^2}{gR} - 1$).

7.1.8 Performances in the runway

After analyzing the performances of an aircraft in the air, we will analyze the performances of an aircraft while taking off and landing.

Take off

The take off is defined as the maneuver covering those phases from the initial acceleration at the runway's head until the aircraft reaches a prescribed altitude and velocity (defined by the aeronavegability norms). This maneuver is performed with maximum thrust, deflected flaps, and landing gear down.

Mechanics of flight

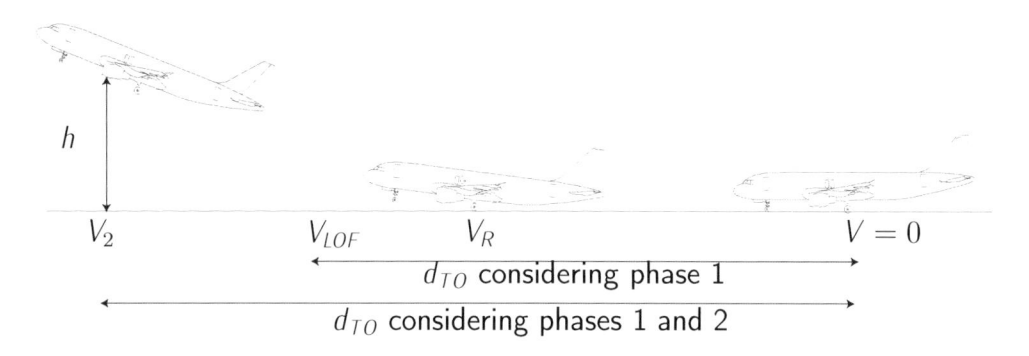

Figure 7.5: Take off distances and velocities.

We can divide the maneuver in two main phases:

1. Rolling in the ground ($0 \leq V \leq V_{LOF}$): From the initial acceleration to the velocity of take off (V_{LOF}), when the aircraft does not touch the runway.

    (a) Rolling with all the wheels in the ground ($0 \leq V \leq V_R$): The aircraft takes off rolling with all the wheels in the ground until it reaches a velocity called rotational velocity, V_R.

    (b) Rolling with the aft wheels in the ground ($V_R \leq V \leq V_{LOF}$): At V_R the nose rotates upwards and the aircraft keep rolling but now just with the aft wheels in the ground.

2. Path in the air ($V_{LOF} \leq V \leq V_2$): From the instant in which the aircraft does not touch the runway to the instant in which the aircraft reaches a velocity V_2 at a given altitude h (such altitude is usually defined as $h = 35\ ft (10.7\ m)$).

    (a) Track of curve transition ($V \approx V_{LOF}$): The aircraft needs a transition until it reaches the desired ascent flight path angle.

    (b) Straight accelerated track ($V_{LOF} \leq V \leq V_2$): The aircraft accelerates with constant flight path angle until it reaches V_2 at a given altitude h.

Of all the above, we are interested on analyzing sub-phase 1.(a). In order to get approximate numbers of taking off distances and times, let us assume the aircraft performs a uniform accelerated movement and the only force is thrust T. According to Newton's second law:

$$ma = T. \qquad (7.20)$$

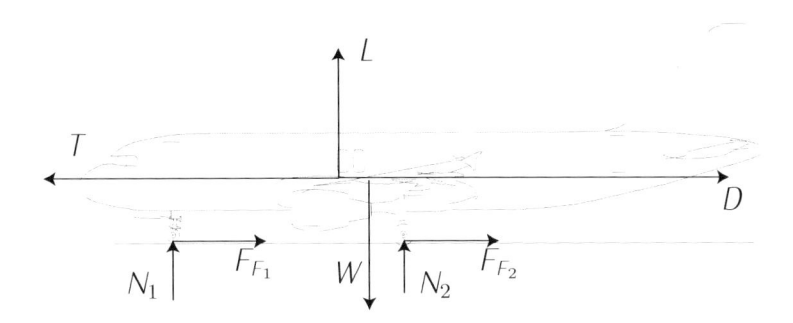

Figure 7.6: Forces during taking off.

The acceleration is $a = \frac{dV}{dt}$. Then:

$$V = \int \frac{T}{m} dt = \frac{T}{m} t. \qquad (7.21)$$

If we make the integral defined between $t = 0$ and $t = t_{TO}$, and $V = 0$ and $V = V_{TO}$:

$$V_{TO} = \int_0^{t_{TO}} \frac{T}{m} dt = \frac{T}{m} t_{TO} \rightarrow t_{TO} = \frac{V_{TO} m}{T}. \qquad (7.22)$$

The velocity is $V = \frac{dx}{dt}$. Then:

$$x = \int \frac{T}{m} t \, dt = \frac{T}{m} \frac{t^2}{2}. \qquad (7.23)$$

If we make the integral defined between $t = 0$ and $t = t_{TO}$ and $x = 0$ and $x = x_{TO}$:

$$x_{TO} = \int_0^{t_{TO}} \frac{T}{m} t \, dt = \frac{T}{m} \frac{t_{TO}^2}{2} \rightarrow x_{TO} = \frac{V_{TO}^2 m}{2T}. \qquad (7.24)$$

Landing

The landing is defined as the maneuver covering those phases starting from a prescribed altitude (defined by the aeronavegability norms, typically $h = 50\ ft$ and $V_A = 1.3 V_S$) until the aircraft stops (to be more precise, when the aircraft reaches a constant taxiing velocity). This maneuver is performed with minimum thrust, deflected flaps, and landing gear down.

We can divide the maneuver in two main phases:

1. Path in the air ($V_A \geq V \leq V_{Touch}$): From the instant in which the aircraft reaches

Mechanics of flight

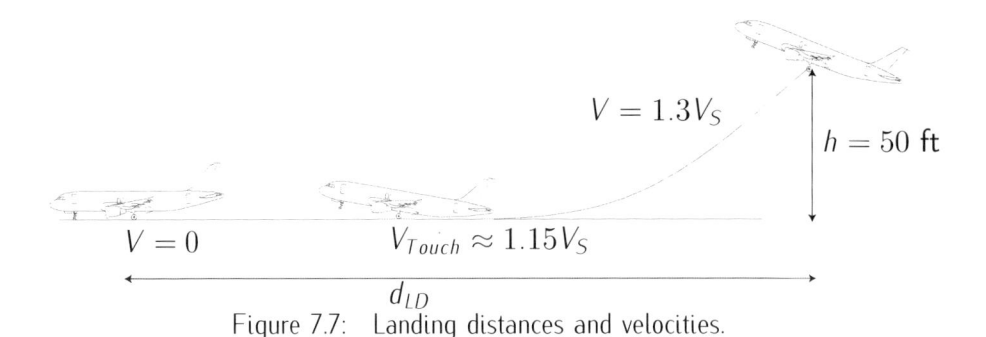

Figure 7.7: Landing distances and velocities.

a prescribed altitude performing a steady descent to the instant the aircraft touches down.

   (a) Final approach: it consist in a steady straight trajectory at a velocity typically 1.3 the stall velocity of the aircraft in the landing configuration.

   (b) Transition: The aircraft performs a transition between the straight trajectory to the horizontal plane of the runway. It can be supposed as a circumference. This transition is performed at a touchdown velocity $V_{Touch} \approx 1.15 V_S$.

2. Rolling in the ground ($V_{Touch} \geq V \geq 0$): From the instant of touchdown to the instant in which stops.

   (a) Rolling with the aft wheels in the ground: the nose rotates downwards and the aircraft rolls but now just with the aft wheels in the ground.

   (b) Rolling with all the wheels in the ground: the aircraft keeps rolling with all the wheels in the ground until it stops.

The equations of sub-phase 2.(b) are basically the same as the equations for sub-phase 1.(a) in taking off. The only differences are that thrust is minimum, zero, o even negative (in reverse gear aircrafts), drag is maximized deflecting the spoiler; the coefficient of friction is much higher due to break and downforce effects. The kinematic analysis for distances and times follow the same patterns as for take off: the movement can be considered herein uniformly decelerated.

7.1.9 Range and endurance

In this section, we study the range and endurance for an aircraft flying a steady, linear-horizontal flight.

- The range is defined as the maximum distance the aircraft can fly given a quantity of fuel.

7.1 Performances

- The endurance is defined as the maximum time the aircraft can be flying given a quantity of fuel.

Considerer the additional hypotheses:

- Consider a symmetric flight in the horizontal plane.
- χ can be considered constant.
- The aircraft performs a leveled wing flight, i.e., $\mu = 0$.
- There is no wind.
- The velocity of the aircraft is constant.

The 3DOF equations governing the motion of the airplane are:

$$T = D, \tag{7.25a}$$
$$L = mg, \tag{7.25b}$$
$$\dot{x}_e = V, \tag{7.25c}$$
$$\dot{m} = -\eta T. \tag{7.25d}$$

Equation (7.25d) means that the aircraft losses weight as the fuel is burt, where η is the specific fuel consumption. Notice that Equation (7.25d) is just valid for jets.

The specific fuel consumption is defined in different ways depending of the type of engines:

- Jets: $\eta_j = \frac{-dm/dt}{T}$.
- Propellers: $\eta_p = \frac{-dm/dt}{P_m} = \frac{-dm/dt}{TV}$, where P_m is the mechanical power.

Focusing on jet engines, operating with Equations (7.25), considering $E = L/D$, and taking into account the initial state $(\cdot)_i$ and the final state $(\cdot)_f$ we obtain the distance and time flown as:

$$x_e = -\int_{m_i}^{m_f} \frac{V}{\eta_j T} dm = -\int_{m_i}^{m_f} \frac{1}{\eta_j g} VE \frac{dm}{m}, \tag{7.26}$$

$$t = -\int_{m_i}^{m_f} \frac{1}{\eta_j T} dm = -\int_{m_i}^{m_f} \frac{1}{\eta_j g} E \frac{dm}{m}. \tag{7.27}$$

In order to integrate such equations we need to make additional assumptions, such for instance consider constant specific fuel consumption and constant aerodynamic efficiency (remember that the velocity has been already assumed to be constant).

Range and endurance (maximum distance and time, respectively) are obtained assuming the aircraft flies with the maximum aerodynamic efficiency (given the weights of the aircraft and given also that for a weight there exists an optimal speed):

$$x_{e\,max} = \frac{1}{\eta_j g} V E_{max} \ln \frac{m_i}{m_f}, \qquad (7.28)$$

$$t_{max} = \frac{1}{\eta_j g} E_{max} \ln \frac{m_i}{m_f}. \qquad (7.29)$$

7.1.10 Payload-range diagram

Weights of the aircraft

Let us start defining the different weights of the aircraft:

- OEW: The **Operating Empty Weight** is the basic weight of an aircraft including the crew, all fluids necessary for operation such as engine oil, engine coolant, water, unusable fuel and all operator items and equipment required for flight but excluding usable fuel and the payload. Also included are certain standard items, personnel, equipment, and supplies necessary for full operation.
- PL: The **Payload** is the load for what the company charges a fee. In transportation aircraft it corresponds to the passenger and its luggage, together with the cargo.
- FW: The **Fuel Weight** of an aircraft is the total weight of fuel carried at take off and it is calculated adding the following two weights:
    1. TF: **Trip Fuel** is the total amount of fuel estimated to be consumed in the trip.
    2. RF: **Reserve Fuel** is the weight of fuel to allow for unforeseen circumstances, such as an inaccurate weather forecast, alternative arrival airports, etc.
- TOW: The **TakeOff Weight** of an aircraft is the weight at which the aircraft takes off. TOW=OEW+PL+FW.
- LW: The **Landing Weight** of an airplane is the total weight of the airplane at destination with no use of reserve fuel. LW=OEW+PL+RF.
- ZFW: The **Zero Fuel Weight** of an airplane is the total weight of the airplane and all its contents, minus the total weight of the fuel on board. ZFW=OEW+PL.

Limitation on the weight of an aircraft

Due to different features, such structural limits, capacity of tanks, or capacity of passengers and cargo, some of the weights have limitations:

1. MPL: The **Maximum PayLoad** of an aircraft is limited due to structural limits and capacity constraints.

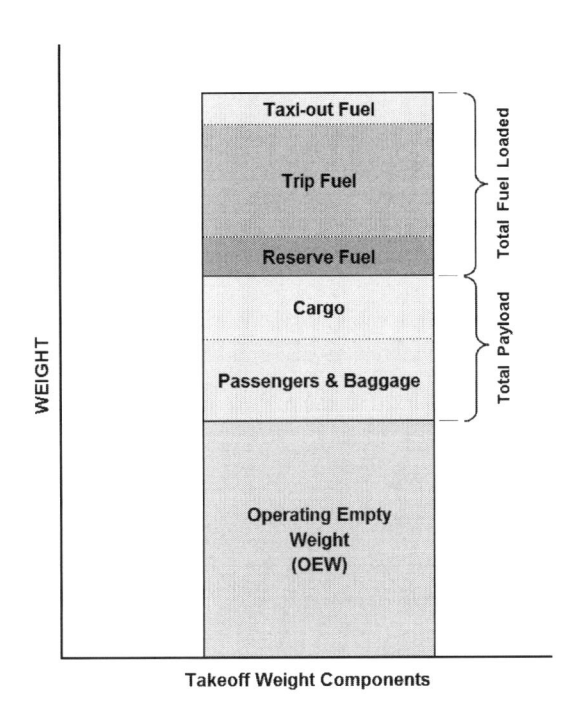

Figure 7.8: Take-off weight components. © Mohsen Alshayef / Wikimedia Commons / CC-BY-SA-3.0.

2. MFW: The **Maximum Fuel Weight** is the maximum weight of fuel to be carried and it is limited by the capacity of tanks.

3. MZFW: The **Maximum Zero Fuel Weight** is the maximum weight allowed before usable fuel and other specified usable agents (engine injection fluid, and other consumable propulsion agents) must be loaded in defined sections of the aircraft as limited by strength and airworthiness requirements. It may include usable fuel in specified tanks when carried instead of payload. The addition of usable and consumable items to the zero fuel weight must be in accordance with the applicable government regulations so that airplane structure and airworthiness requirements are not exceeded.

4. MTOW: The **Maximum Takeoff Weight** of an aircraft is the maximum weight at which the pilot of the aircraft is allowed to attempt to take off due to structural or other limits.

5. MLW: The **Maximum Landing Weight** of an aircraft is the maximum weight at which the pilot of the aircraft is allowed to attempt to land due to structural or other limits. In particular, due to structural limits in the landing gear.

Payload-range diagram

A payload range diagram (also known as the *elbow chart*) illustrates the trade-off between payload and range. The top horizontal line represents the maximum payload. It is limited structurally by maximum zero fuel weight (MZFW) of the aircraft. Maximum payload is the difference between maximum zero-fuel Weight and operational empty weight (OEW). Moving left-to-right along the line shows the constant maximum payload as the range increases. More fuel needs to be added for more range.

Weight in the fuel tanks in the wings does not contribute as significantly to the bending moment in the wing as does weight in the fuselage. So even when the airplane has been loaded with its maximum payload that the wings can support, it can still carry a significant amount of fuel.

The vertical line represents the range at which the combined weight of the aircraft, maximum payload and needed fuel reaches the maximum take-off weight (MTOW) of the aircraft. See point A in Figure 7.17. If the range is increased beyond that point, payload has to be sacrificed for fuel.

The maximum take-off weight is limited by a combination of the maximum net power of the engines and the lift/drag ratio of the wings. The diagonal line after the range-at-maximum-payload point shows how reducing the payload allows increasing the fuel (and range) when taking off with the maximum take-off weight. See point B in Figure 7.17

The second kink in the curve represents the point at which the maximum fuel capacity is reached. Flying further than that point means that the payload has to be reduced further, for an even lesser increase in range. See point C in Figure 7.17. The absolute range is thus the range at which an aircraft can fly with maximum possible fuel without carrying any payload.

In order to relate the ranges with weights we can use to so-called Breguet equation:

$$R = \frac{1}{g\eta_j} VE \ln \frac{TOW}{LW}, \qquad (7.30)$$

For the three marked points, respectively A, B and C:

$$R_A = \frac{1}{g\eta_j} VE \ln \frac{MTOW}{OEW + MPL + RF}, \qquad (7.31)$$

$$R_B = \frac{1}{g\eta_j} VE \ln \frac{MTOW}{MTOW - MFW + RF}, \qquad (7.32)$$

$$R_C = \frac{1}{g\eta_j} VE \ln \frac{OEW + MFW}{OEW + RF}. \qquad (7.33)$$

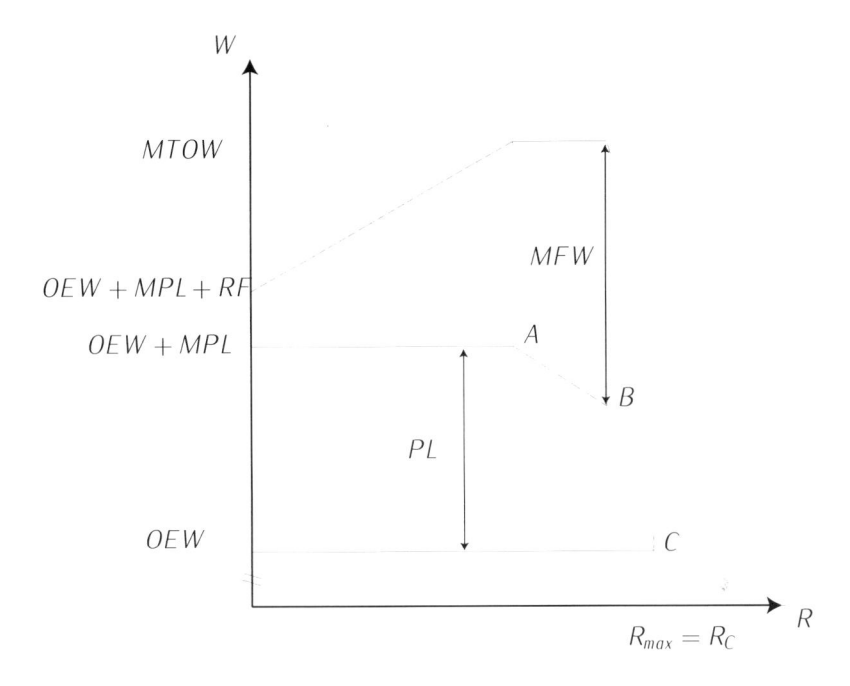

Figure 7.9: Payload-range diagram.

7.2 Stability and control

In Section 7.1 we have studied the performances of the aircraft, modeling the aircraft as a 3DOF solid and studying the point mass model movement due to external actions. In this section we study the fundamentals of stability and control, considering the aircraft as a 6DOF[6] model, so that we must take into consideration the geometric dimensions, the distribution of mass, and thus we study the external forces and torques which define the movement of the center of gravity and the orientation and angular velocity of the aircraft.

7.2.1 Fundamentals of stability

A vehicle is said to be in equilibrium when remains constant or the movement is uniform, that is, both the linear and angular quantity of movement are constant. In the case of an aircraft, the state of equilibrium typically refers to an uniform movement in which the angular velocities are null, and then the movement is simply a translation.

The stability is a property related with the state of equilibrium, which studies the behavior of the aircraft when any of the variables describing its state of equilibrium suffers from a variation (for instance a wind gust). The variation in one of those variables is

---
[6]Please refer to Appendix A for the deduction of the 6-DOF equations.

MECHANICS OF FLIGHT

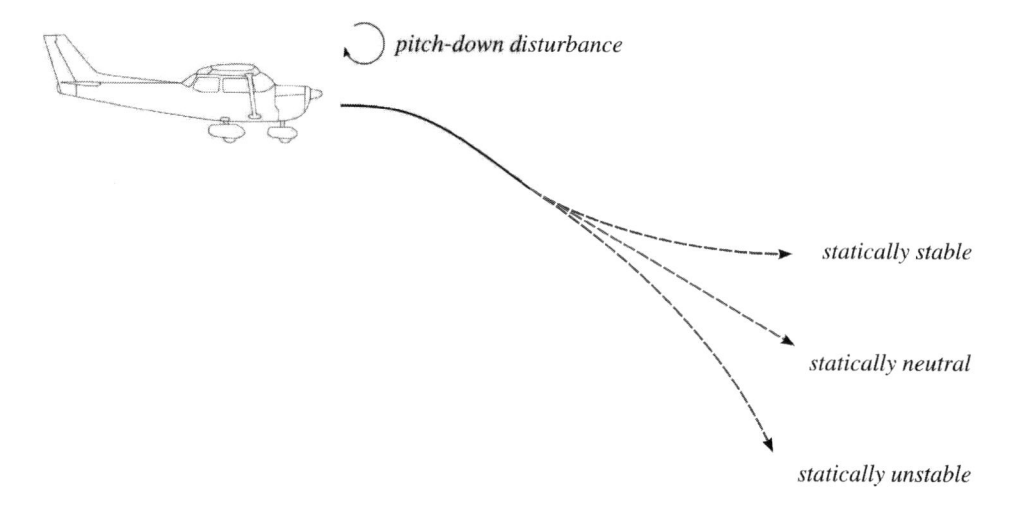

Figure 7.10: Diagram showing the three main cases for aircraft pitch static stability, following a pitch disturbance: Aircraft is statically stable (corrects attitude); Aircraft is statically neutral (does not correct attitude); Aircraft is statically unstable (exacerbates attitude disturbance). © User:Ariadacapo / Wikimedia Commons / CC0-1.0.

typically referred to as perturbation.

The stability can be studied in two different ways depending on the time scale:

- Static stability.
- Dynamic stability.

Static stability

The interest lies on the instant of time immediately after the perturbation takes place. In the static stability, the forces and torques which appear immediately after the perturbation are studied. If the value of the state variables describing the equilibrium tends to increase or amplify, the state of equilibrium is statically unstable. On the contrary, it is statically stable. Figure 7.10 shows a sketch of an aircraft with three different options for the static stability after a pitch down disturbance, e.g., a downwards wind gust.

Dynamic stability

The dynamic stability studies the evolution with time of the different variables of flight (yaw, pitch, and roll angles, velocity, angular velocities, altitude, etc.) when the condition of

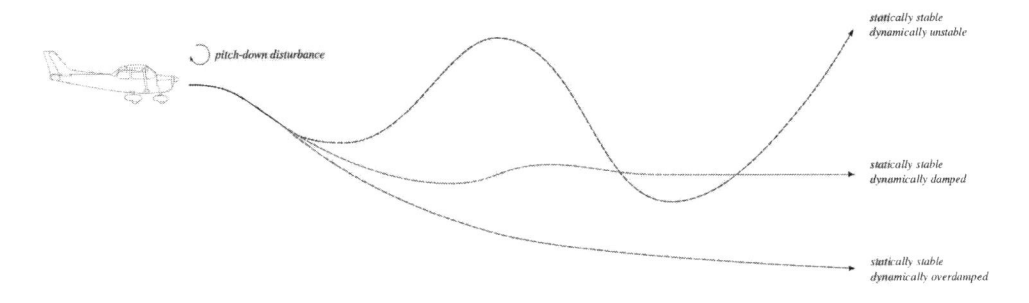

Figure 7.11: Diagram showing the three main cases for aircraft pitch dynamic stability. Here all three cases are for a statically stable aircraft. Following a pitch disturbance, three cases are shown: Aircraft is dynamically unstable (although statically stable); Aircraft is dynamically damped (and statically stable); Aircraft is dynamically overdamped (and statically stable). © User:Ariadacapo / Wikimedia Commons / CC0-1.0.

equilibrium is perturbed. Figure 7.11 shows a sketch of an aircraft with different dynamic-stability behaviors after a pitch down disturbance, e.g., a downwards wind gust. In order to study this evolution we need to solve the system of equation describing the 6DOF movement of the aircraft. As illustrated in Appendix A, the equations are:

$$\vec{F} = m\frac{d\vec{V}}{dt}, \quad (7.34)$$

$$\vec{G} = I\frac{d\vec{\omega}}{dt}. \quad (7.35)$$

An aircraft is said to be dynamically stable when after a perturbation the variables describing the movement of the aircraft tend to a stationary value (either the same point of equilibrium or a new one). On the contrary, if the aircraft does not reach an equilibrium it is said to be dynamically unstable.

7.2.2 Fundamentals of control

Control theory is an interdisciplinary branch of engineering and mathematics that deals with the behavior of dynamical systems. The external input of a system is called the reference. When one or more output variables of a system need to follow a certain reference over time, a controller manipulates the inputs to a system to obtain the desired effect on the output of the system.

The usual objective of control theory is to calculate solutions for the proper corrective action from the controller that result in system stability, that is, the system will hold the reference state values and not oscillate around them.

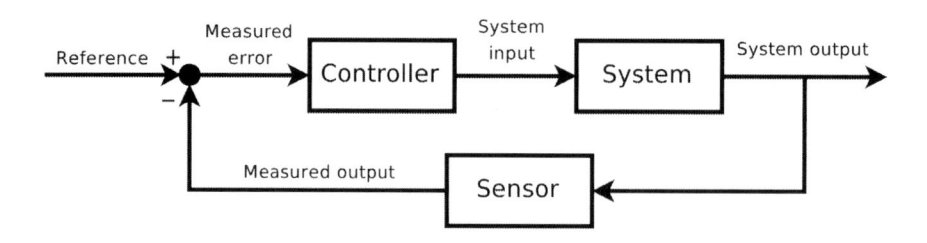

Figure 7.12: Feedback loop to control the dynamic behavior of the system: The sensed value is subtracted from the desired value to create the error signal, which is amplified by the controller. © User:Myself / Wikimedia Commons / CC-BY-SA-3.0.

The input and output of the system are related to each other by what is known as a transfer function (also known as the system function or network function). The transfer function is a mathematical representation, in terms of spatial or temporal frequency, of the relation between the input and output of a linear time-invariant system.

Example

Consider an aircraft's autopilot, which simplifying is a device designed to maintain vehicle speed at a constant desired or reference speed provided by the pilot. The controller is the autopilot, the plant is the aircraft (the equations of motion), and the system is together the aircraft and the autopilot. The system output is the aircraft's speed, and the control itself is the engine's throttle position which determines how much power the engine generates.

A primitive way to implement velocity control is simply to lock the throttle position when the pilot engages autopilot. However, if the velocity control is engaged on an atmosphere with no wind, then the aircraft will travel slower going against wind and faster when having tail wind. This type of controller is called an open-loop controller because no measurement of the system output (the aircraft's speed) is used to alter the control (the throttle position.) As a result, the controller can not compensate for changes acting on the aircraft, like the effect of varying wind.

In a closed-loop control system, a sensor monitors the system output (the aircraft's speed) and feeds the data to a controller which adjusts the control (the throttle position) as necessary to maintain the desired system output (match the aircraft's speed to the reference speed). Now when the aircraft goes against wind the decrease in speed is measured, and the throttle position changes to increase engine power, speeding the vehicle. Feedback from measuring the aircraft's speed has allowed the controller to dynamically compensate for changes to the speed. It is from this feedback that the paradigm of the control loop arises: the control affects the system output, which in turn is measured and looped back to alter the control.

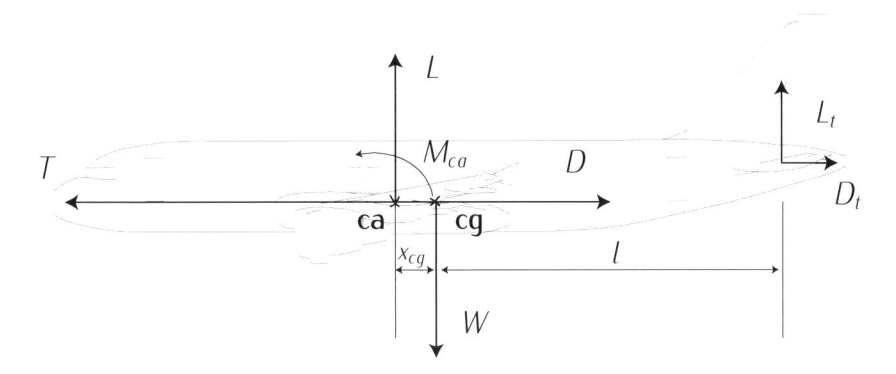

Figure 7.13: Longitudinal equilibrium. Adapted from Franchini and García [2].

7.2.3 Longitudinal balancing

The longitudinal balancing is the problem of determining the state of equilibrium of a longitudinal movement in which the lateral and directional variables are considered uncoupled. For the longitudinal analysis, one must consider forces on z-axis (F_z) and torques around y-axis (M_y). Generally, it is necessary to consider external actions coming from aerodynamics, propulsion, and gravity. However, it is common to consider only the gravity and the lift forces in wing and horizontal stabilizer. Additional hypotheses are:

- There is no wind.

- The mass and the velocity of the aircraft are constant.

The equations to be fulfilled are:

$$\sum F_z = 0, \qquad (7.36)$$

$$\sum M_y = 0. \qquad (7.37)$$

Which results in

$$mg - L - L_t = 0, \qquad (7.38)$$

$$-M_{ca} + L x_{cg} - L_t l = 0, \qquad (7.39)$$

where L_t is the lift generated by the horizontal stabilizer, M_{ca} is the pitch torque with respect to the aerodynamic center, x_{cg} is the distance between the center of gravity and the aerodynamic center, and l is the distance between the center of gravity and the aerodynamic center of the horizontal stabilizer.

MECHANICS OF FLIGHT

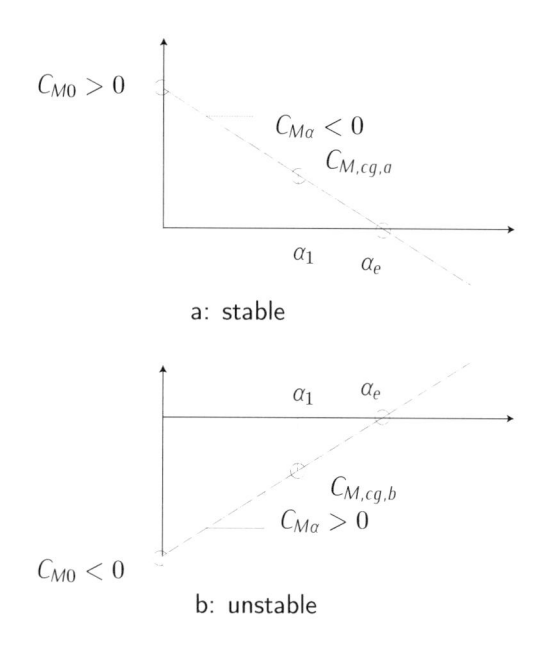

Figure 7.14: Longitudinal stability. Adapted from FRANCHINI and GARCÍA [2].

7.2.4 LONGITUDINAL STABILITY AND CONTROL

Longitudinal static stability

Consider an aircraft in horizontal, steady, linear flight. The aircraft is in equilibrium under equations (7.38)-(7.39).

Consider now that such equilibrium is perturbed by a vertical wind gust, so that the angle of attack increases, that is, there is a perturbation in the angle of attack. In this case, both L and L_t increase according to the lift-angle of attack curves (we assume that the behavior for the horizontal stabilizer is similar to the one for the wing[7]) so that we have $L + \Delta L$ and $L_t + \Delta L_t$. If $\Delta L_t l > \Delta L$, then the angle of attack tends to decrease and the aircraft is statically stable. On the contrary, the aircraft is statically unstable. In other words, the torque generated by the horizontal stabilizer is *stabilizer* (so the name). Obviously, this also depends on the relative position of the aerodynamic centers with respect to the center of gravity of the aircraft [8]. Therefore, the aerodynamic design is also key to determine the static stability.

The external longitudinal moments acting on the center of gravity can be made

---
[7] To be more precise, it is necessary to derive a model which gives us the effective angle of attack of the stabilizer since the wing modifies the incident current. However, this will be studied in posterior courses.
[8] remember that the aerodynamic center is the point at with the pitching moment does not vary with respect the increase in C_L

dimensionless as follows:

$$\frac{\sum M_{cg,y}}{\frac{1}{2}\rho S V_\infty^2 \bar{c}} = c_{M,cg}, \tag{7.40}$$

where $\sum M_{cg,y} = -M_{ca} + L x_{cg} - L_t l$ and $\bar{c}$ is the mean chord of the aircraft. The dimensionless equation is:

$$c_{M,cg} = c_{M0} + c_{M\alpha}\alpha + c_{M\delta_e}\delta_e, \tag{7.41}$$

where $c_{M,cg}$ is the coefficient of moments of the aircraft with respect to its center of gravity, c_{M0} is the coefficient of moments independently of the angle of attack and the deflection of the elevator, $c_{M\alpha}$ is the derivative of the coefficient of moments of the aircraft with respect to the angle of attack, and $c_{M\delta_e}$ is the derivative of the coefficient of moments of the aircraft with respect to the deflection of the elevator. The coefficients of Equation (7.41) c_{M0}, $c_{M\alpha}$, and $c_{M\delta_e}$ depend on the geometry and the aerodynamics of the aircraft.

In Figure 7.14 the coefficient of moments of two aircraft as a function of the angle of attack is shown for a given center of gravity, a given aerodynamic center for wing and horizontal stabilizer, and a given deflection of the elevator. The intersection of the curves with the *abscissa* axis determine the angle of attack of equilibrium α_e. Imagine that a perturbation appears, for instance a wind gust which decreases the angle of attack so that both aircrafts have an angle of attack $\alpha_1 < \alpha_e$. In the case or aircraft (a), the $c_{M,cg,a} > 0$ (the curve for α_1 is above the *abscissa* axis), which means the moment tends to pitch up the aircraft so that it returns to the initial state of equilibrium: the aircraft is statically stable. In the case or aircraft (b), $c_{M,cg,b} < 0$ (the curve for α_1 is below the *abscissa* axis) and the moment tends to pitch down the aircraft, so that it is statically unstable.

From this reasoning, we can conclude that for an aircraft to be statically stable it must be fulfilled:

$$\frac{dc_{M,cg}}{d\alpha} = c_{M\alpha} < 0. \tag{7.42}$$

$c_{M\alpha}$ depends, among other, of the center of gravity of the aircraft. Therefore, one of the key issues in the design of an aircraft is to determine the center of gravity to make the aircraft statically stable. This is not trivial, since the center of gravity varies during the flight (depends on the payload, varies as the fuel is burnt, etc.). Therefore, during a flight the pilot (or the autopilot in control systems) must modify the angle of attack to maintain the flight in the equilibrium (since α_e varies). As the center of gravity makes its way aft the angle of attack of equilibrium increases. There is a point for which $c_{M\alpha} = 0$, the neutral point. The center of gravity can not go back beyond this point by any means because $c_{M\alpha} > 0$ and the aircraft becomes statically unstable.

Longitudinal control

The coefficient $c_{M\delta_e}$ is referred to as the power of the longitudinal control and represents a measure of the capacity that the elevator has to generate a moment and, therefore, to change the angle of attack at which the aircraft can fly in equilibrium (α_e).

An elevator's positive deflection ($\delta_e > 0$) generates an increase in the horizontal stabilizer's lift (ΔL_t), which gives rise to a negative pitch moment $M_{cg} < 0$ ($c_{M\delta_e} < 0$). In the condition of equilibrium, the sum of moments around the center of gravity is null, and therefore it must be fulfilled that

$$c_{M,cg} = c_{M0} + c_{M\alpha}\alpha + c_{M\delta_e}\delta_e = 0. \tag{7.43}$$

Two main problems can be derived in the longitudinal control:

1. Determine the deflection angle of the elevator, δ_e, to be able to fly in equilibrium at a given angle of attack, α_e:

$$\delta_e = \frac{-c_{M0} - c_{M\alpha}\alpha_e}{c_{M\delta_e}} \tag{7.44}$$

2. Determine the angle of attack to fly in equilibrium, α_e, for a known deflection of the elevator, α_e:

$$\alpha_e = \frac{-c_{M0} - c_{M\delta_e}\delta_e}{c_{M\alpha}} \tag{7.45}$$

Figure 7.15 shows the effects of the elevator's deflection in the angle of attack of equilibrium. Simplifying, for a $\delta_e < 0$, the angle of attack of equilibrium at which the aircraft flies increases and so does the coefficient of lift. Since the lift must be equal to weight, the aircraft must fly slower. In other words, the elevator is used to modify the velocity of a steady horizontal flight.

As we have pointed out before, the geometric condition of the aircraft vary during the flight. Therefore it is necessary to re-calculate this conditions and modify the variables continuously. This is made using control systems.

7.2.5 LATERAL-DIRECTIONAL STABILITY AND CONTROL

Consider again an aircraft in horizontal, steady, linear flight. Suppose in this case that the lateral-directional elements (vertical stabilizer, rudder, ailerons) do not produce forces nor moments, so that there not exists a primary problem of balancing (as there was in the longitudinal case) since we have a longitudinal plane of symmetry.

In this case, the lateral-direction control surfaces (rudder and ailerons) fulfill a mission

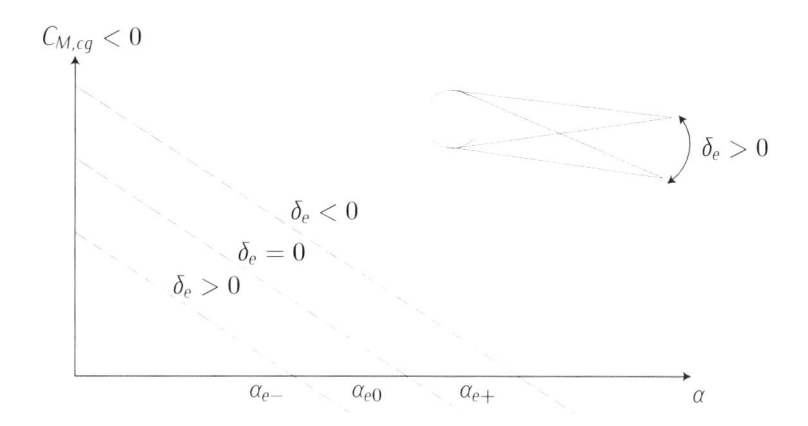

Figure 7.15: Effects of elevator on moments coefficient. Adapted from FRANCHINI and GARCÍA [2].

of secondary balancing since they are used when there exists an asymmetry (propulsive or aerodynamic). For instance, aircraft must be able to fly under engine failure, and thus the asymmetry must be compensated with the rudder. Another instance could be the landing operation under lateral wind, which must be also compensated with the rudder deflection. Notice that the center of gravity lays on the plane of symmetry, so that its position does not affect the lateral-directional control. Further mathematical analysis will be studied in posterior courses.

7.3 Problems

Problem 7.1: Performances

Consider an Airbus A-320 with the following characteristics:

- $m = 64$ tonnes.
- $S_w = 122.6 \ m^2$.
- $C_D = 0.024 + 0.0375 C_L^2$.

1. The aircraft starts an ascent maneuver with uniform velocity at 10.000 feet of altitude (3048 meters). At that flight level, the typical performances of the aircraft indicate a velocity with respect to air of 289 knots (148.67 m/s) and a rate of climb (vertical velocity) of 2760 feet/min (14 m/s). Assuming that $\gamma \ll 1$, calculate:

    (a) The angle of ascent, γ.

    (b) Required thrust at those conditions.

2. The aircraft reaches an altitude of 11000 m, and performs a horizontal, steady, straight flight. Determine:

    (a) The velocity corresponding to the maximum aerodynamic efficiency.

3. The pilot switches off the engines and starts gliding at an altitude of 11000 m. Calculate:

    (a) The minimum descent velocity (vertical velocity), and the corresponding angle of descent, γ_d.

Solution to Problem 7.1:

Besides the data given in the statement, the following data have been used:

- $g = 9.81\ m/s^2$.
- $R = 287\ J/(kgK)$.
- $\alpha_T = 6.5 \cdot 10^{-3}\ K/m$.
- $\rho_0 = 1.225\ kg/m^3$.
- $T_0 = 288.15\ K$.
- ISA: $\rho = \rho_0(1 - \frac{\alpha_T h}{T_0})^{\frac{gR}{\alpha_T}-1}$.

1. *Uniform-ascent under the following flight conditions:*

    - $h = 3048\ m$. Using ISA $\rightarrow \rho = 0.904\ kg/m^3$.
    - $V = 148.67\ m/s$.
    - $\dot{h}_e = 14\ m/s$.

    The system that governs the motion of the aircraft is:

    $$T = D + mg \sin \gamma; \tag{7.46a}$$
    $$L = mg \cos \gamma; \tag{7.46b}$$
    $$\dot{x}_e = V \cos \gamma; \tag{7.46c}$$
    $$\dot{h}_e = V \sin \gamma. \tag{7.46d}$$

    Assuming that $\gamma \ll 1$, and thus that $\cos \gamma \sim 1$ and $\sin \gamma \sim \gamma$, System (7.46) becomes:

    $$T = D + mg\gamma; \tag{7.47a}$$
    $$L = mg; \tag{7.47b}$$
    $$\dot{x}_e = V; \tag{7.47c}$$
    $$\dot{h}_e = V\gamma. \tag{7.47d}$$

    a) *From Equation (7.47d), $\gamma = \frac{\dot{h}_e}{V} = 0.094\ rad\ (5.39°)$.*
    b) *From Equation (7.47a), $T = D + mg\gamma$.*

    $$D = C_D \frac{1}{2}\rho S_w V^2, \tag{7.48}$$

197

MECHANICS OF FLIGHT

where $C_D = 0.024 + 0.0375 C_L^2$, and ρ, S_w, V^2 are known.

$$C_L = \frac{L}{\frac{1}{2}\rho S_w V^2} = 0.512, \tag{7.49}$$

where, according to Equation (7.47b), $L = mg$. With Equation (7.49) in Equation (7.48), $D = 41398$ N.

Finally:

$$T = D + mg\gamma = 100 \text{ kN}.$$

2. **Horizontal, steady, straight flight under the following flight conditions:**

- $h = 11000$ m. Using ISA $\rightarrow \rho = 0.3636$ kg/m^3.

- The aerodynamic efficiency is maximum.

The system that governs the motion of the aircraft is:

$$T = D; \tag{7.50a}$$
$$L = mg \cos \gamma. \tag{7.50b}$$

The maximum Efficiency is $E_{max} = \dfrac{1}{2\sqrt{C_{D_0} C_{D_i}}} = 16.66$.

The optimal coefficient of lift is $C_{L_{opt}} = \sqrt{\dfrac{C_{D_0}}{C_{D_i}}} = 0.8$.

$$C_L = \frac{L}{\frac{1}{2}\rho S_w V^2} \rightarrow V = \sqrt{\frac{L}{\frac{1}{2}\rho S_w C_L}} = 187 \text{ m/s},$$

where, according to Equation (7.50b), $L = mg$, and in order to fly with maximum efficiency: $C_L = C_{L_{opt}}$.

3. **Gliding under the following flight conditions:**

- $h = 11000$ m. Using ISA $\rightarrow \rho = 0.3636$ kg/m^3.

- At the minimum descent velocity.

The system that governs the motion of the aircraft is:

$$D = mg \sin \gamma_d; \quad (7.51a)$$
$$L = mg \cos \gamma_d; \quad (7.51b)$$
$$\dot{x}_e = V \cos \gamma_d; \quad (7.51c)$$
$$\dot{h}_{e_{des}} = V \sin \gamma_d. \quad (7.51d)$$

Notice that $\gamma_d = -\gamma$.

Assuming that $\gamma_d \ll 1$, and thus that $\cos \gamma_d \sim 1$ and $\sin \gamma_d \sim \gamma_d$, System (7.51) becomes:

$$D = mg\gamma_d; \quad (7.52a)$$
$$L = mg; \quad (7.52b)$$
$$\dot{x}_e = V; \quad (7.52c)$$
$$\dot{h}_{e_{des}} = V\gamma_d. \quad (7.52d)$$

In order to fly with maximum descent velocity $\dot{h}_{e_{des}}$ must be maximum. Operating with Equation (7.52a) and Equation (7.52b), $\gamma_d = \frac{D}{L}$.

$$\dot{h}_{e_{des}} = V\gamma = V\frac{D}{L} = V\frac{(0.024 + 0.0375 C_L^2)\frac{1}{2}\rho S_w V^2}{C_L \frac{1}{2}\rho S_w V^2}. \quad (7.53)$$

Knowing that $C_L = \frac{L}{\frac{1}{2}\rho S_w V^2}$, where, according to Equation (7.52b), $L = mg$, Equation (7.53) becomes:

$$\dot{h}_{e_{des}} = \frac{V}{mg}\left(0.024\frac{1}{2}\rho S_w V^2 + \frac{0.0375(mg)^2}{\frac{1}{2}\rho S_w V^2}\right). \quad (7.54)$$

Make $\frac{\partial \dot{h}_{e_{des}}}{\partial V} = 0$.

The velocity with respect to air so that the vertical velocity is minimum is:

$$V = \sqrt[4]{\frac{4}{3}\frac{C_{D_i}}{C_{D_0}}}\sqrt{\frac{mg}{\rho S_w}} = 142.57 \text{ m/s}. \quad (7.55)$$

Substituting V=142.57 m/s in Equation (7.54), $\dot{h}_{e_{des}} = 9.87$ m/s.

Problem 7.2: Runway performances

We want to estimate the take-off distance of an Airbus A-320 taking off at Madrid-Barajas airport. Such aircraft mounts two turbojets, whose thrust can be estimated as: $T = T_0(1 - k \cdot V^2)$, where T is the thrust, T_0 is the nominal thrust, k is a constant and V is the true airspeed.

Considering that:

- $g \cdot \left(\frac{T_0}{m \cdot g} - \mu_r\right) = 1.31725 \; \frac{m}{s^2}$;
- $\frac{\rho S (C_D - \mu_r C_L) + 2 \cdot T_0 \cdot k}{2 \cdot m} = 3.69 \cdot 10^{-5} \; \frac{m}{s^2}$;
- *The velocity of take off is $V_{TO} = 70$ m/s;*

where g is the force due to gravity, m is the mass of the aircraft, μ_r is the friction coefficient, ρ is the density of air, S is the wet surface area of the aircraft, C_D is the coefficient of drag and C_L is the coefficient of lift[9].

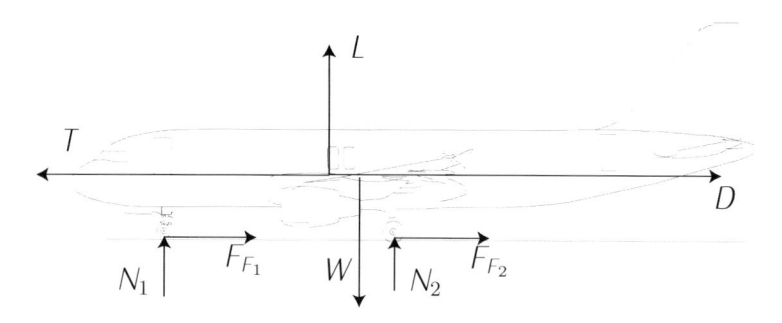

Figure 7.16: Forces during taking off (Problem 7.2).

Calculate:

1. *Take-off distance.*

---

[9]All this variables can be considered constant during take off.

7.3 PROBLEMS

Solution to Problem 7.2:

We apply the 2nd Newton's Law:

$$\sum F_z = 0, \qquad (7.56)$$

$$\sum F_x = m\dot{V}. \qquad (7.57)$$

Regarding Equation (7.56), notice that while rolling on the ground, the aircraft is assumed to be under equilibrium along the vertical axis.

Looking at Figure 7.16, Equations (7.56)-(7.57) become:

$$L + N - mg = 0; \qquad (7.58)$$

$$T - D - F_F = m\dot{V}. \qquad (7.59)$$

being L the lift, N the normal force, mg the weight; T the trust, D the drag and F_F the total friction force.

It is well known that:

$$L = C_L \frac{1}{2} \rho S V^2; \qquad (7.60)$$

$$D = C_D \frac{1}{2} \rho S V^2. \qquad (7.61)$$

It is also well known that:

$$F_F = \mu_r N. \qquad (7.62)$$

Equation (7.58) states that: $N = mg - L$. Therefore:

$$F_F = \mu_r (mg - L). \qquad (7.63)$$

Given that $T = T_0(1 - kV^2)$, with Equation (7.63) and Equations (7.60)-(7.61), Equation (7.59) becomes:

$$\left(\frac{T_0}{m} - \mu_r g\right) + \frac{(\rho S(\mu_r C_L - C_D) - 2T_0 k)}{2m} V^2 = \dot{V}. \qquad (7.64)$$

Now, we have to integrate Equation (7.64).

In order to do so, we know, as it was stated in the statement, that: $T_0, m, \mu_r, g, \rho, S, C_L, C_D$ and k can be considered constant along the take off phase.

We have that:

$$\frac{dV}{dt} = \frac{dV}{dx}\frac{dx}{dt}, \qquad (7.65)$$

201

MECHANICS OF FLIGHT

and knowing that $\frac{dx}{dt} = V$, Equation (7.64) becomes:

$$\frac{(\frac{T_0}{m} - \mu_r g) + \frac{(\rho S(\mu_r C_L - C_D) - 2T_0 k)}{2m} V^2}{V} = \frac{dV}{dx}. \qquad (7.66)$$

In order to simplify Equation (7.66):

- $(\frac{T_0}{m} - \mu_r g) = g(\frac{T_0}{mg} - \mu_r) = A$;

- $\frac{(\rho S(\mu_r C_L - C_D) - 2T_0 k)}{2m} = B$.

We proceed on integrating Equation (7.66) between $x = 0$ and x_{TO} (the distance of take off); $V = 0$ (Assuming the maneuver starts with the aircraft at rest) and the velocity of take off that was given in the statement: $V_{TO} = 70$ m/s. It holds that:

$$\int_0^{x_{TO}} dx = \int_0^{V_{TO}} \frac{V dV}{A + BV^2}. \qquad (7.67)$$

Integrating:

$$\left[x \right]_0^{x_{TO}} = \left[\frac{1}{2B} Ln(A + BV^2) \right]_0^{V_{TO}}. \qquad (7.68)$$

Substituting the upper and lower limits:

$$x_{TO} = \frac{1}{2B} Ln(1 + \frac{B}{A} V_{TO}^2). \qquad (7.69)$$

Substituting the data given in the statement:

- $A = 1.31725 \, \frac{m}{s^2}$;

- $B = -3.69 \cdot 10^{-5} \, \frac{m}{s^2}$;

- $V_{TO} = 70$ m/s.

The distance to take off is $x_{TO} = 2000$ m.

7.3 PROBLEMS

Problem 7.3: Performances

An aircraft has the following characteristics:

- $S_w = 130 \ m^2$.
- $b = 40 \ m$.
- $m = 70000 \ kg$.
- $T_{max,av}(h=0) = 120000 \ N$ (Maximum available thrust at sea level).
- $C_{D_0} = 0.02$.
- Oswald coefficient (wing efficiency coefficient): $e = 0.9$.
- $C_{L_{max}} = 1.5$.

We can consider that the maximum thrust only varies with altitude as follows: $T_{max,av}(h) = T_{max,av}(h=0)\frac{\rho}{\rho_0}$. Consider standard atmosphere ISA and constant gravity $g = 9.8 \ m/s^2$. Determine:

1. The required thrust to fly at an altitude of $h = 11000 \ m$ with $M_\infty = 0.7$ in horizontal, steady, straight flight.

2. The maximum velocity due to propulsive limitations of the aircraft and the corresponding Mach number in horizontal, steady, straight flight at $h = 11000$.

3. The minimum velocity due to aerodynamic limitations (stall speed) in horizontal, steady, straight flight at an altitude of $h = 11000$.

4. The theoretical ceiling (maximum altitude) in horizontal, steady, straight flight.

We want to perform a horizontal turn at an altitude of $h = 11000$ with a load factor $n = 2$, and with the velocity corresponding to the maximum aerodynamic efficiency in horizontal, steady, straight flight. Determine:

5. The required bank angle.
6. The radius of turn.
7. The required thrust. Can the aircraft perform the complete turn?

We want to perform a horizontal turn with the same load factor and the same radius as in the previous case, but at an altitude corresponding to the theoretical ceiling of the aircraft.

8. Can the aircraft perform such turn?

MECHANICS OF FLIGHT

Solution to Problem 7.3:

Besides the data given in the statement, the following data have been used:

- $R = 287$ J/(kgK).
- $\gamma_{air} = 1.4$.
- $\alpha_T = 6.5 \cdot 10^{-3}$ K/m.
- $\rho_0 = 1.225$ kg/m^3.
- $T_0 = 288.15$ K.
- ISA: $\rho = \rho_0 (1 - \frac{\alpha_T h}{T_0})^{\frac{gR}{\alpha_T} - 1}$.

1. **Required Thrust to fly a horizontal, steady, straight flight under the following flight conditions:**

   - $h = 11.000$ m;
   - $M_\infty = 0.7$.

   According to ISA:

   - $\rho(h = 11000) = 0.364$ Kg/m^3;
   - $a(h = 11000) = \sqrt{\gamma_{air} R (T_0 - \alpha_T h)} = 295.04$ m/s.

   where a corresponds to the speed of sound.
   The system that governs the dynamics of the aircraft is:

$$T = D; \qquad (7.70a)$$
$$L = mg; \qquad (7.70b)$$

being L the lift, mg the weight; T the trust and D the drag.
It is well known that:

$$L = C_L \frac{1}{2} \rho S_w V^2; \qquad (7.71)$$
$$D = C_D \frac{1}{2} \rho S_w V^2. \qquad (7.72)$$

It is also well known that the coefficient of drag can be expressed in a parabolic form as follows:

$$C_D = C_{D_0} + C_{D_i} C_L^2, \qquad (7.73)$$

where C_{D_0} is given in the statement and $C_{D_i} = \frac{1}{\pi A e}$. The enlargement, A, can be calculated as $A = \frac{b^2}{S_w} = 12.30$ and therefore: $C_{D_i} = 0.0287$.

According to Equation (7.70b): $L = 686000$ N. The velocity of flight can be calculated as $V = M_\infty a = 206.5$ m/s. Once these values are obtained, with the values of density and wet surface, and entering in Equation (7.71), we obtain that $C_L = 0.68$.

With the values of C_L, C_{D_i} and C_{D_0}, entering in Equation (7.73) we obtain that $C_D = 0.0332$.

Looking now at Equation (7.70a) and using Equation (7.72), we can state that the required thrust is as follows:

$$T = C_D \frac{1}{2}\rho S_w V^2.$$

Since all values are known, the required thrust yields:

$$T = 33567 \text{ N}.$$

Before moving on, we should look wether the required thrust exceeds or not the maximum available thrust at the given altitude. In order to do that, it has been given that the maximum thrust only varies with altitude as follows:

$$T_{max,av}(h) = T_{max,av}(h=0)\frac{\rho}{\rho_0}. \tag{7.74}$$

The maximum available thrust at h=11000 yields:

$$T_{max,av}(h=11000) = 35657.14 \text{ N}. \tag{7.75}$$

Since $T < T_{max,av}$, the flight condition is flyable.

2. **The maximum velocity due to propulsive limitations of the aircraft and the corresponding Mach number in horizontal, steady, straight flight at $h = 11000$:**

The maximum velocity due to propulsive limitation at the given altitude implies flying at the maximum available thrust that was obtained in Equation (7.75).

Looking again at Equation (7.70a) and using Equation (7.72), we can state that:

$$T_{max,av} = C_D \frac{1}{2}\rho S_w V^2. \tag{7.76}$$

Using Equation (7.73) and Equation (7.71), and entering in Equation (7.76) we have that:

$$T_{max,av} = \left(C_{D_0} + C_{D_i}\left(\frac{L}{\frac{1}{2}\rho S_w V^2}\right)^2\right)\frac{1}{2}\rho S_w V^2. \tag{7.77}$$

Mechanics of Flight

Multiplying Equation (7.77) by V^2 we obtain a quadratic equation of the form:

$$ax^2 + bx + c = 0. \qquad (7.78)$$

where $x = V^2$, $a = \frac{1}{2}\rho S_w C_{D_0}$, $b = -T_{max,av}$, and $c = \frac{C_{D_L} L^2}{\frac{1}{2}\rho S_w}$.

Solving the quadratic equation we obtain two different speeds at which the aircraft can fly given the maximum available thrust[10]. Those velocities yield:

$$V_1 = 228 \ m/s;$$
$$V_2 = 151 \ m/s.$$

The maximum corresponds, obviously, to V_1.

3. **The minimum velocity due to aerodynamic limitations (stall speed) in horizontal, steady, straight flight at an altitude of** $h = 11000$.

The stall speed takes place when the coefficient of lift is maximum, therefore, using equation Equation (7.71), we have that:

$$V_{stall} = \sqrt{\frac{L}{\frac{1}{2}\rho S_w C_{L_{max}}}} = 139 \ m/s.$$

4. **The theoretical ceiling (maximum altitude) in horizontal, steady, straight flight.**

In order to obtain the theoretical ceiling of the aircraft the maximum available thrust at that maximum altitude must coincide with the minimum required thrust to fly horizontal, steady, straight flight at that maximum altitude, that is:

$$T_{max,av} = T_{min}. \qquad (7.79)$$

Let us first obtain the minimum required thrust to fly horizontal, steady, straight flight. Multiplying and dividing by L in the second term of Equation (7.70a), and given that the aerodynamic efficiency is $E = \frac{L}{D}$, we have that:

$$T = \frac{D}{L}L = \frac{L}{E}. \qquad (7.80)$$

Since L is constant at those conditions of flight, the minimum required thrust occurs when the efficiency is maximum: $T_{min} \Leftrightarrow E_{max}$.

---

[10] Notice that given an altitude and a thrust setting, the aircraft can theoretically fly at two different velocities meanwhile those velocities lay between the minimum velocity (stall) and a maximum velocity (typically near the divergence velocity).

7.3 PROBLEMS

Let us now proceed deducing the maximum aerodynamic efficiency:

The aerodynamic efficiency is defined as:

$$E = \frac{L}{D} = \frac{C_L}{C_D}. \tag{7.81}$$

Substituting the parabolic polar curve given in Equation (7.73) in Equation (7.81), we obtain:

$$E = \frac{C_L}{C_{D_0} + C_{D_i} C_L^2}. \tag{7.82}$$

In order to seek the values corresponding to the maximum aerodynamic efficiency, one must derivate and make it equal to zero, that is:

$$\frac{dE}{dC_L} = 0 = \frac{C_{D_0} - C_{D_i} C_L^2}{(C_{D_0} + C_{D_i} C_L^2)^2} \rightarrow (C_L)_{E_{max}} = C_{L_{opt}} = \sqrt{\frac{C_{D_0}}{C_{D_i}}}. \tag{7.83}$$

Substituting the value of $C_{L_{opt}}$ into Equation (7.82) and simplifying we obtain that:

$$E_{max} = \frac{1}{2\sqrt{C_{D_0} C_{D_i}}}.$$

E_{max} yields 20.86, and $T_{min} = 32870$ N.

According to Equation (7.75) and based on Equation (7.79) with $T_{min} = 32870$ N, we have that:

$$32870 = T_{max,av}(h = 0) \frac{\rho}{\rho_0}.$$

Given that $T_{max,av}(h = 0)$ was given in the statement and ρ_0 is known according to ISA, we have that $\rho = 0.335$ kg/m³.

Since $\rho_{h_{max}} < \rho_{11000}$ we can easily deduce that the ceiling belongs to the stratosphere. Using the ISA equation corresponding to the stratosphere we have that:

$$\rho_{h_{max}} = \rho_{11} \exp^{-\frac{g}{RT_{11}}(h_{max} - h_{11})}, \tag{7.84}$$

where the subindex 11 corresponds to the values at the tropopause (h=11000 m). Operating in Equation (7.84), the ceiling yields $h_{max} = 11526$ m.

5. **The required bank angle:**

The equations governing the dynamics of the airplane in an horizontal turn are:

$$T = D; \tag{7.85a}$$
$$mV\dot\chi = L\sin\mu; \tag{7.85b}$$
$$L\cos\mu = mg, \tag{7.85c}$$

In a uniform (stationary) circular movement, it is well known that the tangential velocity is equal to the angular velocity ($\dot\chi$) multiplied by the radius of turn (R):

$$V = \dot\chi R. \tag{7.86}$$

Therefore, System (7.85) can be rewritten as:

$$T = D; \tag{7.87a}$$
$$n\sin\mu = \frac{V^2}{gR}; \tag{7.87b}$$
$$n = \frac{1}{\cos\mu}; \tag{7.87c}$$

where $n = \frac{L}{mg}$ is the load factor.

Therefore, looking at Equation (7.87c), it is straightforward to determine that the bank angle of turn is $\mu = 60°$.

6. **The radius of turn.**

First, we need to calculate the velocity corresponding to the maximum efficiency. As we have calculated before in Equation (7.83), the coefficient of lift that generates maximum efficiency is the so-called optimal coefficient of lift, that is, $C_{L_{opt}} = \sqrt{\frac{C_{D_0}}{C_{D_i}}} = 0.834$. The velocity yields then:

$$V = \sqrt{\frac{L}{\frac{1}{2}\rho S_w C_{L_{opt}}}} = 186.4 \text{ m/s}.$$

Entering in Equation (7.87b) with $\mu = 60$ [deg] and $V = 186.4$ m/s: $R = 4093.8$ m.

7.3 PROBLEMS

7. **The required thrust.**

As exposed in Question 3., Equation (7.87a) can be expressed as:

$$T = \frac{1}{2}\rho S_w V^2 C_{D_0} + \frac{L^2}{\frac{1}{2}\rho V^2 S_w} C_{D_i}.$$

Since all values are known: $T = 32451$ N.

Since $T \leq T_{max,av}(h = 110000)$, the aircraft can perform the turn.

8. **We want to perform a horizontal turn with the same load factor and the same radius as in the previous case, but at an altitude corresponding to the theoretical ceiling of the aircraft. Can the aircraft perform the turn?**

If the load factor is the same, $n = 2$, necessarily (according to Equation (7.87c)) the bank angle is the same, $\mu = 60°$. Also, if the radius of turn is the same, $R = 4093.8$ m, necessarily (according to Equation (7.87b)), the velocity of the turn must be the same as in the previous case, $V = 186.4$ m/s. Obviously, since the density will change according to the new altitude ($\rho = 0.335$ Kg/m^3), the turn will not be performed under maximum efficiency conditions.

In order to know wether the turn can be performed or not, we must compare the required thrust with the maximum available thrust at the ceiling altitude:

$$T = \frac{1}{2}\rho S_w V^2 C_{D_0} + \frac{L^2}{\frac{1}{2}\rho V^2 S_w} C_{D_i} = 32983 \text{ N}.$$

$$T_{max,av}(h = 11526) = T_{max,av}(h = 0)\frac{\rho}{\rho_0} = 32816 \text{ N}.$$

Since $T > T_{max,av}$ at the ceiling, the turn can not be performed.

Problem 7.4: Weights

Consider an Airbus A-320. Simplifying, we assume the wing of the aircraft is rectangular and it is composed of NACA 4415 airfoils. The characteristics of the NACA 4415 airfoils as as follows:

- $c_l = 0.2 + 5.92\alpha$. (α in radians).
- $c_d = 6.4 \cdot 10^{-3} - 1.2 \cdot 10^{-3} c_l + 3.5 \cdot 10^{-3} c_l^2$.

Regarding the aircraft, the following data are known:

- *Wing wet surface of 122.6 $[m^2]$.*
- *Wing-span of 34.1 $[m]$.*
- *Oswald efficiency factor of 0.95.*
- *Specific consumption per unity of thrust and time: $\eta_j = 6.8 \cdot 10^{-5} [\frac{Kg}{N \cdot s}]$.*

Calculate:

1. *The lift curve of the wing in its linear range.*
2. *The drag polar curve, assuming it can be expressed as: $C_D = C_{D_0} + C_{D_i} C_L^2$.*
3. *The optimal coefficient of lift of the wing, $C_{L_{opt}}$. Compare it with the airfoil's one.*

On regard of the characteristic weights of the aircraft, the following data are known:

- *Operating empty weight OEW=42.4 $[Ton.]$.*
- *Maximum take-off weight $MTOW = 77000 \cdot g$ $[N]$[11].*
- *Maximum fuel weight $MFW = 29680 \cdot g$ $[N]$.*
- *Maximum zero fuel weight $MZFW = 59000 \cdot g$ $[N]$.*
- *Moreover, the reserve fuel (RF) can be calculated as the 5% of the Trip Fuel (TF).*

---

[11] g represents the force due to gravity.

7.3 Problems

Calculate:

4. *The Payload and Trip Fuel in the following cases:*
   (a) *Initial weight equal to MTOW with the Maximum Payload (MPL).*
   (b) *Initial weight equal to MTOW with the Maximum Fuel Weight (MFW).*
   (c) *Initial weight equal to OEW plus MFW.*

Assuming that in cruise conditions the aircraft flies at a constant altitude of $h = 11500$ [m], constant Mach number M=0.78, and maximum aerodynamic efficiency, considering ISA standard atmosphere, calculate:

5. *Range and autonomy of the aircraft for the three initial weights pointed out above.*
6. *According to the obtained results, draw the payload-range diagram.*

Solution to Problem 7.4:

1. *Wing's lift curve:*

   The lift curve of a wing can be expressed as follows:

   $$C_L = C_{L_0} + C_{L_\alpha}\alpha, \qquad (7.88)$$

   and the slope of the wing's lift curve can be expressed related to the slope of the airfoil's lift curve as:

   $$C_{L_\alpha} = \frac{C_{l_\alpha}}{1 + \frac{C_{l_\alpha}}{\pi A}} e = 4.69 \cdot 1/rad.$$

   In order to calculate the independent term of the wing's lift curve, we must consider the fact that the zero-lift angle of attack of the wing coincides with the zero-lift angle of attack of the airfoil, that is:

   $$\alpha(L = 0) = \alpha(l = 0). \qquad (7.89)$$

   First, notice that the lift curve of an airfoil can be expressed as follows

   $$C_l = C_{l_0} + C_{l_\alpha}\alpha. \qquad (7.90)$$

   Therefore, with Equation (7.88) and Equation (7.89) in Equation (7.90), we have that:

   $$C_{L_0} = C_{l_0}\frac{C_{L_\alpha}}{C_{l_\alpha}} = 0.158.$$

   The required curve yields then:

   $$C_L = 0.158 + 4.69\alpha \quad [\alpha \text{ in } rad].$$

2. *The expression of the parabolic polar of the wing:*

   Notice first that the statement of the problem indicates that the polar should be in the following form:

   $$C_D = C_{D0} + C_{D_i}C_L^2. \qquad (7.91)$$

   For the calculation of the parabolic drag of the wing we can consider the parasite term approximately equal to the parasite term of the airfoil, that is, $C_{D_0} = C_{d_0}$.

212

7.3 Problems

The induced coefficient of drag can be calculated as follows:

$$C_{D_i} = \frac{1}{\pi A e} = 0.035.$$

The expression of the parabolic polar yields then:

$$C_D = 0.0064 + 0.035 C_L^2. \tag{7.92}$$

3. **The optimal coefficient of lift, $C_{L_{opt}}$, for the wing. Compare it with the airfoils's one.**

The optimal coefficient of lift is that making the aerodynamic efficiency maximum. The aerodynamic efficiency is defined as:

$$E = \frac{L}{D} = \frac{C_L}{C_D}. \tag{7.93}$$

Substituting the parabolic polar curve given in Equation (7.91) in Equation (7.93), we obtain:

$$E = \frac{C_L}{C_{D_0} + C_{D_i} C_L^2}. \tag{7.94}$$

In order to seek the values corresponding to the maximum aerodynamic efficiency, one must derivate and make it equal to zero, that is:

$$\frac{dE}{dC_L} = 0 = \frac{C_{D_0} - C_{D_i} C_L^2}{(C_{D_0} + C_{D_i} C_L^2)^2} \rightarrow (C_L)_{E_{max}} = C_{L_{opt}} = \sqrt{\frac{C_{D_0}}{C_{D_i}}}. \tag{7.95}$$

For the case of an airfoil, the aerodynamic efficiency is defined as:

$$E = \frac{l}{d} = \frac{c_l}{c_d}. \tag{7.96}$$

Substituting the parabolic polar curve given in the statement in the form $c_{d_0} + b c_l + k c_l^2$ in Equation (7.96), we obtain:

$$E = \frac{c_l}{c_{d_0} + b c_l + k c_l^2}. \tag{7.97}$$

In order to seek the values corresponding to the maximum aerodynamic efficiency, one must derivate and make it equal to zero, that is:

$$\frac{dE}{dC_l} = 0 = \frac{c_{d_0} - k c_l^2}{(c_{d_0} + b c_l + k c_l^2)^2} \rightarrow (c_l)_{E_{max}} = (c_l)_{opt} = \sqrt{\frac{c_{d_0}}{k}}. \tag{7.98}$$

According to the values previously obtained ($C_{D_0} = 0.0064$ and $C_{D_i} = 0.035$) and the values given in the statement for the airfoil's polar ($c_{d_0} = 0.0064$, $k = 0.0035$), substituting them in Equation (7.95) and Equation (7.98), respectively, we obtain:

- $(C_L)_{opt} = 0.42$;

- $(c_l)_{opt} = 1.35$.

4. **Payload and Trip Fuel for cases a), b), and c):**

Before starting we the particular cases, it is necessary to point out that:

$$TOW = OEW + PL + FW, \qquad (7.99)$$
$$FW = TF + RF, \qquad (7.100)$$
$$MZFW = OEW + MPL. \qquad (7.101)$$

Moreover, according to the statement,

$$RF = 0.05 \cdot TF. \qquad (7.102)$$

Based on the data given in the statement, and using Equation (7.101):

$$MPL = 16.6 \ [Ton.]. \qquad (7.103)$$

Case a) Initial weight[12] $\rightarrow$ MTOW with MPL.

Equation (7.99) becomes:

$$MTOW = OEW + MPL + TF + 0.05 \cdot TF, \qquad (7.104)$$

Isolating in Equation (7.104): $TF = 17.14 \ [Ton]$. The payload is equal to the maximum payload MPL.

Case b) Initial weight $\rightarrow$ MTOW with MFW.

Equation (7.99) becomes:

$$MTOW = OEW + PL + MFW, \qquad (7.105)$$

Isolating in Equation (7.105): $PL = 4.92 \ [Ton]$. In order to calculate the trip fuel, looking at Equation (7.100), we have that

$$MFW = TF + RF = TF + 0.05 \cdot TF. \qquad (7.106)$$

---

[12]Notice that we can convert mass in weight by simply multiplying by the force due to gravity.

Isolating, $TF = 28.266 \, [Ton.]$.

Case c) Initial weight $\to$ OEW + MFW.

Equation (7.99) becomes:

$$TOW = OEW + MFW, \qquad (7.107)$$

That means $PL = 0$. In order to calculate the trip fuel, we proceed exactly as in Case b). Looking at Equation (7.100), we have that

$$MFW = TF + RF = TF + 0.05 \cdot TF. \qquad (7.108)$$

Isolating, $TF = 28.266 \, [Ton.]$

5. **Range and Endurance for cases a), b), and c):** Considering that the aircraft performs a linear, horizontal, steady flight, we have that:

$$L = mg; \qquad (7.109)$$
$$T = D; \qquad (7.110)$$
$$\dot{x} = V; \qquad (7.111)$$
$$\dot{m} = -\eta T. \qquad (7.112)$$

Since $\dot{x} = \frac{dx}{dt}$, it is clear that the Range, R, looking at Equation (7.111), can be expressed as:

$$R = \int_{t_i}^{t_f} V \, dt. \qquad (7.113)$$

Now, since $\dot{m} = \frac{dm}{dt} = -\eta T$, Equation (7.113) yields:

$$R = \int_{m_i}^{m_f} -\frac{V}{\eta T} dm, \qquad (7.114)$$

where m_i is the initial mass and m_f is the final mass.

Using Equation (7.109) and Equation (7.110), Equation (7.114) yields:

$$R = \int_{m_i}^{m_f} -\frac{VE}{\eta g} \frac{dm}{m}. \qquad (7.115)$$

with $E = \frac{L}{D}$, V, η, and g are constant values.

Integrating:

$$R = \frac{VE}{\eta g}\ln\left(\frac{m_i}{m_f}\right). \tag{7.116}$$

For the endurance, we operate analogously, integrating Equation (7.112), which yields

$$t = \int_{m_i}^{m_f} -\frac{1}{\eta T}dm. \tag{7.117}$$

Using Equation (7.109) and Equation (7.110), Equation (7.117) yields:

$$t = \int_{m_i}^{m_f} -\frac{E}{\eta g}\frac{dm}{m}. \tag{7.118}$$

where again $E = \frac{L}{D}$, η, and g are constant values. Integrating:

$$t = \frac{E}{\eta g}\ln\left(\frac{m_i}{m_f}\right). \tag{7.119}$$

Before calculating Range and Endurance for the three cases, we need to calculate the values of velocity and aerodynamic efficiency, which is maximum.

The velocity can be expressed as $V = M \cdot a$, where a is the speed of sound, that can be expressed as

$$a = \sqrt{\gamma R T}, \tag{7.120}$$

where $\gamma = 1.4$ is adiabatic coefficient of air, and $R = 287$ J/KgK is the perfect gas constant. Notice that, using ISA, the temperature at the stratosphere is constant, and thus $T=216.6$ K.

Substituting all terms, it yields $V = 230.1$ $[m/s]$.

The aerodynamic efficiency is maximum. Thus, substituting $C_{L_{opt}}$ obtained in Equation (7.95) into Equation (7.94), it yields:

$$E = \frac{1}{2\sqrt{C_{D_0}C_{D_i}}} = 33.4. \tag{7.121}$$

Now, we should calculate the initial and final mass for of the three cases a), b), c). Notice that the final mass results from subtracting the trip fuel from the take-off mass:

$$m_f = m_i - TF. \tag{7.122}$$

Thus,

a) $m_i = \frac{MTOW}{g}$ *[Kg] and* $m_f = 59860$ *[Kg].*

b) $m_i = \frac{MTOW}{g}$ *[Kg] and* $m_f = 48734$ *[Kg].*

c) $m_i = 72080$ *[Kg] and* $m_f = 43814$ *[Kg].*

Substituting in Equation (7.116) and Equation (7.119), it yields:

a) $R_a = 2900$ *[Km] and* $t_a = 12600$ *[s].*

b) $R_b = 5270$ *[Km] and* $t_b = 22900$ *[s].*

c) $R_c = 5736$ *[Km] and* $t_c = 24928$ *[s].*

6. *Payload-Range diagram cases a), b), and c):*

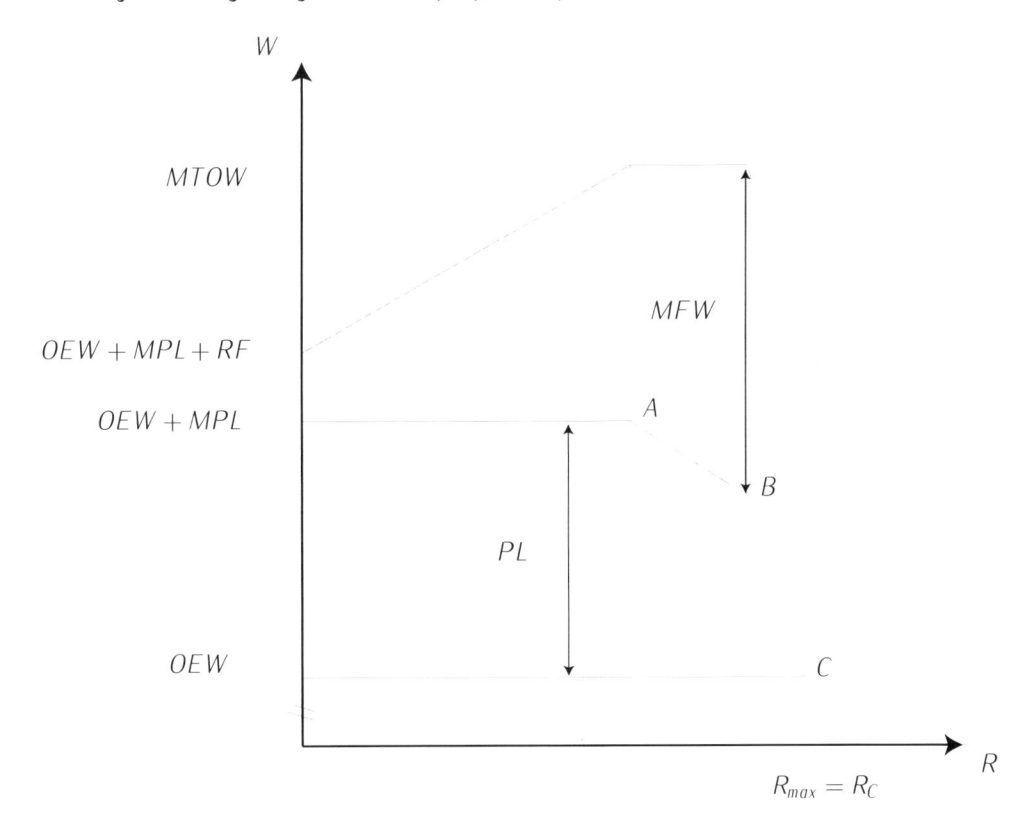

Figure 7.17: Payload-range diagram (Problem 7.4).

Problem 7.5: Performances and stability

An aircraft has the following characteristics:

- $S_w = 130\ m^2$.
- $b = 40\ m$.
- $m = 70000\ kg$.
- $T_{max,av,0} = 130000\ N$ (Maximum available thrust at sea level).
- $C_{D_0} = 0.02$.
- Oswald efficiency factor: $e = 0.9$.

The maximum available thrust can be considered to vary according to the following law:

$$T_{max,av}(h) = T_{max,av,0}\frac{\rho}{\rho_0}.$$

Consider ISA atmosphere and constant gravity $g = 9.8\ m/s^2$. Determine:

1. The required thrust to fly at an altitude of $h = 11250\ m$ at $M_\infty = 0.78$ in steady linear-horizontal flight.

2. The maximum speed of the aircraft due to propulsive limitations in steady linear-horizontal flight at an altitude of $h = 11250\ m$.

3. The theoretical ceiling of the aircraft in steady linear-horizontal flight.

We want now to determine the surface of the horizontal stabilizer and as a design criterion we take the flight conditions of equilibrium at an altitude of $h = 11250\ m\ y\ M_\infty = 0.78$. In those conditions, the coefficient of lift of the stabilizer is equal to 1.4. Moreover, we can assume that the distribution of lift of the wing can be reduced to a resultant force in the aerodynamic center (lift) and a pitching down moment with respect to the aerodynamic center equal to $10000\ N \cdot m$. The distance between the aerodynamic center and the center of gravity (note that the aerodynamic center is closer to the nose of the aircraft) is $x_{cg} = 2\ m$. The distance between the stabilizer and the center of gravity is $l = 20\ m$.

4. Determine the surface of the horizontal stabilizer for those flight conditions.

Solution to Problem 7.5:

Besides the data given in the statement, the following data have been used:

- $R = 287$ J/(kgK).
- $\gamma_{air} = 1.4$.
- $T_{11} = 216.6$ K.
- $\rho_{11} = 0.36$ kg/m^3.
- $T_0 = 288.15$ K.
- ISA: $\rho = \rho_{11} \cdot e^{\frac{-gR}{T_{11}}(h-11000)}$.

1. **Required Thrust to fly a horizontal, steady, straight flight under the following flight conditions:**

   - $h = 11.250$m.
   - $M_\infty = 0.78$.

   According to ISA:

   - $\rho(h = 11250) = 0.3461$ Kg/m^3.
   - $a(h = 11250) = \sqrt{\gamma_{air} R(T_{11})} = 295$ m/s.

where a corresponds to the speed of sound.

The system that governs the dynamics of the aircraft is:

$$T = D, \qquad (7.123a)$$
$$L = mg, \qquad (7.123b)$$

being L the lift, mg the weight; T the trust and D the drag.

It is well known that :

$$L = C_L \frac{1}{2} \rho S_w V^2; \qquad (7.124)$$

$$D = C_D \frac{1}{2} \rho S_w V^2. \qquad (7.125)$$

It is also well known that the coefficient of drag can be expressed in a parabolic form as follows:

$$C_D = C_{D_0} + C_{D_i} C_L^2, \qquad (7.126)$$

where C_{D_0} is given in the statement and $C_{D_i} = \frac{1}{\pi A e}$. The enlargement A can be calculated as $A = \frac{b^2}{S_w} = 12.30$ and therefore: $C_{D_i} = 0.0287$.

According to Equation (7.123b): $L = 686000$ N. The velocity of flight can be calculated as $V = M_\infty a = 230.1$ m/s. Once these values are obtained, with the values of density and wet surface, and entering in Equation (7.124), we obtain that $C_L = 0.5795$.

With the values of C_L, C_{D_i} and C_{D_0}, entering in Equation (7.126) we obtain that $C_D = 0.0295$.

Looking now at Equation (7.123a) and using Equation (7.125), we can state that the required thrust is as follows:

$$T = C_D \frac{1}{2} \rho S_w V^2.$$

Since all values are known, the required thrust yields:

$$T = 35185 \text{ N}.$$

Before moving on, we should look wether the required thrust exceeds or not the maximum available thrust at the given altitude. In order to do that, it has been given that the maximum thrust only varies with altitude as follows:

$$T_{max,av}(h) = T_{max,av,0} \frac{\rho}{\rho_0}. \tag{7.127}$$

The maximum available thrust at $h=11250$ yields:

$$T_{max,av}(h = 11250) = 36729 \text{ N}. \tag{7.128}$$

Since $T < T_{max,av}$, the flight condition is flyable.

2. **The maximum velocity due to propulsive limitations of the aircraft in horizontal, steady, straight flight at $h = 11250$:**

The maximum velocity due to propulsive limitation at the given altitude implies flying at the maximum available thrust that was obtained in Equation (7.128).

Looking again at Equation (7.123a) and using Equation (7.125), we can state that:

$$T_{max,av} = C_D \frac{1}{2} \rho S_w V^2. \tag{7.129}$$

Using Equation (7.126) and Equation (7.124), and entering in Equation (7.129) we have that :

$$T_{max,av} = \left(C_{D_0} + C_{D_i}\left(\frac{L}{\frac{1}{2}\rho S_w V^2}\right)^2\right)\frac{1}{2}\rho S_w V^2. \qquad (7.130)$$

Multiplying Equation (7.130) by V^2 we obtain a quadratic equation of the form:

$$ax^2 + bx + c = 0. \qquad (7.131)$$

where $x = V^2$, $a = \frac{1}{2}\rho S_w C_{D_0}$, $b = -T_{max,av}$ and $c = \frac{C_{D_i}L^2}{\frac{1}{2}\rho S_w}$.

Solving the quadratic equation we obtain two different speeds at which the aircraft can fly given the maximum available thrust[13]. Those velocities yield:

$$V_1 = 218 \ m/s;$$
$$V_2 = 184.06 \ m/s.$$

The maximum speeds corresponds, obviously, to V_1.

3. **The theoretical ceiling (maximum altitude) in horizontal, steady, straight flight.**

In order to obtain the theoretical ceiling of the aircraft the maximum available thrust at that maximum altitude must coincide with the minimum required thrust to fly horizontal, steady, straight flight at that maximum altitude, that is:

$$T_{max,av} = T_{min}. \qquad (7.132)$$

Let us first obtain the minimum required thrust to fly horizontal, steady, straight flight. Multiplying and dividing by L in the second term of Equation (7.123a), and given that the aerodynamic efficiency is $E = \frac{L}{D}$, we have that:

$$T = \frac{D}{L}L = \frac{L}{E}. \qquad (7.133)$$

Since L is constant at those conditions of flight, the minimum required thrust occurs when the efficiency is maximum: $T_{min} \Leftrightarrow E_{max}$.

Let us now proceed deducing the maximum aerodynamic efficiency:

The aerodynamic efficiency is defined as:

$$E = \frac{L}{D} = \frac{C_L}{C_D}. \qquad (7.134)$$

---

[13] Notice that given an altitude and a thrust setting, the aircraft can theoretically fly at two different velocities meanwhile those velocities lay between the minimum velocity (stall) and a maximum velocity (typically near the divergence velocity).

MECHANICS OF FLIGHT

Substituting the parabolic polar curve given in Equation (7.126) in Equation (7.134), we obtain:

$$E = \frac{C_L}{C_{D_0} + C_{D_i} C_L^2}. \tag{7.135}$$

In order to seek the values corresponding to the maximum aerodynamic efficiency, one must derivate and make it equal to zero, that is:

$$\frac{dE}{dC_L} = 0 = \frac{C_{D_0} - C_{D_i} C_L^2}{(C_{D_0} + C_{D_i} C_L^2)^2} \rightarrow (C_L)_{E_{max}} = C_{L_{opt}} = \sqrt{\frac{C_{D_0}}{C_{D_i}}}. \tag{7.136}$$

Substituting the value of $C_{L_{opt}}$ into Equation (7.135) and simplifying we obtain that:

$$E_{max} = \frac{1}{2\sqrt{C_{D_0} C_{D_i}}}.$$

E_{max} yields 20.86, and $T_{min} = 32904$ N.

According to Equation (7.128) and based on Equation (7.132) with $T_{min} = 32904$ N, we have that:

$$32904 = T_{max,av,0} \frac{\rho}{\rho_0}.$$

Given that $T_{max,av,0}$ was given in the statement and ρ_0 is known according to ISA, we have that $\rho = 0.3101$ kg/m^3.

Since $\rho_{h_{max}} < \rho_{11000}$ we can easily deduce that the ceiling belongs to the stratosphere. Using the ISA equation corresponding to the stratosphere we have that:

$$\rho_{h_{max}} = \rho_{11} \exp^{-\frac{g}{RT_{11}}(h_{max}-h_{11})}, \tag{7.137}$$

where the subindex 11 corresponds to the values at the tropopause (h=11000 m). Operating in Equation (7.137), the ceiling yields $h_{max} = 11946$ m.

4. **We want to determine the surface of the horizontal stabilizer:**

   In order to do that, we state the equations for the longitudinal balancing problem:

$$mg - L - L_t = 0; \tag{7.138}$$
$$-M_{ca} + L x_{cg} - L_t l = 0; \tag{7.139}$$

where L_t is the lift generated by the horizontal stabilizer, $M_{ca} = 10000$ Nm is the pitch torque with respect to the aerodynamic center, $x_{cg} = 2$ m is the distance between the

7.3 PROBLEMS

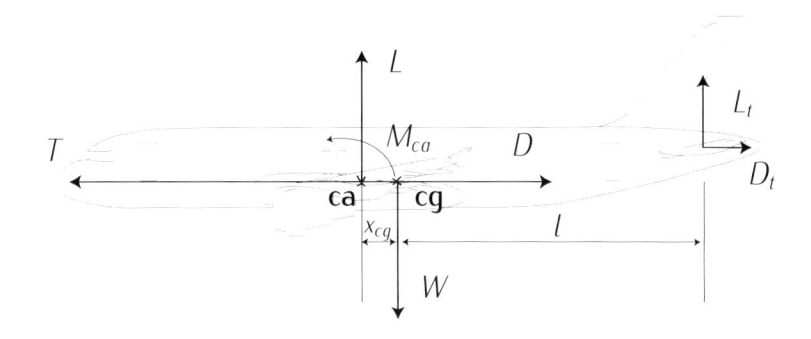

Figure 7.18: Longitudinal equilibrium.

center of gravity and the aerodynamic center, and $l = 20$ m is the distance between the center of gravity and the aerodynamic center of the horizontal stabilizer.

Entering in Equation (7.138), we have that $L = mg - L_t$. Knowing that $L_t = 0.5\rho S_t V^2 C_{L_t}$ and substituting in Equation (7.139), we have that:

$$L_t = \frac{mg \cdot x_{cg} - M_{ca}}{l + x_{cg}} \rightarrow S_t = \frac{1}{0.5\rho V^2 C_{L_t}} \frac{mg \cdot x_{cg} - M_{ca}}{l + x_{cg}} = 4.78 \text{ m}^2. \quad (7.140)$$

Problem 7.6: Performances

Consider an Airbus A-320 with the following characteristics:

- $m = 64$ tonnes.
- $S_w = 122.6\ m^2$.
- $T_{max,av,0} = 130000\ N$ (maximum available thrust at sea level).
- $C_D = 0.024 + 0.0375 C_L^2$.

The maximum available thrust can be considered to vary according to the following law:

$$T_{max,av}(h) = T_{max,av,0} \frac{\rho}{\rho_0}.$$

Consider ISA atmosphere and constant gravity $g = 9.8\ m/s^2$.

1. The aircraft starts an uniform ascent maneuver (constant speed and flight path angle) at an altitude of 10.000 feet (3048 m). At this flight level, the aerodynamic speed is 150 m/s and the vertical speed is 12 m/s. Assuming a small angle of attack $\gamma \ll 1$, calculate:

    (a) The ascent flight path angle, γ.
    (b) The required thrust at these conditions.
    (c) Thrust relation[14] with respect to the maximum available thrust at that altitude.

2. Calculate the maximum angle of ascent (flight path angle) in uniform ascent at an altitude of 10000 feet. In these conditions, determine:

    (a) The aerodynamic speed and the vertical speed.

3. We want now to analyze the turn performances in the horizontal plane. Consider sea level conditions, maximum aerodynamic efficiency, and structural limitations characterized by a maximum load factor of $n_{max} = 2.5$. in these conditions, calculate:

    (a) The required bank angle.
    (b) Speed and radius of turn.
    (c) The required thrust.
    (d) Can the aircraft perform the complete turn?

---

[14]it refers to a value between 0 y 1 associated to the throttle level position.

Solution to Problem 7.6:

Besides the data given in the statement, the following data have been used:

- $g = 9.81 \ m/s^2$.
- $R = 287 \ J/(kgK)$.
- $\alpha_T = 6.5 \cdot 10^{-3} \ K/m$.
- $\rho_0 = 1.225 \ kg/m^3$.
- $T_0 = 288.15 \ K$.
- ISA: $\rho = \rho_0 (1 - \frac{\alpha_T h}{T_0})^{\frac{gR}{\alpha_T} - 1}$.

1. **Uniform-ascent under the following flight conditions:**

   - $h = 3048 \ m$. Using ISA $\rightarrow \rho = 0.904 \ kg/m^3$.
   - $V = 150 \ m/s$.
   - $\dot{h}_e = 12 \ m/s$.

   The system that governs the motion of the aircraft is:

   $$T = D + mg \sin \gamma; \tag{7.141a}$$
   $$L = mg \cos \gamma; \tag{7.141b}$$
   $$\dot{x}_e = V \cos \gamma; \tag{7.141c}$$
   $$\dot{h}_e = V \sin \gamma. \tag{7.141d}$$

   Assuming that $\gamma \ll 1$, and thus that $\cos \gamma \sim 1$ and $\sin \gamma \sim \gamma$, System (7.141) becomes:

   $$T = D + mg\gamma; \tag{7.142a}$$
   $$L = mg; \tag{7.142b}$$
   $$\dot{x}_e = V; \tag{7.142c}$$
   $$\dot{h}_e = V\gamma. \tag{7.142d}$$

   a) *From Equation (7.142d), $\gamma = \frac{\dot{h}_e}{V} = 0.08 \ rad \ (4.58°)$.*
   b) *From Equation (7.142a), $T = D + mg\gamma$.*

   $$D = C_D \frac{1}{2} \rho S_w V^2, \tag{7.143}$$

225

where $C_D = 0.024 + 0.0375 C_L^2$, and ρ, S_w, V^2 are known.

$$C_L = \frac{L}{\frac{1}{2}\rho S_w V^2} = 0.5057, \tag{7.144}$$

where, according to Equation (7.142b), $L = mg$. With Equation (7.144) in Equation (7.143), $D = 41398$ N.

Finally:

$$T = D + mg\gamma = 91886 \text{ N}.$$

c) The maximum thrust at those conditions is

$$T_{max,av}(h = 3048) = T_{max,av,0} \frac{\rho}{\rho_0} = 95510 \text{ N}.$$

The relation is $\Pi = \frac{T}{T_{max,av}} = 0.962$.

2. **Maximum angle of ascent.**

We consider again the set of equations (7.142). Diving Equation (7.142a) by mg, we have that :

$$\gamma = \frac{T}{mg} - \frac{1}{E}. \tag{7.145}$$

In order γ to be maximum:

- $T = T_{max,av} = 95510$ N.

- $E = E_{max}$.

Let us now proceed deducing the maximum aerodynamic efficiency:

The aerodynamic efficiency is defined as:

$$E = \frac{L}{D} = \frac{C_L}{C_D}. \tag{7.146}$$

Substituting the parabolic polar curve given in the statement in Equation (7.146), we obtain:

$$E = \frac{C_L}{C_{D_0} + C_{D_i} C_L^2}. \tag{7.147}$$

In order to seek the values corresponding to the maximum aerodynamic efficiency, one *must derivate and make it equal to zero, that is:*

$$\frac{dE}{dC_L} = 0 = \frac{C_{D_0} - C_{D_i} C_L^2}{(C_{D_0} + C_{D_i} C_L^2)^2} \rightarrow (C_L)_{E_{max}} = C_{L_{opt}} = \sqrt{\frac{C_{D_0}}{C_{D_i}}}. \quad (7.148)$$

Substituting the value of $C_{L_{opt}}$ into Equation (7.147) and simplifying we obtain that:

$$E_{max} = \frac{1}{2\sqrt{C_{D_0} C_{D_i}}}.$$

Substituting: $C_{L_{opt}} = 0.8$, $E_{max} = 16.66$, and $\gamma_{max} = 5.27°$.
The aerodynamic velocity will be given by:

$$V = \sqrt{\frac{m \cdot g}{\frac{1}{2}\rho S_w C_{L_{opt}}}} = 119 \ m/s.$$

From Equation (7.142d), $\dot{h}_e = V \cdot \gamma = 10.95 \ m/s$.

3. **We want to perform a horizontal turn at an altitude of $h = 0$ with a load factor $n = 2.5 = n_{max}$, and with the velocity corresponding to the maximum aerodynamic efficiency in horizontal, steady, straight flight.**

The equations governing the dynamics of the airplane in an horizontal turn are:

$$T = D; \quad (7.149a)$$
$$mV\dot{\chi} = L\sin\mu; \quad (7.149b)$$
$$L\cos\mu = mg. \quad (7.149c)$$

In a uniform (stationary) circular movement, it is well known that the tangential velocity is equal to the angular velocity $\dot{\chi}$ multiplied by the radius of turn R:

$$V = \dot{\chi} R. \quad (7.150)$$

Therefore, System (7.149) can be rewritten as:

$$T = D; \quad (7.151a)$$
$$n\sin\mu = \frac{V^2}{gR}; \quad (7.151b)$$
$$n = \frac{1}{\cos\mu}; \quad (7.151c)$$

where $n = \frac{L}{mg}$ is the load factor.

MECHANICS OF FLIGHT

a. *The required bank angle:*

Therefore, looking at Equation (7.151c), it is straightforward to determine that the bank angle of turn is $\mu = 66.4°$.

b. *Velocity and radius of turn.*

First, we need to calculate the velocity corresponding to the maximum efficiency. As we have calculated before in Equation (7.148), the coefficient of lift that generates maximum efficiency is the so-called optimal coefficient of lift, that is, $C_{L_{opt}} = \sqrt{\frac{C_{D_0}}{C_{D_i}}} = 0.8..$ The velocity yields then:

$$V = \sqrt{\frac{L}{\frac{1}{2}\rho_0 S_w C_{L_{opt}}}} = 161.57 \ m/s.$$

Entering in Equation (7.151b) with $\mu = 66.4$ [deg] and $V = 161.57$ m/s: $R = 1167$ m.

c. *The required thrust.*

Equation (7.151a) can be expressed as:

$$T = \frac{1}{2}\rho_0 S_w V^2 C_{D_0} + \frac{L^2}{\frac{1}{2}\rho_0 V^2 S_w} C_{D_i}.$$

Since all values are known: $T = 94093$ N.

d. Since $T \leq T_{max,av}(h=0)$, the aircraft can perform the turn.

References

[1] ANDERSON, J. (2012). *Introduction to flight, seventh edition*. McGraw-Hill.

[2] FRANCHINI, S. and GARCÍA, O. (2008). *Introducción a la ingeniería aeroespacial*. Escuela Universitaria de Ingeniería Técnica Aeronáutica, Universidad Politécnica de Madrid.

Part III

Air Transportation, Airports, and Air Navigation

AIR TRANSPORTATION

Contents

8.1	Regulatory framework	234
	8.1.1 ICAO	234
	8.1.2 IATA	239
8.2	The market of aircraft for commercial air transportation	240
	8.2.1 Manufacturers in the current market of aircraft	241
	8.2.2 Types of aircraft	243
	8.2.3 Future market of aircraft	245
8.3	Airlines' cost strucutre	247
	8.3.1 Operational costs	248
8.4	Environmental impact	256
	8.4.1 Sources of environmental impact	256
	8.4.2 Aircraft operations' environmental fingerprint	257
References		268

Air transportation plays an integral role in our way of life. Commercial airlines allow millions of people every year to attend business conventions or take vacations around the globe. Air transportation also represents the fastest way to ship most types of cargo over long distances. Air transportation must be seen both as a business and as a technical and operational activity. Therefore, how an aircraft is operated and exploited, and what is the cost of operating and maintaining an aircraft are questions to be assessed.

First, we need to understand the complex regulatory framework needed for reliable and safe air transportation. ICAO and IATA will be studied in Section 8.1.1 and Section 8.1.2, respectively. Second, for air transport economy we need to consider the performances of the aircraft studied in Chapter 7 and the particular characteristics of air transportation. Thus, this chapter will briefly focus on the types of aircraft and manufactures in Section 8.2, on the structure of costs of a typical airline in Section 8.3, and on aviation's environmental fingerprint in Section 8.4. A good introductory reference is NAVARRO [8]. Thorough overviews are given, for instance, in PINDADO CARRIÓN [11] and BELOBABA et al. [2].

8.1 Regulatory framework

8.1.1 ICAO[2]

The International Civil Aviation Organization (ICAO) is a specialized agency of the United Nations that was created in 1944 to promote the safe and orderly development of international civil aviation throughout the world. It sets standards and regulations necessary for aviation safety, security, efficiency, and regularity, as well as for aviation environmental protection.

Why are Standards Necessary?

Civil aviation is a powerful force for progress in our modern global society. A healthy and growing air transport system creates and supports millions of jobs worldwide. It forms part of the economic lifeline of many countries. It is a catalyst for travel and tourism, the world's largest industry. Beyond economics, air transport enriches the social and cultural fabric of society and contributes to the attainment of peace and prosperity throughout the world.

Twenty four hours a day, 365 days of the year, an airplane takes off or lands every few seconds somewhere on the face of the earth. Every one of these flights is handled in the same, uniform manner, whether by air traffic control, airport authorities, or pilots at the controls of their aircraft. Behind the scenes are millions of employees involved in manufacturing, maintenance, and monitoring of the products and services required in the never-ending cycle of flights. In fact, modern aviation is one of the most complex systems of interaction between human beings and machines ever created.

This clock-work precision in procedures and systems is made possible by the existence of universally accepted standards known as Standards and Recommended Practices, or SARPs. SARPs cover all technical and operational aspects of international civil aviation, such as safety, personnel licensing, operation of aircraft, aerodromes, air traffic services, accident investigation, and the environment. Without SARPs, our aviation system would be at best chaotic and at worst unsafe.

How it works

The constitution of ICAO is the Convention on International Civil Aviation, drawn up by a conference in Chicago in November and December 1944, and to which each ICAO Contracting State is a party. According to the terms of the Convention, the Organization is

---

[2] The information included in this section has been retrieved from ICAO's website @ http://www.icao.int/Pages/icao-in-brief.aspx.

made up of an Assembly, a Council of limited membership with various subordinate bodies and a Secretariat.

The Assembly, composed of representatives from all Contracting States, is the sovereign body of ICAO. It meets every three years, reviewing in detail the work of the Organization and setting policy for the coming years.

Foundation of ICAO

The consequence of the studies initiated by the US and subsequent consultations between the Major Allies was that the US government extended an invitation to 55 states or authorities to attend, in November 1944, an International Civil Aviation Conference in Chicago. Fifty four states attended this conference end of which a Convention on International Civil Aviation was signed by 52 States set up the permanent International Civil Aviation Organization (ICAO) as a mean to secure international cooperation an highest possible degree of uniformity in regulations and standards, procedures, and organization regarding civil aviation matters.

The most important work accomplished by the Chicago Conference was in the technical field because the Conference laid the foundation for a set of rules and regulations regarding air navigation as a whole which brought safety in flying a great step forward and paved the way for the application of a common air navigation system throughout the world.

From the very assumption of activities of ICAO, it was realized that the work of the Secretariat, especially in the technical field, would have to cover the following major activities: those which covered generally applicable rules and regulations concerning training and licensing of aeronautical personnel both in the air and on the ground, communication systems and procedures, rules for the air and air traffic control systems and practices, airworthiness requirements for aircraft engaged in international air navigation as well as their registration and identification, aeronautical meteorology, and maps and charts. For obvious reasons, these aspects required uniformity on a world-wide scale if truly international air navigation was to become a possibility.

Chicagos's convention

In response to the invitation of the United States Government, representatives of 54 nations met at Chicago from November 1 to December 7, 1944, to *make arrangements for the immediate establishment of provisional world air routes and services* and *to set up an interim council to collect, record and study data concerning international aviation and to make recommendations for its improvement.* The Conference was also invited to *discuss the principles and methods to be followed in the adoption of a new aviation convention.*

Convention on International Civil Aviation (also known as Chicago Convention), was signed on 7 December 1944 by 52 States. Pending ratification of the Convention by 26

States, the Provisional International Civil Aviation Organization (PICAO) was established. It functioned from 6 June 1945 until 4 April 1947. By 5 March 1947 the 26th ratification was received. ICAO came into being on 4 April 1947. In October of the same year, ICAO became a specialized agency of the United Nations linked to Economic and Social Council (ECOSOC). The Convention on International Civil Aviation set forth the purpose of ICAO:

> *WHEREAS the future development of international civil aviation can greatly help to create and preserve friendship and understanding among the nations and peoples of the world, yet its abuse can become a threat to the general security; and*
>
> *WHEREAS it is desirable to avoid friction and to promote that co-operation between nations and peoples upon which the peace of the world depends;*
>
> *THEREFORE, the undersigned governments having agreed on certain principles and arrangements in order that international civil aviation may be developed in a safe and orderly manner and that international air transport services may be established on the basis of equality of opportunity and operated soundly and economically;*
>
> *Have accordingly concluded this Convention to that end.*

The Convention has since been revised eight times (in 1959, 1963, 1969, 1975, 1980, 1997, 2000 and 2006). It is constituted by a preamble and 4 parts:

- Air navigation.
- Organization of the international civil aviation.
- International air transport.
- Final dispositions.

Some important articles are:

- Article 1: Every state has complete and exclusive sovereignty over airspace above its territory.
- Article 5: (Non-scheduled flights over state's territory): The aircraft of states, other than scheduled international air services, have the right to make flights across state's territories and to make stops without obtaining prior permission. However, the state may require the aircraft to make a landing.
- Article 6: (Scheduled air services) No scheduled international air service may be operated over or into the territory of a contracting State, except with the special permission or other authorization of that State.

- Article 10: (Landing at customs airports): The state can require that landing to be at a designated customs airport and similarly departure from the territory can be required to be from a designated customs airport.

- Article 12: Each state shall keep its own rules of the air as uniform as possible with those established under the convention, the duty to ensure compliance with these rules rests with the contracting state.

- Article 13: (Entry and clearance regulations) A state's laws and regulations regarding the admission and departure of passengers, crew or cargo from aircraft shall be complied with on arrival, upon departure and whilst within the territory of that state.

- Article 16: The authorities of each state shall have the right to search the aircraft of other states on landing or departure, without unreasonable delay...

- Article 24: Aircraft flying to, from or across, the territory of a state shall be admitted temporarily free of duty. Fuel, oil, spare parts, regular equipment and aircraft stores retained on board are also exempt custom duty, inspection fees or similar charges.

- Article 29: Before an international flight, the pilot in command must ensure that the aircraft is airworthy, duly registered and that the relevant certificates are on board the aircraft. The required documents are: certificate of registration; certificate of airworthiness; passenger names; place of boarding and destination; crew licenses ; journey logbook; radio license; cargo manifest.

- Article 30: The aircraft of a state flying in or over the territory of another state shall only carry radios licensed and used in accordance with the regulations of the state in which the aircraft is registered. The radios may only be used by members of the flight crew suitably licensed by the state in which the aircraft is registered.

- Article 32: the pilot and crew of every aircraft engaged in international aviation must have certificates of competency and licenses issued or validated by the state in which the aircraft is registered.

- Article 33: (Recognition of certificates and licenses) Certificates of Airworthiness, certificates of competency and licenses issued or validated by the state in which the aircraft is registered, shall be recognized as valid by other states. The requirements for issue of those Certificates or Airworthiness, certificates of competency or licenses must be equal to or above the minimum standards established by the Convention.

- Article 40: No aircraft or personnel with endorsed licenses or certificate will engage in international navigation except with the permission of the state or states whose territory is entered. Any license holder who does not satisfy international standard relating to that license or certificate shall have attached to or endorsed on that license information regarding the particulars in which he does not satisfy those standards.

The Convention is supported by eighteen annexes containing standards and recommended practices (SARPs). The annexes are amended regularly by ICAO and are as follows:

- Annex 1: Personnel Licensing
- Annex 2: Rules of the Air
- Annex 3: Meteorological Service for International Air Navigation
    - Vol I: Core SARPs
    - Vol II: Appendices and Attachments
- Annex 4: Aeronautical Charts
- Annex 5: Units of Measurement to be used in Air and Ground Operations
- Annex 6: Operation of Aircraft
    - Part I: International Commercial Air Transport: Aeroplanes
    - Part II: International General Aviation: Aeroplanes
    - Part III: International Operations: Helicopters
- Annex 7: Aircraft Nationality and Registration Marks
- Annex 8: Airworthiness of Aircraft
- Annex 9: Facilitation
- Annex 10: Aeronautical Telecommunications
    - Vol I: Radio Navigation Aids
    - Vol II: Communication Procedures including those with PANS status
    - Vol III: Communication Systems
        * Part I: Digital Data Communication Systems
        * Part II: Voice Communication Systems
    - Vol IV: Surveillance Radar and Collision Avoidance Systems
    - Vol V: Aeronautical Radio Frequency Spectrum Utilization
- Annex 11: Air Traffic Services: Air Traffic Control Service, Flight Information Service and Alerting Service
- Annex 12: Search and Rescue
- Annex 13: Aircraft Accident and Incident Investigation
- Annex 14: Aerodromes
    - Vol I: Aerodrome Design and Operations
    - Vol II: Heliports
- Annex 15: Aeronautical Information Services
- Annex 16: Environmental Protection

- Vol I: Aircraft Noise
- Vol II: Aircraft Engine Emissions

- Annex 17: Security: Safeguarding International Civil Aviation Against Acts of Unlawful Interference
- Annex 18: The Safe Transport of Dangerous Goods by Air

8.1.2 IATA[4]

Air transport is one of the most dynamic industries in the world. The International Air Transport Association (IATA) is its global trade organization. Over 60 years, IATA has developed the commercial standards that built a global industry. Today, IATA's mission is to represent, lead, and serve the airline industry. Its members comprise some 240 airlines representing 84% of total air traffic.

Mission

Representing: IATA seeks to improve understanding of the industry among decision makers and increase awareness of the benefits that aviation brings to national and global economies. It fights for the interests of airlines across the globe, challenging unreasonable rules and charges, holding regulators and governments to account, and striving for sensible regulation.

Leading: IATA's aim is to help airlines help themselves by simplifying processes and increasing passenger convenience while reducing costs and improving efficiency. The ground-breaking simplifying the business initiative is crucial in this area. Moreover, safety is IATA's number one priority, and IATA's goal is to continually improve safety standards, notably through IATA's Operational Safety Audit (IOSA). Another main concern is to minimize the impact of air transport on the environment.

Serving: IATA ensures that people and goods can move around the global airline network as easily as if they were on a single airline in a single country. In addition, it provides essential professional support to all industry stakeholders with a wide range of products and expert services, such as publications, training and consulting. IATA's financial systems also help carriers and the travel industry maximize revenues.

---

[4] The information included in this section has been retrieved from IATA's website @ http://www.iata.org/about/Pages/index.aspx.

For the benefit for all parties involved: For consumers, IATA simplifies the travel and shipping processes, while keeping costs down. Passengers can make one telephone call to reserve a ticket, pay in one currency and then use the ticket on several airlines in several countries. IATA allows airlines to operate safely, securely, efficiently and economically under clearly defined rules. IATA serves as an intermediary between airlines and passenger as well as cargo agents via neutrally applied agency service standards and centralized financial systems. A large network of industry suppliers and service providers gathered by IATA provides solid expertise to airlines in a variety of industry solutions. For governments, IATA seeks to ensure they are well informed about the complexities of the aviation industry to ensure better, long-term decisions.

History

IATA was founded in Havana, Cuba, in April 1945. It was the prime vehicle for inter-airline cooperation in promoting safe, reliable, secure, and economical air services. At the founding, its main goals were:

- to promote safe, regular, and economical air transport for the benefit of the peoples of the world, to foster air commerce, and to study the problems connected therewith;
- to provide means for collaboration among the air transport enterprises engaged directly or indirectly in international air transport service; and
- to cooperate with the newly created International Civil Aviation Organization (ICAO – the specialized United Nations agency for civil aviation) and other international organizations.

The international scheduled air transport industry is now more than 100 times larger than it was in 1945. Few industries can match the dynamism of that growth, which would have been much less spectacular without the standards, practices, and procedures developed within IATA.

In order to provide a quantitative measure of the importance of the aviation industry, a snapshot of the industry in 2010 is provided BISIGNANI [4]: 2.4000 million passengers; 43 million tones of cargo; 32 million jobs; just 1 accident for every 1.4 million flights; 2% of global carbon emissions; $545.000 million in revenue; etc.

8.2 THE MARKET OF AIRCRAFT FOR COMMERCIAL AIR TRANSPORTATION

As it has been exposed in the two previous sections, ICAO and IATA are the two fundamental international organizations that provide a juridic framework to all stakeholders in order them to carry out the air transportation business under reliable, safe, and efficient standards. Intuitively, one can think straightforward that one of the key elements of air transportation are aircraft. Therefore, we move on to analyze the market of aircraft for commercial air transportation. We will analyze the main manufactures, the fundamental aircraft, its prices, and future trends in the industry.

8.2.1 Manufacturers in the current market of aircraft

In a wide sense, the current market of commercial aircraft is dominated by four major manufacturers of civil transportation aircraft:

- Airbus, based in Europe.
- Boeing, based in the United States.
- Bombardier, based in Canada.
- Embraer, based in Brazil.

Airbus [5]

Airbus is an aircraft manufacturing subsidiary of EADS, a European aerospace company. Based in Toulouse, and with significant activity across Europe, the company produces more than half of the world's jet airliners. Airbus began as a consortium of aerospace manufacturers, Airbus Industrie. The consolidation of European defence and aerospace companies in 1999 and 2000 allowed the establishment of a simplified joint-stock company in 2001, owned by EADS (80%) and British Aerospace (BAE) Systems (20%). In 1006, BAE sold its shareholding to EADS. Airbus employs around 52,000 people at sixteen sites in four European Union countries: France, Germany, the United Kingdom, and Spain. Final assembly production is at Toulouse (France), Hamburg (Germany), Seville (Spain) and, since 2009, Tianjin (People's Republic of China). The company produced and markets the first commercially viable fly-by-wire airliner, the Airbus A320, and the world's largest airliner, the A380.

Boeing [6]

The Boeing Company is an American multinational aerospace and defense corporation, founded in 1916 in Seattle, Washington. Boeing has expanded over the years, merging with McDonnell Douglas in 1997. Boeing is made up of multiple business units, which are Boeing Commercial Airplanes (BCA); Boeing Defense, Space & Security (BDS); Engineering, Operations & Technology; Boeing Capital; and Boeing Shared Services Group. Boeing is among the largest global aircraft manufacturers by revenue, orders and deliveries, and the third largest aerospace and defense contractor in the world based on defense-related revenue. Boeing is the largest exporter by value in the United States. Its Boeing 737 has resulted (counting with the different versions and evolutions) the all times most sold aircraft type.

---

[5] Data retrieved from http://en.wikipedia.org/wiki/Airbus.
[6] Data retrieved from http://en.wikipedia.org/wiki/Boeing.

Air transportation

(a) Boeing 787 dreamliner. © MilborneOne / Wikimedia Commons / CC-BY-SA-3.0.

(b) Airbus A380, the largest airliner. © Axel Peju / Wikimedia Commons / CC-BY-2.0.

(c) Embraer 190, regional jet. © AntŹnio Milena / Wikimedia Commons / CC-BY-SA-3.0.

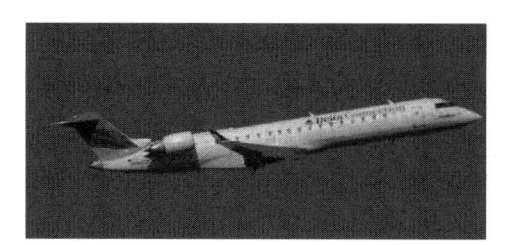

(d) A Bombardier CRJ-700 regional jet in Delta Connection colors. © Mark Wagner / Wikimedia Commons / CC-BY-2.5.

Figure 8.1: Aircraft manufacturers.

Embraer [7]

Embraer is a Brazilian aerospace conglomerate that produces commercial, military, and executive aircraft and provides aeronautical services. Embraer is the third-largest commercial aircraft manufacturing in the world, and the forth-largest aircraft manufacturing when including business jets into account, and it is Brazil's top exporter of industrial products.

Bombardier [8]

Bombardier is a Canadian conglomerate. Over the years it has been a large manufacturer of regional aircraft, business jets, mass transportation equipment, recreational equipment and a financial services provider. Its headquarters are in Montreal, Canada.

---

[7] Data retrieved from http://en.wikipedia.org/wiki/Embraer.
[8] Data retrieved from http://en.wikipedia.org/wiki/Bombardier_Inc..

Long-haul

Aircraft	Pax	MTOW	Mach	Range	long	b
B747-400	416-524	397	0.85-0.88	13450	70.7	64.4
B777-300	368-550	297	0.84-	11000	74	61
A340-600	372 (3-class)	365	0.83-	13890	75.3	63.70
A380-800	555 (3-class)	560	0.85-0.88	14800	72.75	79.8
A350	253-300	245	0.82	13900-16300	82.8-65.2	61.1
B787-8	210-250	228	0.86-0.9	15200	56.70	60

Table 8.1: Long-haul aircraft specifications. MTOW in tones, Range in kilometers, longitude and wing-span in meters. The Mach number corresponds to long-range operating Mach and maximum operating Mach, respectively. Data retrieved from http://www.airliners.net.

8.2.2 Types of aircraft

Boeing and Airbus concentrate on wide-body and narrow-body jet airliners, while Bombardier and Embraer concentrate on regional airliners. Large networks of specialized parts suppliers from around the world support these manufacturers, who sometimes provide only the initial design and final assembly in their own plants.

Jet aircraft can be generally divided into:

- Medium/long-haul (> 100 seats)
    - Wide body (two decks): A380; B787; A350; B747; A340 family.
    - Narrow body (one deck): B737 family and A320 family.
- Short haul (< 100 seats)
    - Bombardier CRJ700/900.
    - Embraer 170-175-190-195.

Regional propellers are also short range with typically 30-80 seats. Table 8.1, Table 8.2, and Table 8.3 show the specifications of different aircraft.

Table 8.4 and Table 8.5 show the average prices of Airbus manufactured aircraft. As it can be observed, medium-haul aircraft have a price of an order of 100 million USD, while the cost of the new A380 is around 400 million USD. Obviously, these prices depend upon many factors, such for instance, the configuration selected by the airline, commercial agreements, the currency exchange rates (note that aircraft are sold in USD currency), etc. Therefore, due to such high value, the policies of aircraft acquisition are key for the viability of the airline. In particular, airlines might acquire the aircraft or simply use it within a leasing or renting formula. Also, after some years of use, old aircraft are sold in the second hand market. In this way, airlines might rotate quickly their fleet in order to count with the newest technological advances.

Medium-haul

Aircraft	Pax	MTOW	Mach	Range	long	b
B737-900	177-189	79	0.785-	3815-5083	42.1	34
A320-200	150 (2-class)	73.5-77.4	0.82-0.85	4843-5676	37.57	34.09
A321-100	186 (2-class)	83-85	0.82-0.85	4352	44.51	34.09

Table 8.2: Medium-haul aircraft specifications. MTOW in tones, Range in kilometers, longitude and wing-span in meters. The Mach number corresponds to long-range operating Mach and maximum operating Mach, respectively. Data retrieved from http://www.airliners.net.

Regional aircraft

Jets

Aircraft	Pax	MTOW	Mach	Range	long	b
CRJ-900	86	36.5	0.83	2700	36.4	23.24
EMBRAER 175	78-86	37.5	0.82	3334	31.68	26
EMBRAER 195	108-118	50	0.82	3300	38.65	28.72

Turboprops

Aircraft	Pax	MTOW	Velocity	Range	long	b
Q100	37-39	16.5	496	1900	22.3	25.9
Q400	38-78	29.3	667	2500	32.8	28.42
ATR 42	46-50	16.7	480	1700	22.67	24.57
ATR 72	66-72	21.5	500	1780	27.1	27

Table 8.3: Regional aircraft specifications. MTOW in tones, range in kilometers with typical pax, longitude and wing-span in meters. Mach number corresponds to maximum cruising speed. The velocity is given in km/h and corresponds to the maximum cruising velocity. Data retrieved from http://www.airliners.net.

Aircraft type	A318	A319	A320	A321	A319neo	A320neo	A321neo
Price	67.7	80.7	88.3	103.6	88.8	96.7	113.3

Table 8.4: Airbus medium-haul aircraft 2012 average prices list (million USD). Data retrieved from Airbus.

Aircraft type	A330-200	A330-300	A350-800	A350-900	A350-1000	A380-800
Price	208.6	231.1	245.5	277.7	320.6	389.9

Table 8.5: Airbus long-haul aircraft 2012 average prices list (million USD). Data retrieved from Airbus.

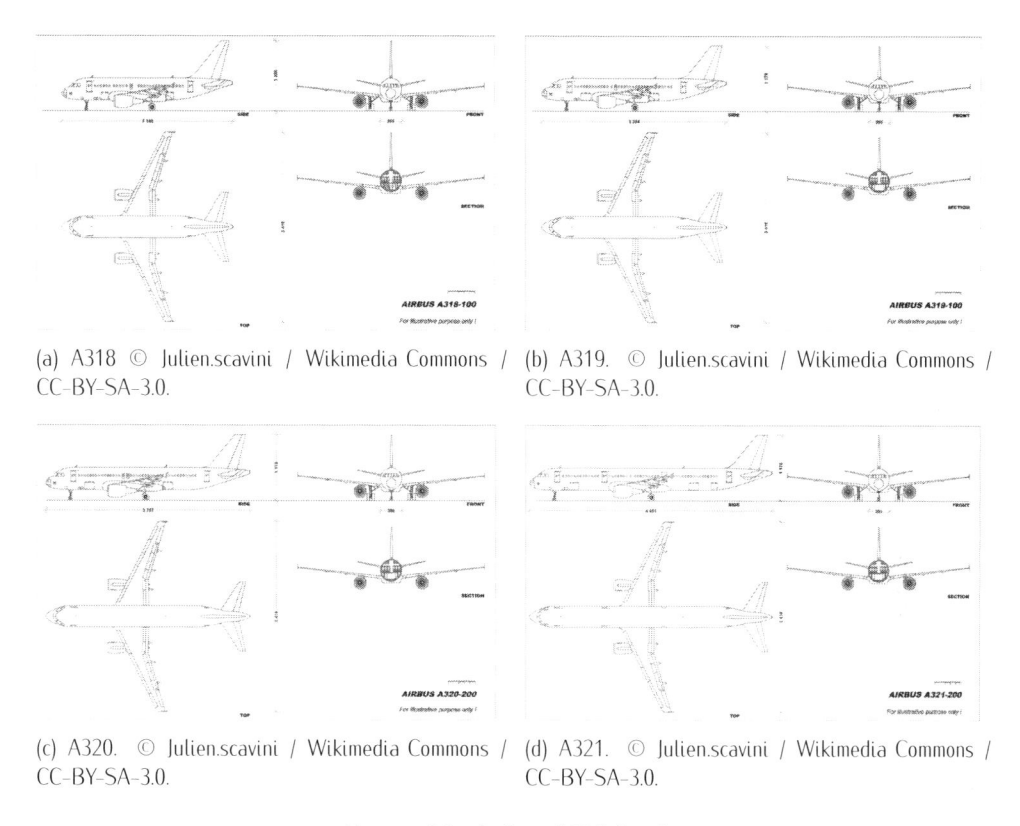

8.2 THE MARKET OF AIRCRAFT FOR COMMERCIAL AIR TRANSPORTATION

(a) A318 © Julien.scavini / Wikimedia Commons / CC-BY-SA-3.0.

(b) A319. © Julien.scavini / Wikimedia Commons / CC-BY-SA-3.0.

(c) A320. © Julien.scavini / Wikimedia Commons / CC-BY-SA-3.0.

(d) A321. © Julien.scavini / Wikimedia Commons / CC-BY-SA-3.0.

Figure 8.2: Airbus A320 family.

8.2.3 FUTURE MARKET OF AIRCRAFT

Future, short-term trends point towards a next generation of medium-haul aircraft. Both Boeing and Airbus are going to relaunch evolved versions of their successful families A320 and B737, respectively. Also, China is trying to get into this niche with the design of its first medium-haul airliner.

A320neo family [9]

The Airbus A320neo is a series of enhanced versions of A320 family under development by Airbus to be served up to 2015. The letters "neo" stand for "New Engine Option". The main change is the use of the larger and more efficient engines which results in 15% less fuel consumption, 8% lower operating costs, less noise production, and a reduction of NOx by at least 10% compared to the A320 series according to Airbus. Two power plants will be available: either the CFM International LEAP-X or the Pratt & Whitney PW1000G. The airframe will also receive some modifications, including the addition of "Sharklet" wingtips

---
[9] Data retrieved from http://en.wikipedia.org/wiki/A320_neo

Airbus A320neo

	A318neo	A320neo	A321neo
Seating capacity	158 (1-class, max) 134 (1-class, typ) 124 (2-class, typ)	180 (1-class, max) 164 (1-class, typ) 150 (1-class, typ)	220 (1-class, max) 199 (1-class, typ) 185 (1-class, typ)
Cruising speed	Mach 0.78		
Maximum speed	Mach 0.82		
Maximum range	7800 km	6900 km	6760 km

Table 8.6: A320neo family specifications. *max* refers to the maximum capacity layout; *typ* refers to the typical seats layout. Maximum range refers to fully loaded. Data retrieved from Wikipedia A320neo.

Boeing 737 MAX

	737 MAX 7	737 MAX 8	737 MAX 8
Seating capacity	126 (2-class, typ)	162 (2-class, typ)	180 (2-class, typ)
Cruising speed	Mach 0.79		
Maximum range	7038 km	6704 km	6658 km

Table 8.7: B737 MAX family specifications. *2-class* refers to two different classes (Business and tourist); *typ* refers to the typical seats layout. Maximum range refers to fully loaded. Data retrieved from Wikipedia B737 MAX.

to reduce drag and interior modifications for the passengers comfort such as larger luggage spaces and an improved air purification system. The A320neo family specifications can be consulted in Table 8.6.

B737 MAX [10]

The Boeing 737 MAX is a new family of aircraft being developed by Boeing in order to replace the current Boeing 737 generation family. The primary change will be the use of the larger and more efficient CFM International LEAP-1B engines. The airframe is to receive some modifications as well. The 737 MAX is scheduled for first delivery in 2017, 50 years after the 737 first flew. The three variants of the new family are the 737 MAX 7, the 737 MAX 8 and the 737 MAX 9, which are based on the 737-700, 737-800 and 737-900ER, respectively, which are the best selling versions of the current 737 generation family. Boeing claims the 737 MAX will provide a 16% lower fuel burn than the current Airbus A320, and 4% lower than the Airbus A320neo. The B737 MAX family specifications can be consulted in Table 8.7.

---

[10]Data retrieved from http://en.wikipedia.org/wiki/Boeing_737_MAX

Comac C919 [11]

The Comac C919 is a planned family of 168–190 seat narrow-body airliners to be built by the Commercial Aircraft Corporation of China (Comac). It will be the largest commercial airliner designed and built in China since the defunct Shanghai Y-10. Its first flight is expected to take place in 2014, with deliveries scheduled for 2016. The C919 forms part of China's long-term goal to break Airbus and Boeing's duopoly, and is intended to compete against Airbus A320neo and the Boeing 737 MAX.

Dimensions of the C919 are very similar to the Airbus A320. Its fuselage will be 3.96 meters wide, and 4.166 meters high. The wingspan will be 33.6 meters. Its cruise speed will be Mach 0.785 and it will have a maximum altitude of 12,100 meters. There will be two variants. The standard version will have a range of 4,075 km, with the extended-range version able to fly 5,555 km. The capacity will go from 156 passengers (with two classes) to 174 passengers (1 class and maximum density of seats).

8.3 Airlines' cost strucutre

The calculus of economic costs constitutes a necessity within every single enterprise. It is valuable for measuring the efficiency of the different areas, deciding on new investments, and obviously to set the prices for the supplied products (in the case of airlines, services) is based on desired profits and estimated forecasts. Two important references in airlines' economics are DOGANIS [6] and DOGANIS [5].

Focusing on costs, the breaking down taxonomy varies depending on the company. However they are typically adjusted to the cost classification stablished by ICAO. A fundamental division arises when dividing operational costs and non operational costs (also refereed to as operative and non-operative costs):

- Operational costs: Expenses associated with administering a business on a daily basis. Operating costs include both fixed costs and variable costs. According to a somehow canonical definition, fixed costs, such as infrastructures or advertising, remain the same regardless of the number of products produced; variable costs, such as materials or labour, can vary according to how much product is produced. In airlines terminology, they are referred to as Direct Operational Costs (DOC) and Indirect Operational Costs (IOC):
    - DOC are related to the operation of the aircraft
    - IOC are related to the running of the airline company and, therefore, regardless of the aircraft operation.
- Non-operational costs: associated to expenses not related to day to day operations, typically financial costs.

---
[11]Data retrieved from http://en.wikipedia.org/wiki/Comac_C919

Operational costs		Non-operational costs
DOC	IOC	-
Salary (crew and cabin)	Station and ground exprenses	Loans amortization
Fuel and oil	Ticketing, sales and promotion	Capital interests
Airport and En-route fees	General and administration	-
Manteinance	Depreciation	-
Handling	Renting, leasing	-
Others	Insurances	-

Table 8.8: Cost structure of a typical airline.

We could keep breaking down the different costs, however with this general framework we can expose a typical taxonomy of the cost structure of an airline company as illustrated in Table 8.8.

The non-operational costs are also refereed to as capital costs or simply financial costs. As pointed out above, they can be divided into:

- Loans amortization.
- Capital interests.

The concept *loans amortizations* refers to the distribution of an acquisition in different payment periods. This fact implies typically interests.

8.3.1 Operational costs

As pointed out above, it is under common agreement to establish a classification based in fixed costs (c_f) and variable costs (c_v):

$$c = c_f + c_v BT, \qquad (8.1)$$

where c are the operational costs, and BT refers to Block Time[12]. The difference between BT and flight time is small in long flights, but it can be important in short flights. BT can be expressed as a function of the aircraft range, R, as follows:

$$BT = A + B \cdot R + C \cdot \ln R, \qquad (8.2)$$

where A, B and C are parameters set by the airline. A linear approximation is usually adopted:

$$BT = A + B \cdot R. \qquad (8.3)$$

---
[12]The time in block hours is the time between the instant in which the aircraft is pulled out in the platform and the instant in which the aircraft parks at the destination airport. It includes, therefore, taxi out and taxi in.

Therefore the operation costs can be expressed either as a function of block time or range:

$$c = c_f + c_v BT = c_f + c_v(A + B \cdot R) = c'_f + c'_v \cdot R. \tag{8.4}$$

Using these expressions one can generate the so called cost indicators: cost per unity of time, range, payload, or any other parameter to be taken into account at accounting level. Three of these cost indicators are:

The hourly costs:

$$\frac{c}{BT} = c_v + \frac{c_f}{A + B \cdot R}. \tag{8.5}$$

The kilometric costs:

$$\frac{c}{R} = \frac{c'_f}{R} + c'_v = \frac{c'_f}{R} + Bc_v. \tag{8.6}$$

The cost per offered ton and kilometer (OTK):

$$\frac{c}{(MPL) \cdot R} = \frac{c'_f}{(MPL) \cdot R} + \frac{c_v B}{MPL}, \tag{8.7}$$

where MPL refers to maximum payload.

Besides this cost ratios, there are also some other ratios in which airlines generally measure their operations. These are Passenger Kilometer Carried (PKC), the passenger carried per kilometer flown; Seat Kilometer Offered (SKO), seats offered per kilometer flown; the Factor of Occupancy (FO=PKC/SKO):

$$PKC = OTK \cdot FO. \tag{8.8}$$

These three ratios, together with the above described cost ratios represent the best metric to analyze the competitiveness of an airline.

Structure of the operational costs

We can identify four main groups:

- Labour;
- Fuel;
- Aircraft rentals and depreciation & amortization; and
- Others: Maintenance, landing and air navigation fees, handling, ticketing, etc.

	North America		Europe		Asia Pacipic		Major Airlines	
	2001	2008	2001	2008	2001	2008	2001	2008
Labour	36.2%	21.5%	27.2%	24.8%	17.2%	14.7%	28.3%	20.1%
Fuel	13.4%	34.2%	12.2%	25.3%	15.7%	36.7%	13.6%	32.3%
Aircraft rentals	5.5%	3.0%	2.9%	2.5%	6.3%	4.5%	5.0%	3.5%
Depreciation & Amortization	6.0%	4.5%	7.1%	5.7%	7.4%	7.8%	6.7%	5.9%
Other	38.9%	36.9%	50.7%	41.8%	53.4%	36.3%	46.4%	38.2%

Table 8.9: Evolution of airlines' operational costs 2001-2008. Source IATA.

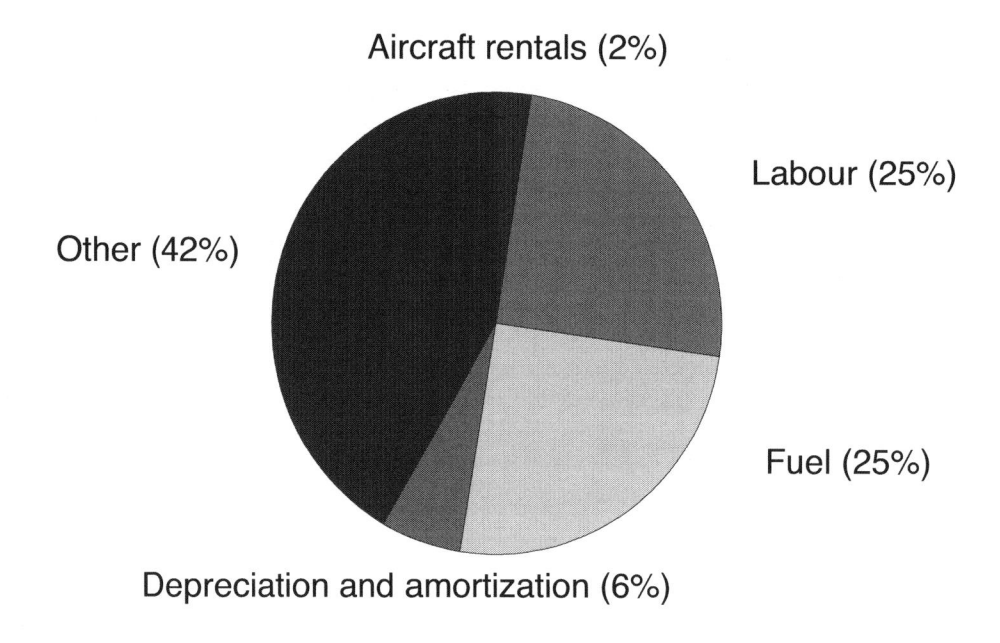

Figure 8.3: European percentage share of airline operational costs in 2008. Data retrieved from IATA.

Labour costs Labour has been traditionally the main budget item not only for airline companies, but also for any other company. However, the world has shifted to one more deregulated and globalized. This fact has lead to a less stable labour market, in which airlines companies are not a exception.

Therefore, meanwhile traditional flag companies had a very consolidated labour staff, the airline deregulation in the 70s brought the appearance of fierce competitors as the low cost companies were (and still are). In that sense all companies have been making big efforts in reducing their labour cost as we can see in Table 8.9 and Figure 8.3 and this tendency will continue growing.

8.3 Airlines' cost strucutre

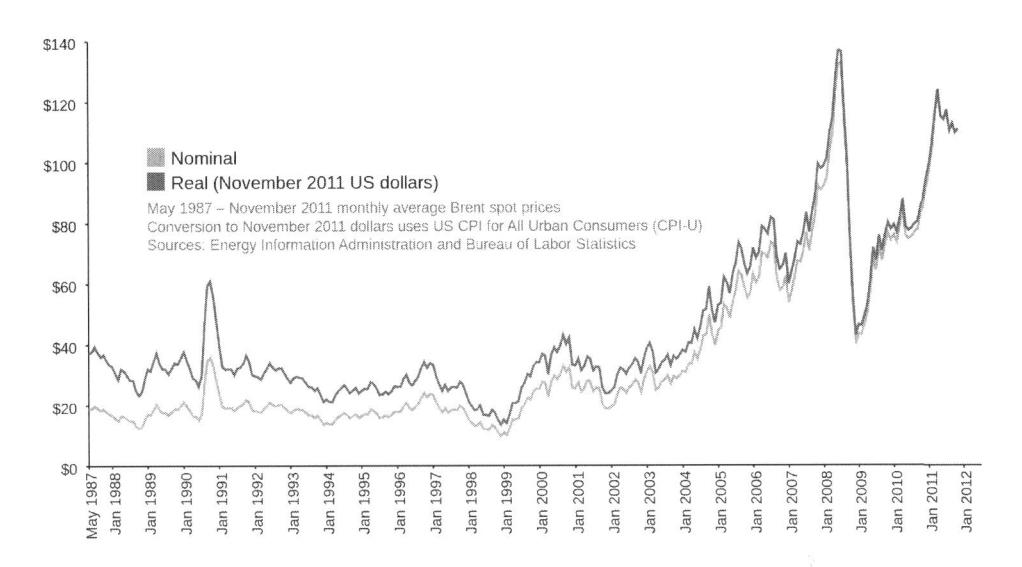

Figure 8.4: Evolution of the price of petroleum 1987–2012. © TomTheHand / Wikimedia Commons / CC-BY-SA-3.0.

Nowadays one can find staff with the same qualification and responsibilities in very different contractual conditions, or even people that works for free (as it is the case of many pilots pursuing the aircraft type habilitation).

Fuel costs: The propellant used in aviation are typically kerosenes, which it is produced derived from the crude oil (or petroleum). The crude oil is a natural, limited resource which is under high demand. It is also a focus for geo-politic conflicts. Therefore its price is highly volatile in the short term and this logically affects airline companies in their operating cost structures. However, as Figure 8.4 illustrates, the long-term evolution of petroleum price has shown since 2000 a clear increasing tendency in both real and nominal value. This increase in the price of fuel has modified substantially the operational cost structure of airline companies; in some cases fuel expenses represent 30-40% of the total operating costs. Notice that in Europe the impact was mitigated due to the strength of Euro with respect to the Dollar, but most likely today in European companies the weight of fuel costs is as well in the 30-40% of the total operating costs.

Since the crude oil is limited as a natural resource and day after day the extractions are more expensive, and since the demand is increasing due to the rapid evolution of countries such China, India, Brazil, etc., forecasts predict that this tendency or price increase will continue. Therefore, airlines will have to either increase tickets (as they have done); reduce other costs items; and encourage research on the direction of alternative fuels.

Maintenance: The maintenance costs depend on the maintenance program approved by the company, the complexity of the aircraft design (number of pieces), the reliability of the aircraft, and the price of the spare pieces.

The maintenance is carried out at different levels after the dictation of an inspection:

- Routine check: It is performed at the gate after every single flight (before first flight or at each stop when in transit). It consists of visual inspection; fluid levels; tyres and brakes; and emergency equipment. The standard duration is around 45 to 1 hour, however the minimum required time to perform it is 20 minutes.
- Check A: It is performed at the gate every 500 flight hours. It consists of routine light maintenance and engine inspection. The standard duration is around 8-10 hours (a night).
- Check B: It is performed at the gate every 1500 flight hours. It is similar to A check but with different tasks and may occur between consecutive A checks. The duration is between 10 hours and 1 day.
- Check C: It is performed in the hangar every 15-18 moths on service. It consists of a structural inspection of airframe and opening access panels; routine and non routine maintenance; and run-in tests. The duration ranges between 3 days and 1 week.
- Check D: It is performed in the hangar after some years (around 8) of service. This last level requires a complete revision (**overhaul**). It consists of major structural inspection of airframe after paint removal; engines, landing gear, and flaps removed; instruments, electronic, and electrical equipment removed; interior fittings (seats and panels) removed; hydraulic and pneumatic components removed. The duration is around 1 month with the aircraft out of service.

The traditional companies used to have internal maintenance services, typically hosted in their hubs. However, this strategy is shifting due to different flight strategies (point to point) and also to take advantage of reduced cost in determined geographic zones. Therefore, nowadays many flight companies externalize these services. This is the last aspect in which the low cost strategies have modified the air transportation industry.

Regarding spare pieces, meanwhile in other industries do not exist a clear regulation in regard of spare pieces, the aeronautical industry and the american government have been pioneers regulating the market of spare pieces. Both original manufactures and spare companies can provide spare pieces. In order private companies to be allowed to commerce sparse pieces they need an habilitation named PMA (Parts Manufacturer Approval), while the original pieces referred to as OEMs (Original Equipment Manufactures).

Handling: The handling services consist of the assistance on the ground given to aircraft, passenger, and freight, so that a stopover in any airport is carried out properly. Handling include devices, vehicles, and services such fuel refilling, aircraft guidance, luggage management, or cabin cleaning.

The handling services can be grouped into:

- Aircraft handling.
- Operational handling.
- Payload handling.

Aircraft handling defines the servicing of an aircraft while it is on the ground, usually parked at a terminal gate of an airport. It includes:

- Operations: includes communications, download and load of cargo, passenger transportation, assistance for turn on, pushback, etc.;
- Cleaning: exterior cleaning, cabin clean up, restrooms, ice or snow, etc.; and
- Fill in and out: Fuel, oil, electricity, air conditioning, etc.;

Operational handling assits in:

- Administrative assistance.
- Flight operations: includes dispatch preparation and modifications on the flight plan;
- Line maintenance: includes the maintenance prior departure, spare services and reservation of parking lot or hangar;
- Catering: includes the unloading of unused food and drink from the aircraft, and the loading of fresh food and drink for passengers and crew. Airline meals are typically delivered in trolleys. Empty or trash-filled trolley from the previous flight are replaced with fresh ones. Meals are prepared mostly on the ground in order to minimize the amount of preparation (apart from chilling or reheating) required in the air.

Payload handling refers to:

- Passenger handling: includes assistance in departure, arrival and transit, tickets and passport control, check-in, and luggage transportation towards the classification area;
- Classification, load, and download of luggage;
- Freight and main services;
- Transportation of passengers, crew and payload between different airport terminals.

As in the case of maintenance, the handling services used to be handled by flag companies. However, this activity was liberalized (in Europe, in the 90s) and it is being increasingly outsourced. As a result, many independent company have arisen in past 10-15 years.

Landing and air navigation fees: According to IATA, the landing fees and air navigation costs represent around 10% of the operating cost for airline companies. Each nation establishes navigation fees due to services provided when overflying an airspace region under its sovereignty. Moreover, each airport establishes landing fees for the services provided to the aircraft when approaching and landing. In Spain, AENA charges for approaching (notice that can be interpreted as a landing fee).

The landing fee is established taking into consideration the MTOW of the aircraft and the type of flight (Schengen, International, etc.). AENA gives therefore a unitary fee that must be multiplied by the aircraft MTOW. The formula is as follows:

$$L_{fee} = u_l \cdot \left(\frac{MTOW}{50}\right)^{0.9}, \tag{8.9}$$

where L_{fee} is the total landing fee, u_l is the unitary landing fee and MTOW is the maximum take off weight of the aircraft. For instance, this unitary fee depends on the airport and ranges 12 to 17 €.

On the other hand, the air navigation fees in Europe are invoiced and charged by Eurocontrol by means of the following formula:

$$\text{Navigation fee} = \text{unit rate} \cdot \text{distance coef} \cdot \text{weight coef}, \tag{8.10}$$

where the unit rate is established in the different European FIR/UIR[13]. For instance, in Spain, the unit rates for FIR Madrid, FIR Barcelona and FIR Canarias are, respectively, 71.84 €, 71.84 € and 58.52 €. The distance coefficient is the orthodromic distance over 100. The weight coefficient is $\sqrt{MTOW/50}$. Therefore, the navigation fee results in:

$$\text{Navigation fee} = \text{unit rate} \cdot \frac{d}{100} \cdot \sqrt{\frac{MTOW}{50}}. \tag{8.11}$$

Depreciation: The depreciation of an aircraft (the most important good airline companies have in their accounting) is typically imposed by the national (or international) accounting regulations, and it is typically associated to the following factors:

- Use.
- The course of time.
- Technological obsolescence.

The *Use*, corresponding to flight hour (also referred to as block hour), could be included somehow into the DOC. The depreciation due to the course of time is due to the loss

---

[13]Flying Information Region and Upper Information Region. The meanings of these regions will be studied in Chapter 10.

of efficiency with respect to more modern aircraft. Last but not least, the incorporation by the competitors of a technological breakthrough (such, for instance, Airbus with the fly-by-wire) implies that the aircraft get depreciated immediately in the market.

The estimation of depreciation set the pace for fleet renovation, and it is obviously association to the utility life of the aircraft. A wide-body jet is typically depreciated to a residual value of (0-10%) in 14-20 years. The utility life of an aircraft is set to 30 years.

Aircraft acquisition: The acquisition of an aircraft is a costly financial operation, indeed the companies typically acquire several aircraft at the same time, not only one. These investments must financed by means of bank loans (also by increases of capital, emissions of bonds and obligations, etc.), which imply interests.

Other forms of aircraft disposition are the *operational leasing* (a simply renting) and the *financial leasing* (renting with the right of formal acquisition). Among the operational leasing there exist different types:

- Dry leasing: the aircraft is all set to be operated, but it does not include crew, maintenance, nor fuel. Sometimes the insurance is neither included.
- Wet leasing: like dry leasing but including crew.
- ACMI leasing: includes Aircraft, Crew, Maintenance and Insurance.
- Charter leasing: Includes everything, even airport and air navigation fees.

There exist important leasing companies, such GECAS, ILFC, Boeing Capital Corp or CIT group. Notice that approximately half of the total orders made to the manufactures correspond to leasing companies. The financial leasing is also very extended. The only difference with operational leasing is that they include a policy with the operator's right to acquire the aircraft at a preset date and price.

Insurances: An insurance is a practice or arrangement by which a company or government agency provides a guarantee of compensation for specified loss, damage, illness, or death in return for payment of a premium. The characteristic of an insurance contract is the displacement of a risk by means of paying a price.

The aeronautical insurance, when compared to maritime or terrestrial, have some peculiarities: The reality of air traffic proofs that air accidents occur with rather low regularity, which makes difficult to apply the rules of *big numbers*. Moreover, exceptionally an accident produces partial damage, but, on the other hand, catastrophic damages including death of crew and passengers, resulting in high compensatory payments. In these circumstances, the insurance companies have agreed to subscribe common insurances so that the risk is hold by a pool of insurance companies.

8.4 Environmental impact

A simple description of air transportation could be based on three elements: the aircraft, the airport, and the air navigation system. The first transports people and goods; airports allow passengers and goods to change transportation mode; the later provides services to ensure air operations are performed in a safe way. The activities of each of these three elements have a characteristic environmental impact, including construction, life-cycle, and reposition. A introductory reference on the topic is BENITO and BENITO [3]. Even though the different environmental impact sources will be briefly described, the focus will be on aircraft operations' environmental fingerprint, in particular to its contribution to climate change. The reader is referred to SCHUMANN [15] for a recent, thorough overview.

8.4.1 Sources of environmental impact

Attending to its geographical range, the different impact sources can be classified into BENITO and BENITO [3]:

- Local effects

    - Noise.
    - Local air pollution.
    - Use of surrounding areas.

- Global effects

    - Consumption of non-recyclable materials.
    - Use of airspace and radio-electric spectrum.
    - Contribution to climate change.

Local effects referred to those effects that are only perceptible in the vicinity of airports. This includes noise nuisance due to aircraft operations (mainly take off and landing), the air pollution due to the airport activity, and also the use of areas for the purpose of airport activity, e.g, areas with bird colonies or natural interest.

On the other hand, global effects refer to those effects that affect the sustainability of the planet worldwide. Among this, we can cite the consumption of non-recyclable materials that are finite and, moreover, need to be stored somewhere after the life-cycle, the use of airspace by aircraft and electromagnetic waves emitted by navigation services and aircraft to provide communication, navigation, and surveillance services, and the contribution of the industry to global warming.

In the sequel, the focus will be on aircraft operations' environmental impact, namely noise and emissions (CO_2, NO_x, etc.) that contribute to climate change.

8.4.2 Aircraft operations' environmental fingerprint.

Noise

Noise nuisance is an important environmental impact in the vicinity of airports. The problem is not related to an isolated take-off or landing operation, but as a consequence of the total set of departures and arrivals taking place in the airport daily. In order to understand the problem, try to empathize with a neighborhood (including hospitals, schools, houses, etc.) that has to bear systematically with an important amount of noise. To quantify it, the decibel [Db] is used. In order to provide a qualitative reference, it is worth mentioning that a typical commercial aircraft during take off emits 130 dB; the pain threshold is 140 dB; a launcher during take off is 180 dB; a concert is 110 dB; a train 80 dB; a conversation 40 dB; etc.

The most important noise emission sources within an aircraft are due to the engines, which work at high power settings during take off and initial climb. The fundamental contribution to this noise is due to rotatory elements (compressor, turbine, fans, etc.). The second fundamental contribution is due to the exhausted jet in case of turbojets. Moreover, there is also a so-called aerodynamic noise, coming form the wing, the fuselage, the empennage, and the landing gear as the aircraft flies. Sound waves propagate in the air at the speed of sound. The intensity that an observer *suffers* is proportional to the intensity in the source (the aircraft in this case) and inversely proportional to the square distance between source and receiver, i.e., the closer the aircraft is, the more intense the noise suffered by the observer.

Noise mitigation strategies: There are four fundamental strategies to mitigate noise:

- Reduction of noise in the source (airframe and engines).
- Urban management and planning.
- Take-off and landing noise abatement procedures.
- Operative restrictions.

The continuous development of more and more modern aircraft and jet engines has led in the past to substantial reduction of noise (among other improvements) emissions. This is expected to continue in the future, since noise emissions are regulated by authorities. It obviously requires the application of new technologies coming from research and innovation.

Urban managing and planning refers to limiting the urban areas next to current limits of the airport, but also to potential future enlargements.

If it happens that there is a neighborhood next to an airport, and the neighborhood is suffering from noise, an interesting strategy is to design the so-called noise abatement procedures both for departure and arrival. These are typically continuous climb or continuous descent procedures that modify the flight path to avoid overflying certain areas. A good reference on this issue is PRATS-MENÉNDEZ [12].

Last, if any of the previous strategies has not been developed, one can always restrict operations, for instance, at night hours. This is not desirable in terms of the economy of the industry, but it might be mandatory due to local legislation.

Climate change impact.

Aviation is one of the transport sectors with currently moderate climate impact. Air transportation contributes a small but growing share of global anthropogenic climate change impact. As aviation grows to meet increasing demand, the United Nations Intergovernmental Panel on Climate Change (IPCC) forecasted in 1999 that its share of global man made CO_2 emissions will increase to around 3% to 5% in 2050 (in 1999 it was estimated to be 2%) PENNER [10]. Moreover, the Royal Commission of Environmental Pollution (RCEP) has estimated that the aviation sector will be responsible for 6% or the total anthropogenic radiative forcing by 2050 ROYAL-COMMISSION [13]. The development of mitigation methods for this purpose is in line with aviation visions and research programs, such as ACARE [1], the European aeronautics projects Clean Sky[14] and SESAR CONSORTIUM [16], and the U.S. Next Generation[15] strategy.

The climate impact of aviation results from CO_2 and non-CO_2 emissions PENNER [10]. While CO_2 is the most widely perceived greenhouse gas agent in aviation, mainly because its long lifetimes in the atmosphere and because of its considerable contribution to radiative forcing, emissions from aircraft engines include other constituents that contribute, via the formation or destruction of atmospheric constituents, to climate change. The non-CO_2 emissions (nitrogen oxides, water vapor, aerosols, etc.) have shorter lifetimes but contribute a large share to aviation climate impact, having a higher climate impact when emitted at cruise than at ground levels. One of these non-CO_2 contributors to climate change is the formation of contrails, which have received significant attention PENNER [10].

The relative importance of CO_2 and non-CO_2 depends strongly on the time horizon for evaluation of climate impacts and scenarios, e.g., future air traffic development. The non-CO_2 effects are more important for short time horizons than for long horizons.

Aviation NOx Climate Impact: Nitrogen oxides (NOx, i.e., NO and NO_2) are one of the major non-CO_2 emissions. NOx emissions in the troposphere and lower stratosphere contribute to ozone (O_3) formation and methane (CH_4) reduction. Both are important greenhouse gases. On average, the O_3 impact of aviation NOx is expected to be stronger than the impact on CH_4, which increases the greenhouse effect, though the precise amounts

---
[14] http://www.cleansky.eu
[15] http://www.faa.gov/nextgen/

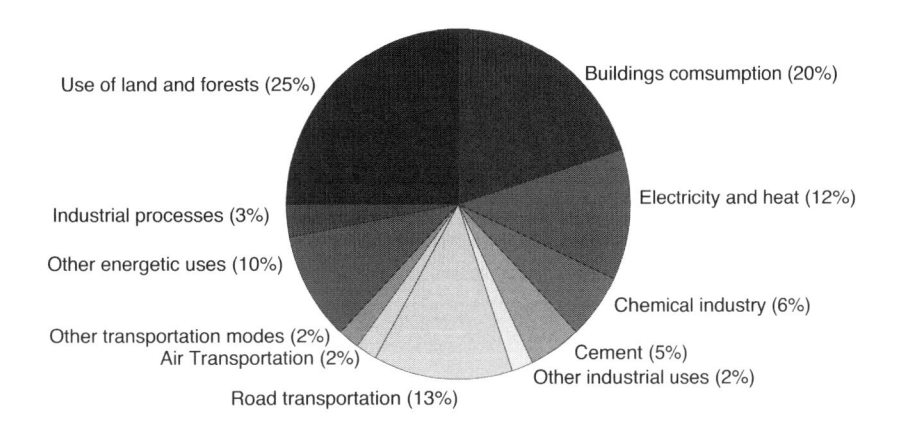

(a) CO_2 emissions by source. Data retrieved from BENITO and BENITO [3].

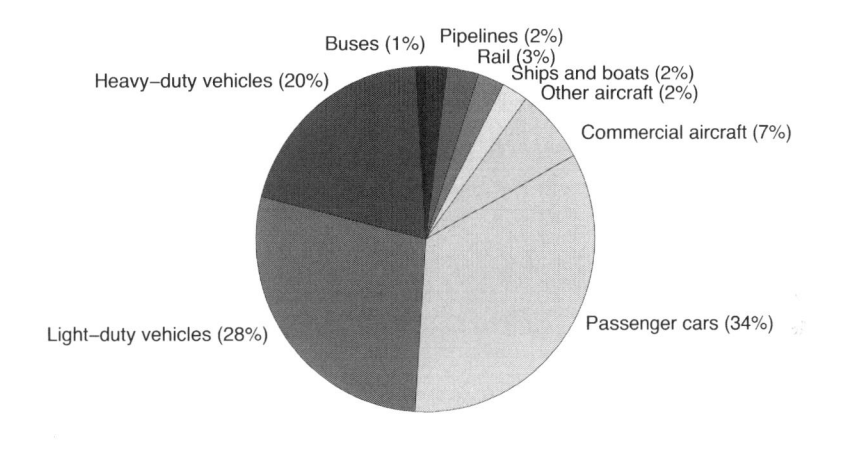

(b) U.S. Transportation Greenhouse Gas Emissions by Source, 2006 (all gases, in Tg CO2 equivalent). Data retrieved from US Department of Transportation.

Figure 8.5: CO_2 and global warming emissions.

are uncertain. The amount of NOx emissions depends on fuel consumption and the engine's type-specific emission index. The emission index for NOx depends on the engine and combustor architecture, power setting, flight speed, ambient pressure, temperature, and humidity. This dependence has to be taken into account when considering changes in aircraft design and operations.

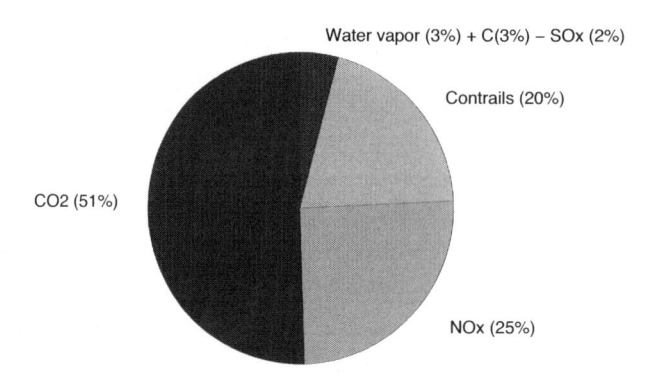

(a) Non considering a (still undetermined) quantity due to the artificial generation of cirrus clouds

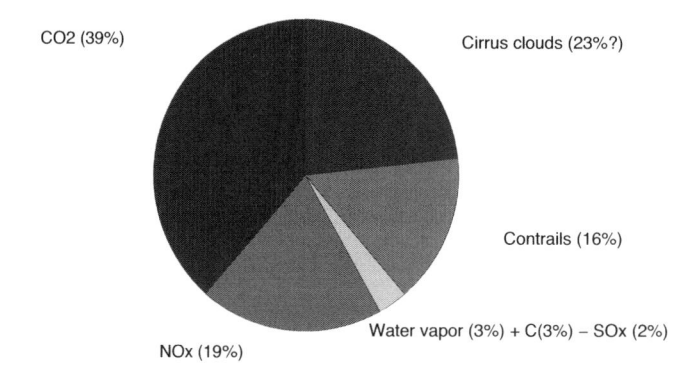

(b) Considering a (still undetermined) quantity due to the artificial generation of cirrus clouds

Figure 8.6: Aircraft emissions contributing to global warming. Data retrieved from BENITO and BENITO [3].

Aviation Water Vapor Climate Impact: The climate impact of water vapor emissions without contrail formation is relatively small for subsonic aviation. The relative impact increases with altitude because of longer lifetimes and lower background concentrations at higher altitudes in the stratosphere, and would be more important for supersonic aircraft; water vapor would also become more important when using hydrogen-powered aircraft. The total route time in the stratosphere can be used as an indicator for water vapor climate impact.

8.4 Environmental impact

(a) Linear contrail in the sunset. Photo taken in Alcañiz, Teruel.

(b) Example of persistent contrail cirrus. Photo taken in Av Via Lusitana, Madrid.

Figure 8.7: Contrails.

Contrails: Contrails (short for condensation trails) are thin, linear ice particle clouds often visible behind cruising aircraft. They form because, under appropriate atmospheric conditions, the exhausted water vapor resulting from combustion inside aircraft engines mixes with cold ambient air, leading to local liquid saturation, condensation of water vapor, and subsequent freezing. A comprehensive analysis of the conditions for persistent contrail formation from aircraft exhausts is given in SCHUMANN [14].

Linear contrails may persist for hours and may eventually evolve into diffuse cirrus clouds, modifying thus the natural cloudiness. As a consequence persistent contrails modify the radiation balance of the Earth–Atmosphere system, resulting into a net increase of earth's surface warming.

Contrails form when a mixture of warm engine exhaust gases and cold ambient air reaches saturation with respect to water, forming liquid drops that quickly freeze. Contrails form in the regions of airspace that have ambient relative humidity with respect to water (RH_w) greater than a critical value r_{contr}. Regions with RH_w greater or equal than 100% are excluded because clouds are already present. Contrails can persist when the environmental relative humidity with respect to ice (RH_i) is greater than 100%. Thus, persistent contrail favorable regions are defined as the regions of airspace that have: $r_{contr} \leq RH_w < 100\%$ and $RH_i \geq 100\%$.

The estimated critical relative humidity for contrail formation at a given temperature T (in degrees Celsius) can be calculated as:

$$r_{contr} = \frac{G(T - T_{contr}) + e_{sat}^{liq}(T_{contr})}{e_{sat}^{liq}(T)}, \qquad (8.12)$$

where $e_{sat}^{liq}(T)$ is the saturation vapor pressure over water at a given temperature. The estimated threshold temperature (in degrees Celsius) for contrail formation at liquid saturation is:

$$T_{contr} = -46.46 + 9.43 \log(G - 0.053) + 0.72 \log^2(G - 0.053), \quad (8.13)$$

where

$$G = \frac{EI_{H_2O} C_p P}{\epsilon Q(1 - \eta)}. \quad (8.14)$$

In equation (8.14), EI_{H_2O} is the emission index of water vapor, C_p is the isobaric heat capacity of air, P is the ambient air pressure, ϵ is the ratio of molecular masses of water and dry air, Q is the specific heat combustion, and η is the average propulsion efficiency of the jet engine.

RH_i is calculated by temperature and relative humidity using the following formula:

$$RH_i = RH_w \frac{6.0612 \exp \frac{18.102 T}{249.52 + T}}{6.1162 \exp \frac{22.577 T}{237.78 + T}}, \quad (8.15)$$

where T is the temperature in degrees Celsius.

Climate impact mitigation Options

Strategies for minimizing the climate impact of air traffic include identifying the most efficient options for airframe and propulsion technology, air traffic management, and alternative route network concepts. Economic measures and market-based incentives may also contribute, but these are out of the scope of this chapter. Minimizing the climate impact of aviation would require addressing all climate impact components. In the following, due to its relative importance, only the reduction of emissions of CO_2 and a contrail mitigation strategy are considered.

Minimizing CO_2 Emissions: Minimum fuel consumption is of primary interest for the aviation industry because it reduces costs. However, fuel is not the only cost driver and various constraints cause fuel penalties. Although fuel reduction below the current state is challenging, further reduction of fuel consumption and hence of fossil CO_2 climate impact is nevertheless feasible. As exposed in Section 1.3, the aim is to reduced the CO_2 emissions by 50% due to 2050 when compared to 2010 emissions. The reduction in CO_2 will require contributions from new technologies in aircraft design (engines, airframe materials, and aerodynamics), alternative fuels (bio fuels), and improved ATM and operational efficiency (mission and trajectory management). See Figure 1.3.

Contrail impact mitigation strategies: Several strategies for persistent contrail mitigation have been studied. See for instance GIERENS et al. [7]. As illustration, a flight planning contrail mitigation strategy is herein presented. Further mitigation potential can be achieved by developing optimized aircraft and jets for these alternative trajectories.

Example 8.1. *In this example the aim is at showing a contrail mitigation strategy based on modifying the vertical profile of the flight. More information about this example can be consulted in SOLER et al. [17].*

More specifically, we optimize the trajectory of a B757-200 BADA 3.6 NUIC [9] model aircraft performing the en-route part of a flight San Francisco (SFO) - New York (JFK) between the waypoint[16] PEONS as initial fix and the waypoint MAGIO as final fix. The route is composed by waypoints given in Table 8.10.

Name	Type	Longitude	Latitude
PEONS	WAYPOINT (RNAV)	−119.1674°	38.5035°
INSLO	WAYPOINT (RNAV)	−117.2981°	38.6791°
DTA	VOR-TAC (NAVAID)	−112.5055°	39.6791°
MTU	VOR-DME (NAVAID)	−110.1270°	40.1490°
CHE	VOR-DME (NAVAID)	−107.3049°	40.5200°
HANKI	REPORTING POINT	−102.9301°	41.6319°
KATES	REPORTING POINT	−96.7746°	42.5525°
FOD	VOR-TAC (NAVAID)	−94.2947°	42.6111°
KG75M	NRS-WAYPOINT	−88°	42.5°
DAFLU	REPORTING POINT	−82.7055°	42.3791°
JHW	VOR-DME (NAVAID)	−79.1213°	42.1886°
MAGIO	REPORTING POINT	−76.5964°	41.5373°

Table 8.10: Route's waypoints, navaids, and fixes

We assume all pairs formed by two consecutive waypoints are connected by bi-directional airways. On an airway, aircraft fly at different flight levels to avoid collisions. The different flight levels are vertically separated 1000 feet. On a bi-directional airway, each direction has its own set of flight levels according to the course. In east direction flights, aircraft are assigned odd flight levels separated 2000 feet. We then assume the aircraft can flight the route in any (if only one) of the following flight levels:

$$\{FL270, FL290, FL310, FL330, FL350, FL370, FL390, FL410\} \quad (8.16)$$

The flight we are analyzing is inspired in DAL30, with scheduled departure from SFO at 06:30 a.m. on June the 30th, 2012. Data of air temperature and relative humidity for June the 30th, 2012 at time 18.00 Z[17] (10.00 a.m. PST) have been retrieved from the

---

[16]Waypoints may be a simple named point in space or may be associated with existing navigational aids, intersections, or fixes.

[17]Z-hour corresponds to Universal Time Coordinates (UTC). The Pacific Standard Time (PST) is given by UTC − 8 hours.

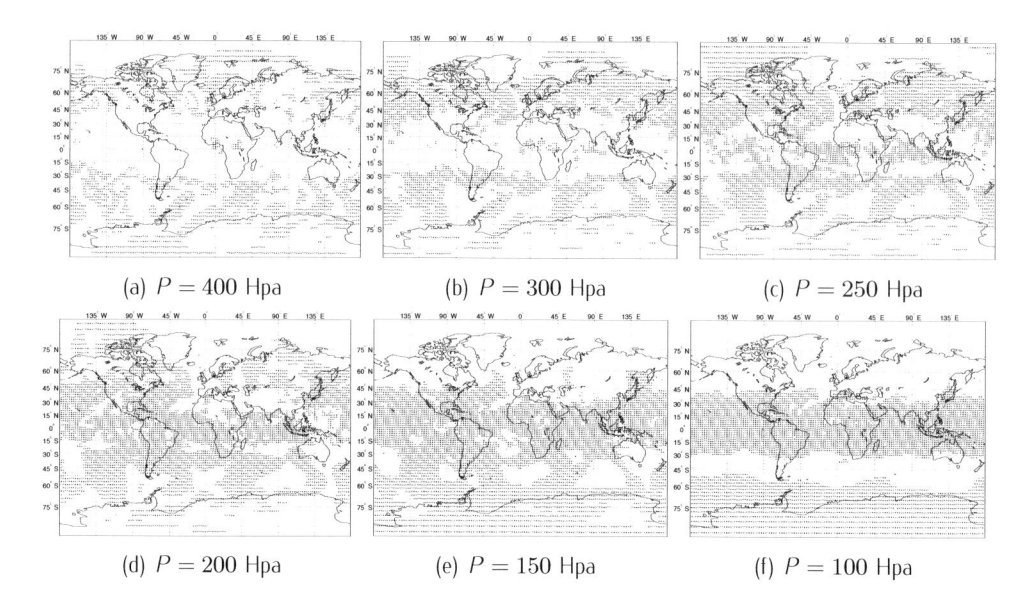

Figure 8.8: Longitude-latitude grid points that present favorable conditions for persistent contrail formation for different barometric altitudes.

(a) $P = 400$ Hpa
(b) $P = 300$ Hpa
(c) $P = 250$ Hpa
(d) $P = 200$ Hpa
(e) $P = 150$ Hpa
(f) $P = 100$ Hpa

NCEP/DOE AMIP-II Reanalysis data provided by the System Research Laboratory at the National Oceanic & Atmospheric Administration (NOAA)[18]. The data have a global spatial coverage with different grid resolutions. Our data have a global longitude-latitude grid resolution of $2.5° \times 2.5°$. Regarding the vertical resolution, the data are provided in 17 pressure levels (hPa): 1000, 925, 850, 700, 600, 500, 400, 300, 250, 200, 150, 100, 70, 50, 30, 20, 10.

According to what has been exposed in Section 8.4.2, we compute the latitude-longitude grid points that are favorable to persistent contrail formation at different barometric altitudes (which defines the pressure). We do so based on gathered data of air temperature and relative humidity, and using equations (8.12)-(8.15) with the following: $EI_{H_2O} = 1.25$; $C_p = 1004$ $[J/KgK]$; $\epsilon = 0.6222$; $Q = 43 \cdot 10^6$ $[J/Kg]$; and $\eta = 0.15$. The longitude-latitude grid points with favorable conditions for persistent contrail formation are represented as red dots in Figure 8.8 for different barometric altitudes.

In order to analyze the regions of persistent contrail formation in our case study, we first need to estimate the values of temperature and relative humidity for the flight levels given in set (8.16). In order to do that, we use the International Standard Atmosphere (ISA) equations to convert altitude into barometric altitude, and then run a linear interpolation between the data of air temperature and relative humidity corresponding to the 17 pressure levels and the desired flight levels (already converted into barometric altitude). Once we have the values of temperature and relative humidity at the desired flight levels, we proceed

---

[18] The data have been downloaded from NOAA website @ http://www.esrl.noaa.gov/psd/

8.4 ENVIRONMENTAL IMPACT

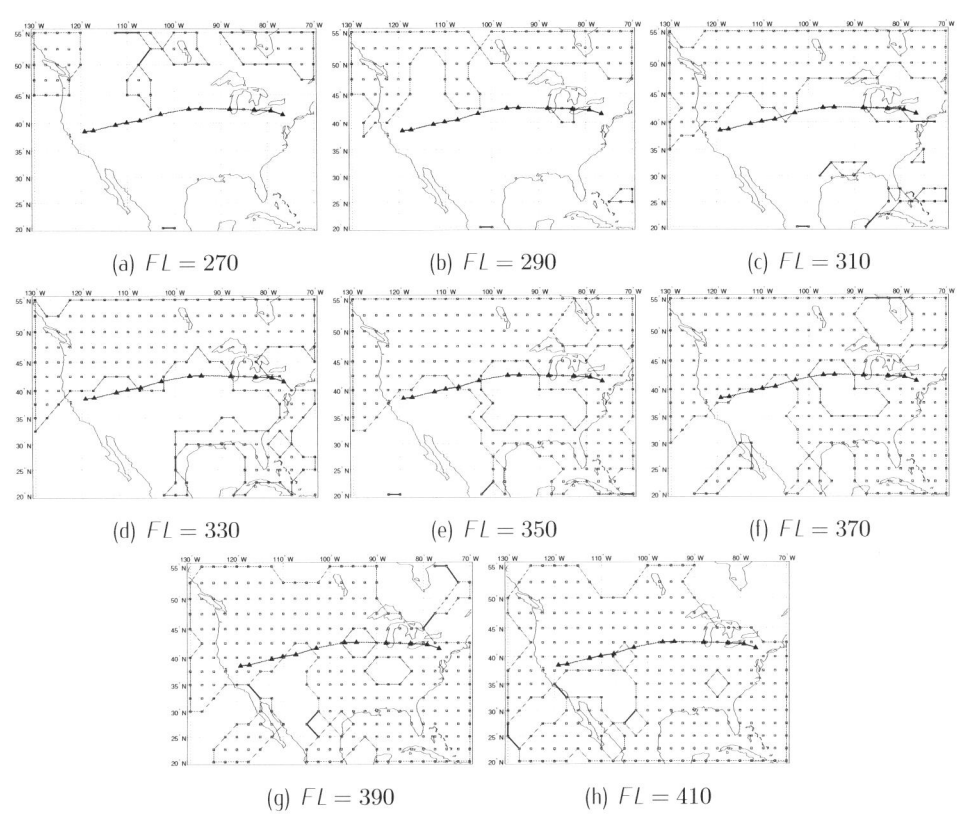

(a) $FL = 270$ (b) $FL = 290$ (c) $FL = 310$
(d) $FL = 330$ (e) $FL = 350$ (f) $FL = 370$
(g) $FL = 390$ (h) $FL = 410$

Figure 8.9: Favorable regions of contrail formation over USA at different flight levels. The horizontal route is depicted to illustrate how the same horizontal route under different flight levels might increase/reduce potential persistent contrail generation.

on using equations (8.12)-(8.15) as exposed above. The favorable regions of persistent contrail formation over the USA at the different flight level can be consulted in Figure 8.9.

It can be observed that flying at high flight levels, e.g., FL370, FL390, and FL410, implies overflying regions of persistent contrail generation. On the contrary, flying at low flight levels, e.g., FL270, FL290, and FL310, implies non overflying regions of persistent contrail generation and thus minimizes environmental impact. However, it is obviously more efficient in terms of fuel burned to fly higher, which is actually what airlines do. Indeed, the flight in which this example is based on flow a flight plan at FL390 and FL410. Concluding, trade-off strategies (fuel-environmental impact) must be found.

REFERENCES

[1] ACARE (2010). Beyond Vision 2020 (Towards 2050). Technical report, European Commission. The Advisory Council for Aeronautics Research in Europe.

[2] BELOBABA, P., ODONI, A., and BARNHART, C. (2009). *The global airline industry*. Wiley.

[3] BENITO, A. and BENITO, E. (2012). *Descubrir el transporte aéreo y el medio ambiente*. Centro de Documentación y Publicaciones Aena.

[4] BISIGNANI, G. (7 June 2010). State of the Air Transport Industry. In *66th IATA Annual General Meeting and World Air Transport Summit*. Director General and CEO of International Air Transport Association (IATA).

[5] DOGANIS, R. (2002). *Flying off course: The economics of international airlines*. Psychology Press.

[6] DOGANIS, R. (2006). *The airline business*. Psychology Press.

[7] GIERENS, K., LIM, L., and ELEFTHERATOS, K. (2008). *The Open Atmospheric Science Journal* 2, 1–7.

[8] NAVARRO, L. U. (2003). *Descubrir el transporte aéreo*. Centro de Documentación y Publicaciones de AENA.

[9] NUIC, A. (2005). *User Manual for the base of Aircraft Data (BADA) Revision 3.6*. Eurocontrol Experimental Center.

[10] PENNER, J. (1999). Aviation and the global atmosphere: a special report of IPCC working groups I and III in collaboration with the scientific assessment panel to the Montreal protocol on substances that deplete the ozone layer. Technical report, International Panel of Climate Change (IPCC).

[11] PINDADO CARRIÓN, S. (2006). *ETSI Aeronáuticos*. Universidad Politécnica de Madrid .

References

[12] PRATS-MENÉNDEZ, X. (2010). *Contributions to the Optimisation of aircraft noise abatement procedures*. PhD thesis, Universitat Politècnica de Catalunya.

[13] ROYAL-COMMISSION (2002). The environmental effects of civil aircraft in flight. Technical report, Royal Commission of Environmental Pollution, TR, London, England, UK.

[14] SCHUMANN, U. (1996). *Meteorologische Zeitschrift-Berlin-* **5**, 4–23.

[15] SCHUMANN, U., editor (2012). *Atmospheric Physics: Background—Methods—Trends*. Research Topics in Aerospace. Springer.

[16] SESAR CONSORTIUM (April 2008). SESAR Master Plan, SESAR Definition Phase Milestone Deliverable 5.

[17] SOLER, M., ZOU, B., and HANSEN, M. (2013). Contrail Sensitive 4D Trajectory Planning with Flight Level Allocation using Multiphase Mixed-Integer Optimal Control. In *Proceedings of the AIAA Guidance, Navigation, and Control Conference*. Boston, Massachusetts, USA.

9 AIRPORTS

Contents

- 9.1 Introduction ... 270
  - 9.1.1 Airport designation and naming ... 270
  - 9.1.2 The demand of air transportation ... 271
  - 9.1.3 The master plan ... 273
- 9.2 Airport configuration ... 273
  - 9.2.1 Airport description ... 273
  - 9.2.2 The runway ... 276
  - 9.2.3 The terminal ... 282
  - 9.2.4 Airport services ... 287
- 9.3 Airport operations ... 288
  - 9.3.1 Air Traffic Management (ATM) services ... 288
  - 9.3.2 Airport navigational aids ... 289
  - 9.3.3 Safety management ... 295
  - 9.3.4 Environmental concerns ... 295
- References ... 297

The aim of this chapter is to give a brief overview of airports, as they are a fundamental infrastructure to facilitate intermodal transportation and ensure that flights are performed in a safe way. Section 9.1 is devoted to define what airports are, providing a brief overview of their history, introducing their naming nomenclature, describing the variables that potentially affect the demand of air transportation, and thus the necessity of building an airport in a determined location, and finally giving a description of the master plan, the set of official documents for the design and construction of an airport. Section 9.2 is devoted to provide a description of the configuration of a modern airport, including air-side and land-side elements. Finally, Section 9.3 analyzes airport operations. Some introductory aspects suitable for this type of course can be consulted in FRANCHINI *et al.* [2]. Two thorough references on the matter are DE NEUFVILLE and ODONI [1], GARCÍA CRUZADO [3].

9.1 Introduction

Over 100 years ago, it arose the necessity of using existing terrains to carry out the first flights. Those terrains were named *airfields*. Later on, airfields evolved to what is referred to as *aerodrome*. ICAO, in its Annex 14 ICAO [4] defines aerodrome as:

Definition 9.1 (*Aerodrome*). *A defined area on land or water (including any buildings, installations, and equipment) intended to be used either wholly or in part for the arrival, departure, and surface movement of aircraft.*

After World War II, when commercial aviation reached its maturity, the term *airport* was generalized. The term airport refers to an aerodrome that is licensed by the responsible government organization (FAA in the United States; AESA in Spain). Airports have to be maintained to higher safety standards according to ICAO standards.

An airport is an intermodal transportation facility where passengers connect from/to ground transportation to/from air transportation. As it will be described in detail in Section 9.2, airports can be divided into land-side and air-side. The land-side embraces all facilities of the airport in which passengers arrive/depart the airport terminal building and move through the terminal building to clear security controls. Air-side embraces those infrastructures devised for movement of the airplanes on the airports surface, but also the boarding lounges. Roughly speaking, land-side corresponds to those facilities in which both passengers and companions (not passengers such as family, friends, etc.) cohabit. On the contrary, air-side infrastructure include all those areas in which only passengers with tickets (and obviously also airport employees) are allowed to be, including those infrastructures made for aircraft parking, taxiing, and landing/taking-off.

The most simple airport consists of one runway (or helipad), but other common components are hangars and terminal buildings. Apart from these, an airport may have a variety of facilities and infrastructures, including airline's services, e.g., hangars; air traffic control infrastructures and services, e.g, the control tower; passenger facilities, e.g., restaurants and lounges; and emergency services, e.g., fire extinction unit.

9.1.1 Airport designation and naming

Airports are uniquely represented by their IATA airport code and ICAO airport code. IATA 3-letter airport codes are typically abbreviated of their names, such as MAD for Madrid Barajas International Airport. Exceptions to this rule are, for instance, O'Hare International Airport in Chicago (retains the code ORD from its former name of Orchard Field) and some named after a prominent national celebrity, e.g., John F. Kennedy, Paris Charles de Gaulle, Istanbul Atartuk , etc. The ICAO 4-letter airport identifier codes uniquely identify individual airports worldwide. Usually, the first letter of ICAO codes identify the country. In the continental USA, the first letter is *K*. In Spain, the first letter is *L*.

Airports by passengers 2012

Airport	Code (IATA/ICAO)	Passengers	Rank	% change
Atlanta Hartsfield-Jackson	ATL/KATL	95462867	1	+3.3%
Beijing Capital	PEZ/ZBAA	81929689	2	+4.5%
London Heathrow	LHR/EGLL	70038857	3	+0.9%
Tokio	HND/RJTT	67788722	4	+8.4%
Chicago O'Hare	ORD/KORD	67091391	5	-0.4%
Los Angeles	LAX/KLAX	63687544	6	+3%
Paris Charles de Gaulle	CDG/LFPG	61611934	7	+1.1%
Dallas Fort Worth	DFW/KDFW	58591842	8	+1.4%
Soekarno-Hatta (Indonesia)	CGK/WIII	57730732	9	+14.4%
Dubai	DXB/OMDB	57684550	10	+13.2%
Frankfurt	FRA/EDDF	57520001	11	+1.9%
Hong Kong	HKG/VHHH	56064248	12	+5.2%
Denver	DEN/KDEN	53156278	13	+0.6%
Suvarnabhumi (Thailand)	BKK/VTBS	53002328	14	+10.6%
Singapure Changi (Singapore)	SIN/WSSS	51181804	15	+10%
Amsterdam Chiphol	AMS/EHAM	51035590	16	+2.6%
Ney York John F Kennedy	JFK/KJFK	49293587	17	+3.1%
Guangzhou Baiyun (China)	CAN/ZGGG	48548430	18	+7.8%
Madrid Barajas	MAD/LEMD	45175501	19	-9%
Istanbul Atartuk	IST/LTBA	44992420	20	+20.1%

Table 9.1: List of biggest airports in 2012 in volume of passengers. % of change refers to the increase of traffic with respect to 2011. Data retrieved from Wikipedia.

9.1.2 THE DEMAND OF AIR TRANSPORTATION

The variables that influence in the potential demand of air transportation in a determined airport can be itemized as follows:

- Historical tendency of geographical related airports.
- Demographic variables of the population under the region of influence of the airport.
- The economical character (industrial, technological, financial, touristic, etc.) of the region.
- Intermodal transportation network.
- Urban and regional strategic development plan.
- Competitors prices.
- Sociocultural changes.

Indeed, all these items can be reduced to one: the Gross Domestic Product (GDP) per capita of the region. GDP per capita and demand of air transportation are strongly correlated.

Airports by movements 2011

Airport	Code (IATA/ICAO)	Passengers	Rank	% change
Atlanta Hartsfield-Jackson	ATL/KATL	923996	1	-2.7%
Chicago O'Hare	ORD/KORD	878798	2	-0.4%
Los Angeles	LAX/KLAX	702895	3	5.4%
Dallas Fort Worth	DFW/KDFW	646803	4	-0.8%
Denver	DEN/KDEN	628796	5	-0.2%
Charlotte	CLT/KCLT	539842	6	2%
Beijing Capital	PEZ/ZBAA	533257	7	3%
Las Vegas McCarran	LAS/KLAS	531538	8	5.1%
Houston George Bush	IAH/KIAH	517262	9	-2.7%
Paris Charles de Gaulle	CDG/LFPG	514059	10	+2.8%
Frankfurt	FRA/EDDF	487162	11	+4.9%
London Heathrow	LHR/EGLL	480931	12	+5.7%
Phoenix Sky Harbor	PHX/KPHX	461989	13	+2.8%
Philadelphia	PHL/KPHL	448129	14	-2.7%
Detroit Metropolitan Wayne County	DTW/KDTW	443028	15	-2.1%
Amsterdam Chiphol	AMS/EHAM	437083	16	+8.6%
Minneapolis Saint Paul	MSP/KMSP	436506	17	0%
Madrid Barajas	MAD/LEMD	429390	18	-1.0%
Toronto Pearson	YYZ/CYYZ	428764	19	+2.5%
Newark Liberty (New Jersey)	EWR/KEWR	410024	20	+2.9%

Table 9.2: List of biggest airports in 2011 in volume of passengers. % of change refers to the increase of movements with respect to 2010. Data retrieved from Wikipedia.

Therefore, according to the long-term estimation of GDP growth, the demand of air transportation is also expected to increase as a worldwide average rate of 5%. Thus, existing airports should be enlarged to absorb increasing demand, but also new airports should be opened in the future.

Table 9.1 and Table 9.2 give a quantitative measure of the busiest airports worldwide.

When analyzing the biggest airports by number of passenger, obviously the big cities appear in the first positions, i.e., Beijing, London, Tokio, Chicago, Los Angles, Paris, etc. Notice however that Atlanta, not such an important city, is the world's busiest airport. This is due to the fact that Atlanta is Delta's hub. Also Dallas, American Airlines' hub, appears in the first positions. Other important issues to notice are: the increasing presence of east and south east asian airports, with very important inter-annual growths; and the fact that Madrid Barajas has dropped 9% in 2012. The later can be explained due to two different phenomena: the very important crisis that Europe, particularly the mediterranean countries, are suffering; and the acquisition of Iberia, Spanish flag company whose hub was Barajas, by British Airways, which has shifted a little bit the South America's connexion demand towards United Kingdom.

In terms of movements, USA's airports cope the first positions. This is due to the fact that many cities in the United States act as hubs. Many connections between american cities are done on a daily basis with medium-haul aircraft types (transporting less people). Also, oversees flights that arrive at the United States typically go first to the airline's hub and then transit to a domestic flight. On the contrary, asian companies have recently started an strategy towards buying big airplanes (A380), transporting thus more people with less movements.

9.1.3 THE MASTER PLAN

The master plan is an official set of documents with information, studies, methodologies, and performances to be carried out in the design and construction of a new airport or an important enlargement into an existing one.

The master plan includes first a study of the current situation, including physical data (topography, meteorology, etc.), socioeconomic data (demography, GPD, etc.), comparative studies with nearby airports, etc. Aeronautical data such forecasted flights and types of aircraft, including a complete long-term demand forecast, are also needed. With the forecasted demand, the necessities in terms of runways, taxiways, platform positions, terminal buildings, etc. are calculated. After analyzing demand and necessities, the emplacement must be selected. For the selection of the emplacement one must take into account:

- The climatology (wind, fog, temperature, etc.).
- The topography (unevenness and slopes in the terrain).
- Obstacles in the surroundings (for safe taking off and landing).
- Intermodal connexions.
- Availability of terrains.
- Environmental impact.

Indeed, the master plan includes an environmental impact report of the operations in the selected emplacement. Economic studies in terms of operations and future development are also included.

9.2 AIRPORT CONFIGURATION

9.2.1 AIRPORT DESCRIPTION

Airports are divided into land-side and air-side areas. Figure 9.1 illustrates and schematic flow in an airport. Figure 9.2 shows a layout of a medium size airport.

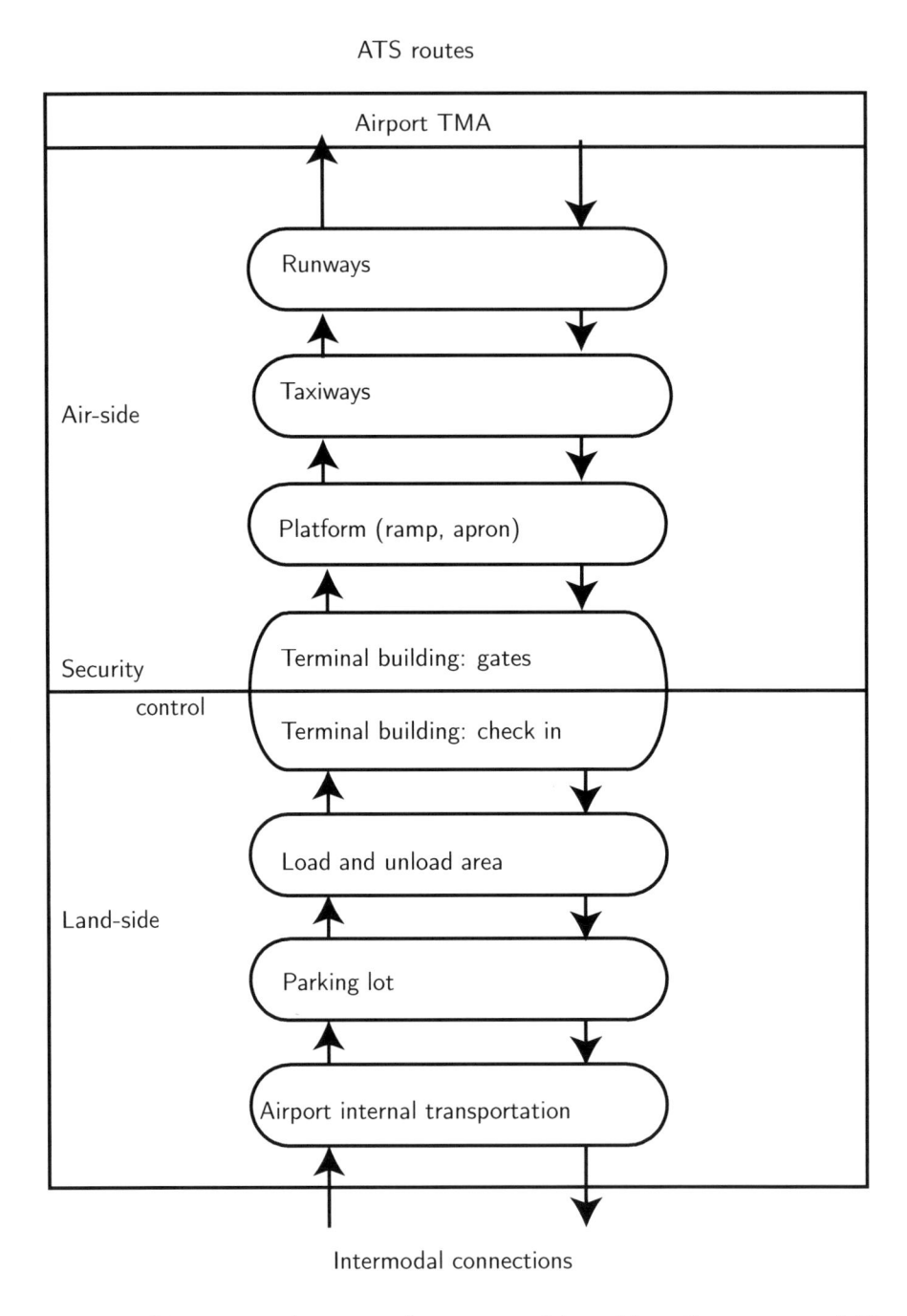

Figure 9.1: Schematic configuration of an airport. Adapted from FRANCHINI et al. [2].

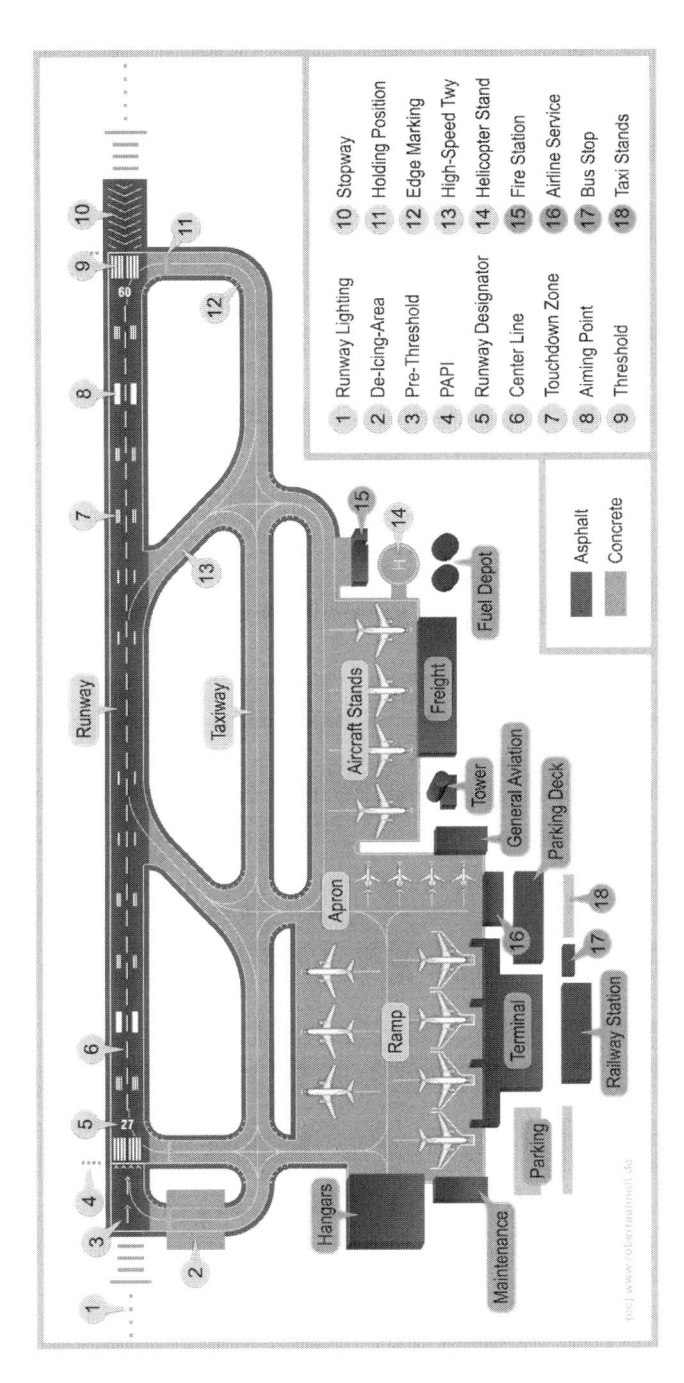

Figure 9.2: Typical airport infrastructure. © Robert Aehnelt. / Wikimedia Commons / CC-BY-SA-3.0.

land-side areas

Land-side areas include parking lots, fuel tank farms, and access roads. Access from land-side areas to air-side areas is controlled at most airports by security systems and personal. Passengers on commercial flights access air-side areas through terminals, where they can purchase tickets, check luggage in, and clear security. One security has been cleared, the passenger is in the air-side areas.

Air-side areas

The air-side is partially composed by a set of infrastructures formed by the runway (or runways), taxiway (or taxiways) and high-speed taxiways, together with the ramp and the apron. Also the waiting areas, which provide passenger access to aircraft and typically include duty free shops and restaurants, are considered air-side and referred to as concourses[1]. Due to their high capacity and busy airspace, most international airports have air traffic control located on site. This is also considered air-side infrastructure. Notice that minor airports might not necessarily have a control tower, instead some air traffic control services would be allocated within the airport facilities.

The area where aircraft park next to a terminal to load passengers and baggage is known as a ramp or platform. Parking areas for aircraft away from terminals are generally called aprons. The difference between ramp and apron is that the ramp is typically connected to the terminal with fingers.

A taxiway is a path on an airport connecting runways with ramps, hangars, terminals, and other facilities. They are typically build on asphalt (or more rarely concrete). Busy airports typically construct high-speed or rapid-exit taxiways in order to allow aircraft to leave the runway at higher speeds. This allows the aircraft to exit the runway quicker, permitting another one to land or depart in a shorter space of time, increasing thus the capacity of the airport as it will be mentioned later on.

9.2.2 THE RUNWAY

According to ICAO [4] a runway is:

Definition 9.2 (*Runway*). *defined as a rectangular area on a land aerodrome prepared for the landing and takeoff of aircraft.*

Runways are typically build based on asphalt or concrete over a previously leveled and compacted surface. Runways are defined together with safety areas, that might be of compacted natural terrain. On both sides of the runway, there is the strip. In the heads of the runway we find a Stop-Way (SWY) area and a Clear-Way (CWY) area. All these

---

[1] this term is often used interchangeably with terminal waiting lounges

areas are due to safety reasons and their specifications are stated by ICAO and can be consulted in ICAO [4].

Runway orientation

The orientation of a runway is typically determined by two main issues:

- The non existence of obstacles involved in the maneuvers of take-off and landing.
- The direction of the dominant winds.

As previously described in Section 9.1, the construction of a new airport (or the enlargement of an existing one) is stated in a set of documents referred to as Master Plan. One of the very first steps is to select the emplacement of a new airport, which is driven by the location and orientation of the runway. Notice that a surrounding area with no obstacles is key, first to fulfill with ICAO regulations and recommendations, second to facilitate the design of more efficient and safer departure and arrival procedures. Moreover, take offs and landings are desirably performed with head/tail wind (head if possible). Lateral wind might be dangerous and sometimes operations would be cancelled if it exceeds determined safety thresholds (which depend on the aircraft type). Therefore, starting 10-15 years prior the construction of the airport, a detailed study on the wind patterns is carried out to statistically select the most appropriate runways configuration.

Runway categories

ICAO has established different categories for the runways according to the size of the aircraft that can operate in such runways. The identifiers are two: a letter associated to the wing-span of the aircraft; and a number designating the longitude of the runway. Table 9.3 shows these categories. The width of the runways is obviously related with the category. It is shown in Table 9.4.

Even though Category 4 establishes a longitude greater that 1800 m, runways can be much longer up to 4500 m. Every single aircraft type must provide a document named *airport planning* to the authorities pertaining a set of data related to airport operations. Data include airplane performance, ground maneuvering, terminal servicing, operating conditions, and pavement data. Within aircraft performance, a take-off distance is provided. This distance is standardized to some conditions (temperature and density according to ISA at sea level, no slope in the runway, etc.). Therefore, at the real operation conditions, the distances that aircraft really need must be compensated with the temperature of the airport, elevation of the airport, and slope of the runway. For instance, the same aircraft would need much less distance to take off in an airport located at sea level that in an airport located at 3000 [m].

Airports

Num. Code	Runway Longitude L[m]	alph. Code	Wing-span b[m]
1	$L < 800$	A	$b < 15$
2	$800 < L < 1200$	B	$15 < b < 24$
3	$1200 < L < 1800$	C	$24 < b < 36$
4	$L > 1800$	D	$36 < b < 52$
–	–	E	$52 < b < 65$
–	–	F	$65 < b < 80$

Table 9.3: Runway ICAO categories. Data retrieved from ICAO [4].

Identifier	A	B	C	D	E	F
1	18	18	23	–	–	–
2	23	23	30	–	–	–
3	30	30	30	45	–	–
4	–	–	45	45	45	60

Table 9.4: Minimum width [m] ICAO identifiers. Data retrieved from ICAO [4].

Runways configuration

The most commonly used configurations are:

- Unique runway configuration.
- Cross runways configuration.
- V runways configuration.
- Parallel runways configuration.
- Double parallel runways configuration.

Notice that these configurations are made attending at the design requirements. One key indicator is the forecasted demand, which determined whether is enough with one runway to cope with all expected demand. Another factor could be related to the dominant winds. If there are two dominant directions, we might be forced to design two runways in different directions not to cancel operation on a regular basis. An overview of different spanish airports layout can be consulted in AIP AENA. See Figure 9.3 and Figure 9.4.

Runway identifiers

Besides the category, runways are identified attending at its geographical orientation, starting from the north and running clockwise, rounding to the closest tens of grades. Therefore, a runway which is being approach with a course of 89° (that is, approaching from the east) will be designated 09. Obviously, in the other head of the runway there is a

9.2 Airport configuration

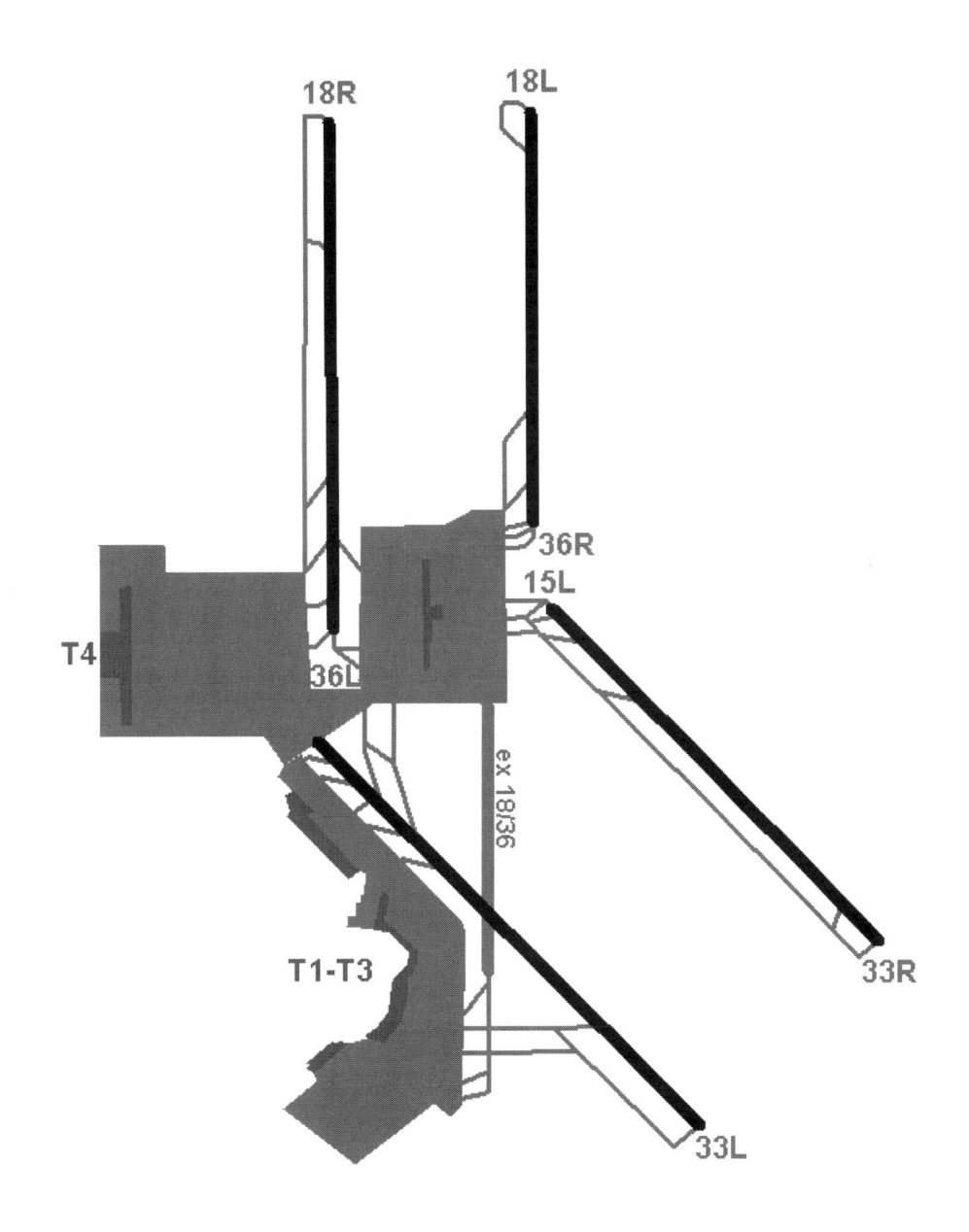

Figure 9.3: Madrid Barajas layout sketch. © user:Mxmxmx. / Wikimedia Commons / CC-BY-SA-2.0.

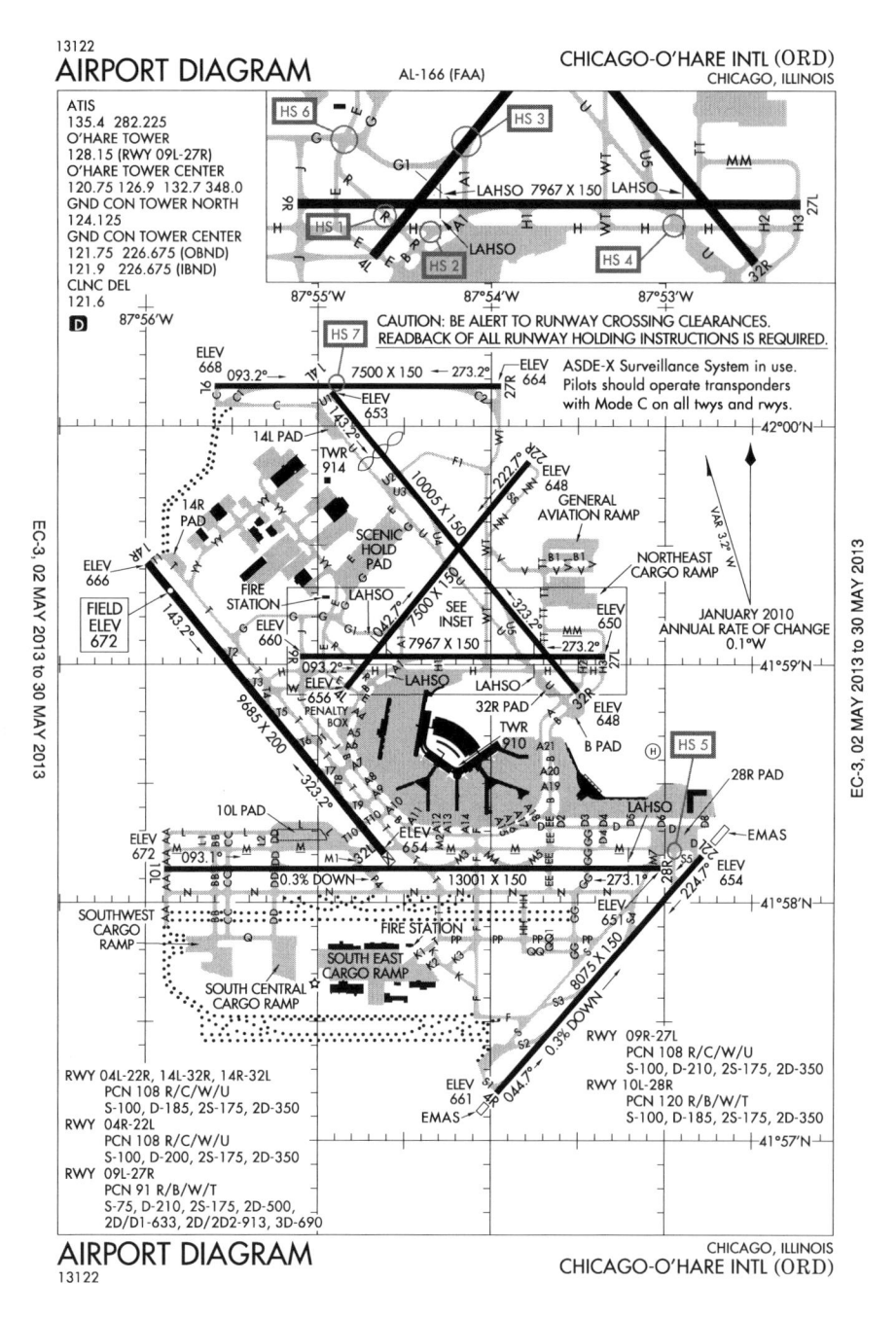

Figure 9.4: FAA airport diagram of O'Hare International Airport. © FAA. / Wikimedia Commons / Public domain.

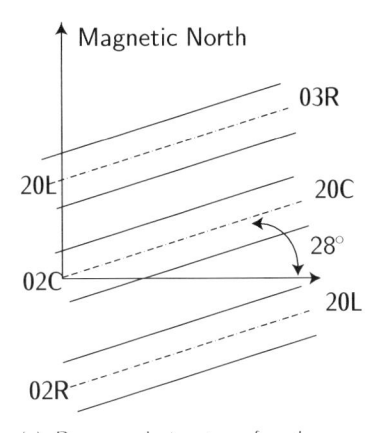

(a) Runway designators for three parallel runways.

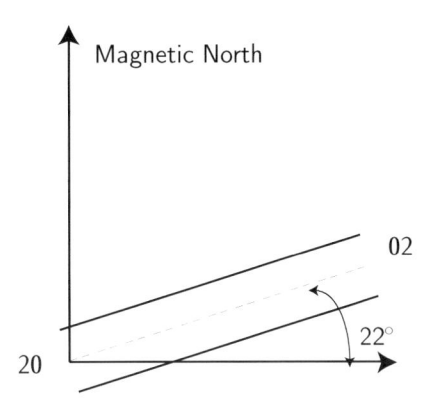

(b) Runway designators for a single runway.

Figure 9.5: Runway designators.

difference of 180°, that is, the designator will be 27. See Figure 9.2. The heads in courses near the north are identified as 36 instead of using 0. See for instance Figure 9.5.b, where a runway is orientated with an angle of 22 degrees with respect to the magnetic north.

When there exist parallel runways, as is the case of big airports such Madrid Barajas, a letter L (left) or right (R)[2] is added prior to the number. L or R is added attending at what the pilot is seeing when approaching on his/her right and left. See for instance Figure 9.5.a, which shows an airport layout with three parallel runways with an orientation of 28 degrees with respect to the magnetic north.

Capacity of the runway

As it has been studied in Chapter 3, an aircraft generates two vortexes in the tips of the wing that travel backwards behind the aircraft. Such trails remain a long distance behind the aircraft and can disturb aircraft flying behind, becoming a potential danger. In order to prevent such danger, ICAO has established a required minimum separation both in flight and in airport operations. Table 9.5 shows these distances in airport operations, being Aircraft 1 the preceding aircraft.

These separations are the key factor which determine the capacity of a runway in nominal conditions. A single runway configuration might have a maximum capacity of approximately 50-60 movements per hour. In configurations with more than one runway, the capacity increases.

In general, the capacity of a runways depends on different factors, some based on the available infrastructure, and other related with airport operations:

---

[2] There might also exist C (center) in case of three parallel runways.

Airports

Aircraft 1	Aircraft 2	Distance (NM)	Time (s)
Heavy	Heavy	4	106
Heavy	Medium	5	133
Heavy	Light	6	159
Medium	Light	5	133
Rest		3	79

Table 9.5: ICAO minimum distance in airport operations. *Heavy* refers to aircraft with $MTOW > 136000$ [kg]. *Medium* refers to aircraft with $7000 < MTOW < 136000$ [kg]. Any other aircraft with $MTOW < 7000$ [kg] are *light*. Data retrieved from ICAO [4].

- The conditions of Air Traffic Control (ATC) in approach and take-off.
- Longitude, orientation, and number of runways.
- The use of the system of runways for different operations (take-off and landing).
- The number, location, and characteristics of the rapid exit taxiways.
- The number of taxiways and the waiting points to runways heads.
- Mix of aircraft.
- Atmospheric conditions (wind, rain, fog, etc.)
- Conditions of the pavement.
- Type of visual aids.
- Approach and take off procedures.
- Interferences of the Terminal Maneuvering Area (TMA) with nearby airports or other flights (military, training, general aviation, etc.)

9.2.3 The terminal

In general, the terminal area designates the set of infrastructures inside the airport different from the aircraft movement area (apron, taxiways, runways). We can distinguish:

- Auxiliary aeronautical buildings (control tower, fire extinction building, electrical building, etc.).
- Freight processing areas (freight terminals).
- Aircraft processing areas (hangars, etc.).
- Industrial and commercial areas (pilot schools, catering services, mail services, etc.).
- Passenger processing and attention infrastructures, also referred to as passenger terminal.

We will focus in what follows on the passenger terminal. An airport passenger terminal is a building at an airport whose main functions are:

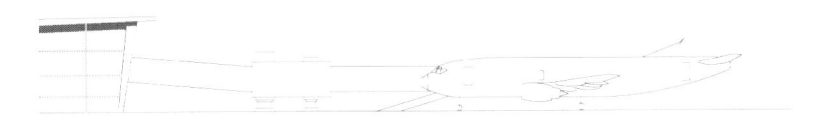

Figure 9.6: Aaircraft fed by a finger.

- The interchange of transportation mode terrestrial-aerial.

- The processing of the passenger before boarding: check-in, security controls, shopping, etc.

- The processing of the passenger after disembarking: luggage claim, customs and security, facilities (car rental, for instance), etc.

- It also fulfills a function of distributing the flows of passengers. Typically passenger reach check in in a bunch, but then they walk alone in small groups, and they reach the gate again as a bunch. Therefore, big lounges and long decks or walkways are needed to distribute the flows.

- Give room to aircraft parking positions fed by fingers, as illustrated in Figure 9.6.

Terminal layout

The terminal layout depends on many factors and typically differs from one airport to another. However, there are some patterns that are typically followed:

- Arrival and departure flows are separated, typically in different levels.
- Domestic and international flow are separated, typically in the same level.

Figure 9.7 shows a typical layout of a medium size airport. We can observe how arrivals and departure flows are separated (typically in two different floors). We can observe that the departure passenger process starts by queuing to check-in, clearing security, waiting in the concourse and proceed to gate, clear the boarding security control, and finally embarking. Notice that aircraft are fed by fingers. On the other hand, the arrival passenger will disembark and go directly to claim luggage (notice that there is no customs on arrivals, so we assume this is the domestic part of the airport).

Also, Figure 9.8 shows an air-side airport's departure terminal layout. Notice that the same basic elements can be identified, i.e., the security controls give access to the concourses that feed aircraft via fingers.

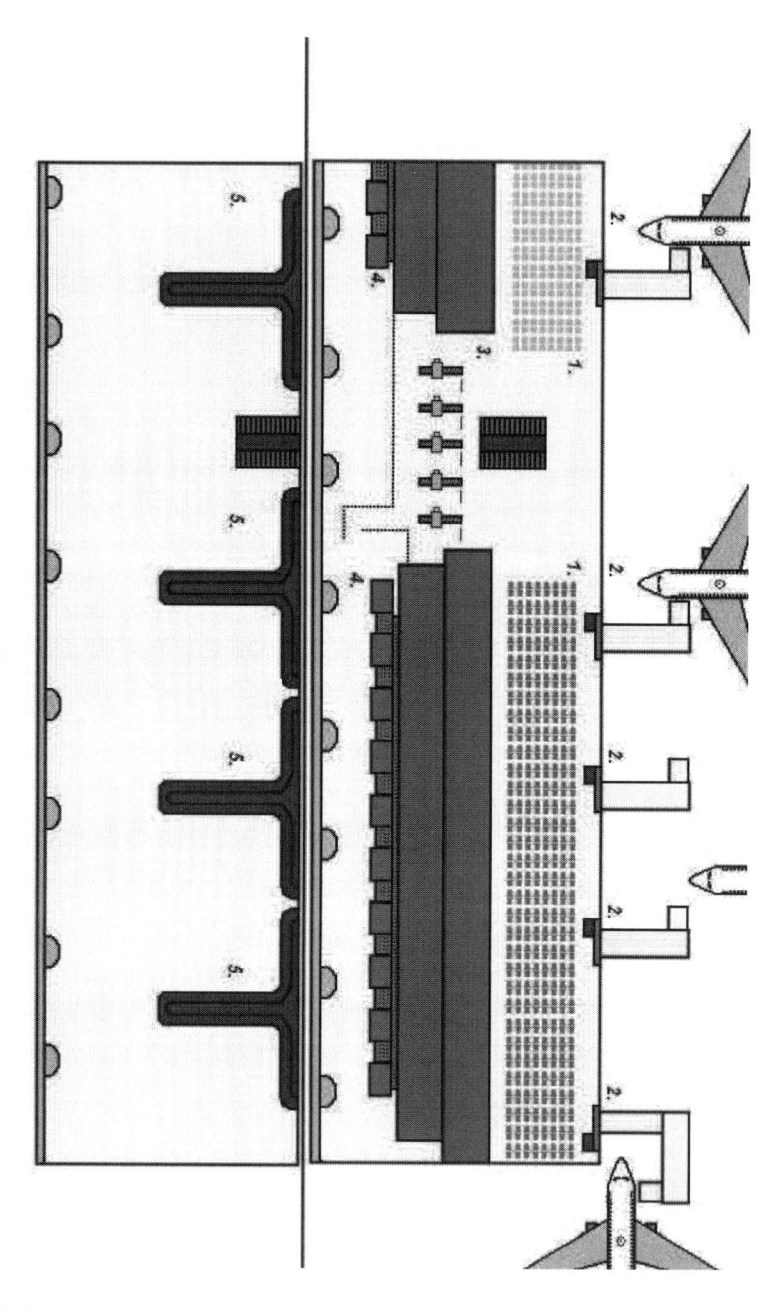

Figure 9.7: Typical design of a terminal, showing departure (right half of page) and arrival levels (left half): 1. Departures lounge; 2. Gates and fingers; 3. Security clearance gates; 4 Baggage check-in; 5. Baggage carousels. © Ohyeh. / Wikimedia Commons / CC-BY-SA-3.0

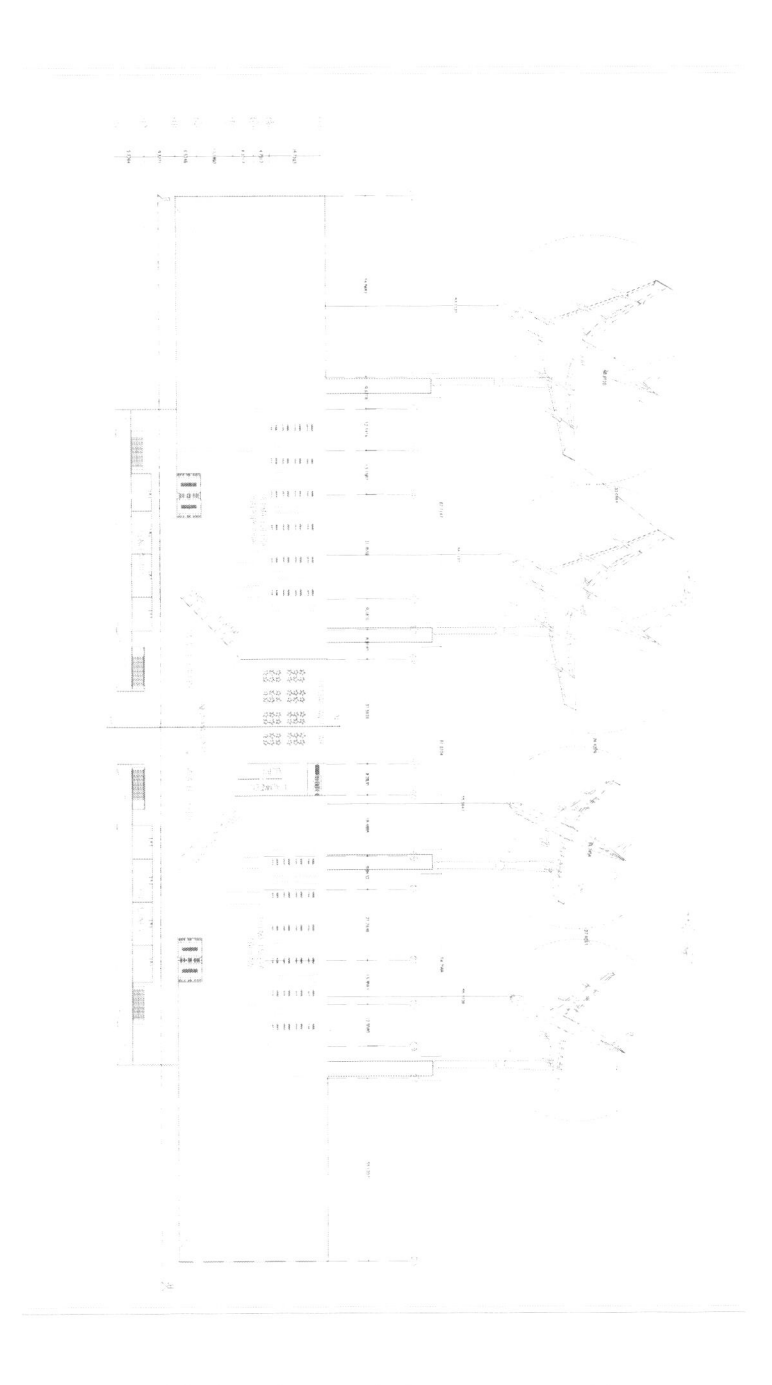

Figure 9.8: Air-side airport's departure terminal layout.

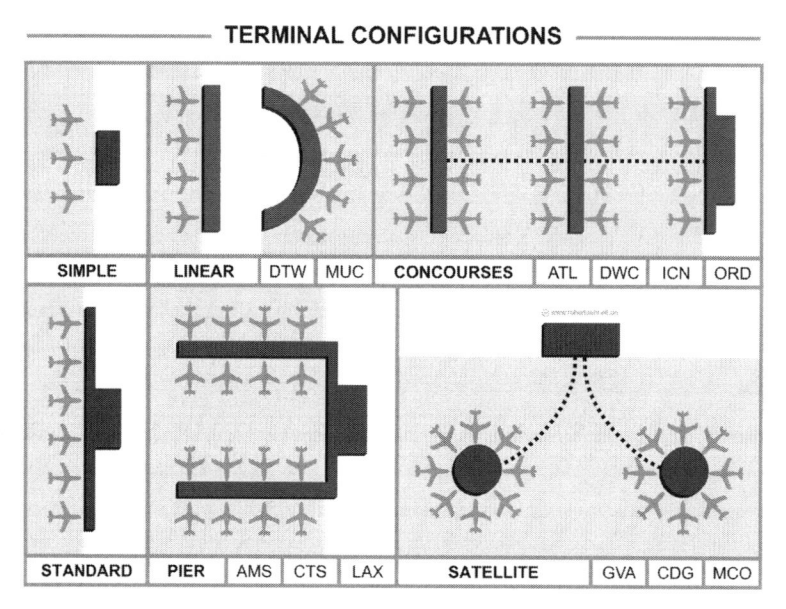

Figure 9.9: Typical terminal configurations. © Robert Aehnelt. / Wikimedia Commons / CC-BY-SA-3.0.

Terminal configuration

The configuration of the terminal is determined by the number and type of aircraft we want to directly feed with fingers. Figure 9.9 shows some of the typical configurations.

The standard configuration allows many aircraft, but the terminal must be long and therefore the distance to be walked by passenger. Another strategy is to use piers. A pier design uses a long, narrow building with aircraft parked on both sides. Piers offer high aircraft capacity and simplicity of design, but often result in a long distance from the check-in counter to the gate and might create problems of capacity due to aircraft parking maneuvers.

Another typical configuration is to build one or more satellite associated to a main passenger processor terminal. The main difference between a satellite and a passenger processor terminal is that the satellite does not allow check-in, nor security controls, is just to give access to gate with walkways, concourses, and maybe duty free shops. The main advantage is that aircraft can park around its entire perimeter. The main disadvantage is that they are expensive: a subway transportation infrastructure in typically needed, also luggages must be transported to the main terminal building. Think for instance in Madrid Barajas with a standard linear terminal (T1-T2-T3) and a Terminal + Satellite (T4 + T4S), in which the satellite is not a processor. In this case a subway transportation infrastructure together with an automated system for luggages were needed and thus constructed.

9.2.4 Airport services

International customs

Any international airport must necessarily have customs facilities, and often require a more perceptible level of physical security. This includes national police and custom agents, drug inspections, and, in general, any inspection to ensure migration and commerce regulations.

Security

Airports are required to have security services in most countries. These services might be sublet to a private security company or carried out by the national security servies of the country (sometimes, one would find a mixture between these two). Airport security normally requires baggage checks, metal screenings of individual persons, and rules against any object that could be used as a weapon. Since the September 11, 2001 attacks, airport security has been dramatically increased worldwide.

Intermodal connections

Airports, specially the largest international airports located in big cities, are often located next to highways or are served by their own highways. Traffic is fed into two access roads (loops) one sitting on top of the other to feed both departures and arrivals (typically in two different levels). Also, many airports have the urban rail system directly connecting the main terminals with the inner city. Very recently, to facilitate connections with medium distance cities (up to 500-600 [km]), there are projects (if is not a reality already) to incorporate the high speed train in the airport facility, connecting big capitals with other important cities.

Shop and food services

Every single airport, even the smallest ones, have shops and food courts (at least one little shop to buy a snack and soda). These services provide passengers food and drinks before they board their flights. If we move to large international airports, these resemble more like a shopping mall, with many franchise food places and the most well-known retail branches (specially clothes stores). International areas usually have a duty-free shop where travelers are not required to pay the usual duty fees on items. Larger airlines often operate member-only lounges for premium passengers (VIP lounges). The key of this business is that airports have a captive audience, sometimes with hours of layover in connections, and consequently the prices charged for food are generally much higher than elsewhere in the region.

Cargo and freight services

Airports are also facilities where large volumes of cargo are continuously moved throughout the entire globe. Cargo airlines carry out this business, and often have their own adjacent infrastructure to rapidly transfer freight items between ground and air modes of transportation.

Support Services

Other services that provide support to airlines are aircraft maintenance, pilot services, aircraft rental, and hangar rental. At major airports, particularly those used as hubs by major airlines, airlines may operate their own support facilities. If this is not the case, every single company operating an airport must have access to the above mentioned services, which are typically rented on demand.

9.3 Airport operations

The main function of an airport, besides facilitating the passenger intermodal connection, is to ensure that aircraft can land, take off, and move around in an efficient and safe manner. Thus, many systems and subsystems are needed to facilitate achieving this end, encompassing many protocols and processes. These processes are most of the times hardly visible to passengers, but have extraordinary complexity, specially at large international airports. We will focus herein on the airport operations duties of Air Traffic Management (ATM) and, fundamentally, on the airport navigational aids that must be available to ensure safe operations. Last, we will briefly point out some issues regarding safety management and environmental concerns in airport operations.

9.3.1 Air Traffic Management (ATM) services

ATM will be deeply described in Chapter 10. As a rough definition, we can say that ATM is about the processes, procedures, and resources which come into play to make sure that aircraft are safely guided in the skies and on the ground. Therefore, it plays an important role in airport operations.

Air traffic control: ATC (to be also studied in Chapter 10) is the tactical part within the Air Traffic Management (ATM) system. It is on charge of separating aircraft safely in the sky as flying at the airports where arriving and departing. These duties are carried out by air traffic controllers, who direct aircraft movements, usually via VHF radio. Air traffic control responsibilities at airports are usually divided into two main areas: ground control and tower control.

Ground control is responsible for directing all ground traffic in designated movement areas, except for the case of traffic on runways, i.e, ground control in on charge of aircraft movements in aprons and taxiways, but also all service vehicles movements (fuel trucks, push-back vehicles, luggage trollies, etc.). Ground Control commands these vehicles on which taxiways to use, which runway to proceed (only for aircraft in this case), where to park, when to cross runways, etc. When a plane is ready to take off, it must wait in the runway head and turned over to tower control, who is responsible to authorize take-off and surveil the operation thereafter. After a plane has landed, it exits the runway and then is automatically turned over to ground control.

Tower Control controls aircraft on the runway and in the controlled airspace immediately surrounding the airport, the so-called Terminal Maneuvering Area (TMA). They coordinate the sequencing and spacing of aircraft and direct aircraft on how to safely join and leave the TMA circuit of arrivals and departures.

Communication services: Together with ATC services, the ATM provides an information system that apply both for airport operations and en-route operations. In regard of airport operations, pilots check before take off the so-called Automatic Terminal Information Service (ATIS), which provides information about airport conditions. The ATIS contains information about weather, which runway and traffic patterns are in use, and other information that pilots should be aware of before boarding the aircraft and entering the movement area and the airspace.

9.3.2 Airport navigational aids

The maneuvers of approach and landing are assisted from the airport by means of radio-electric and visual aids. The flight that is assisted with radio-electric aids is said to be under Instrumental Flight Rules (IFR flight); the flight that is assisted only with visual rules is said to be under Visual Flight Rules (VFR flight).

Visual aids

When flying, there are a number of visual aids available to pilots:

- Signaling devices (such for instance, windsock indicator).
- Guidance signs (information, compulsory instructions, etc.).
- Signs painted over the pavement (runway, taxiways, aprons).
- Lights (runway, taxiway).

Airports

(a) Airport's winsock / © User:Saperaud / Wikimedia Commons / CC-BY-SA-3.0.

(b) Example of a holding sign on an airport runway surface. Wikimedia Commons / Public Domain.

Figure 9.10: Airport visual aids.

Windsock: Planes take-off and land in the presence of head/tail wind in order to achieve maximum performance. Wind speed and direction information is available through the ATIS or ATC, but pilots need instantaneous information during landing. For this purpose, a windsock is kept in view of the runway. As already pointed out before in order to justify the fact that airports might have runways in different directions, the presence of wind is dangerous and limiting in terms of operations. The aeronautical authorities have established a maximum crosswind of 15-40 knots depending of the aircraft (for instance, a B777 has a limit of 38 knots) for landing and take-off. If these values are exceeded the runway can not be operated.

Guidance signs: Airport guidance signs provide moving directions and information to aircraft operating in the airport, but also to airport vehicles. There are two classes of guidance signs at airports, with several types of each:

- Location signs (yellow colored on black background): Identifies the runway or taxiway in which the aircraft is or is about to enter.
- Direction/runway exit signs (black colored on yellow background): Identifies the intersecting taxiways the aircraft is approaching when rolling on the runway right after having landed. They also have an arrow indicating the direction to turn.
- Other: Many airports use conventional traffic signs such for instance stop signs throughout the airport.

Mandatory instruction signs: They show entrances to runways or critical areas. Vehicles and aircraft are required to stop at these signs until the control tower provides clearance to proceed on.

- Runway signs (white on red): These signs simply identify a runway intersection ahead of the aircraft.

9.3 AIRPORT OPERATIONS

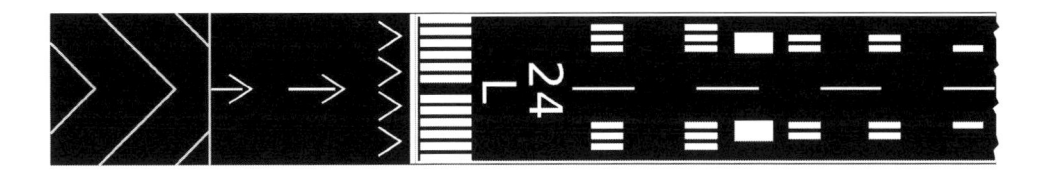

Figure 9.11: Runway pavement signs. © User:Mormegil / Wikimedia Commons / CC-BY-SA-3.0.

- Frequency change signs: Typically consists of a stop sign and an instruction to change to another frequency. These signs are used at airports with different areas of ground control, where different communication frequencies might be used.
- Holding position signs: A single solid yellow bar across a taxiway (painted over the pavement) indicates a position where ground control may require a stop. If a two solid yellow bars and two dashed yellow bars are encountered (painted over the pavement), it indicates a holding position for a runway intersection ahead. Runway holding lines must never be crossed without ATC permission.

Signs painted over the pavement: There are three main sets of signs painted over the pavement:

- Runway signs (white).
- Taxiway signs (yellow).
- Apron signs (red).

The runway signs, white colored, can be consulted in Figure 9.2 and Figure 9.11, namely: threshold, aiming point, touchdown zone, center line, runway designator, edge lines, etc. The taxiway signs, yellow colored, can also be consulted in Figure 9.2, namely: strip and axis lines, holding positions, crossing points, etc. The apron signs, red colored, are typically the so-called envelopes where aircraft park.

Lighting: Many airports have lighting devices that help guide planes using the runways and taxiways at night or in rain or fog. There are two main sets of lighting painted in the movement area:

- Runway lights (green, red, and white).
- Taxiway signs (blue and green).

On runways, green lights indicate the beginning of the runway for landing, while red lights indicate the end of the runway. Runway edge lights are white spaced out on both sides

AIRPORTS

Figure 9.12: Runway lighting. © Hansueli Krapf / Wikimedia Commons / CC-BY-SA-3.0.

Figure 9.13: PAPI: The greater number of red lights visible compared with the number of white lights visible in the picture means that the aircraft is flying below the glide slope. Wikimedia Commons / Public domain.

of the runway. Some airports have more sophisticated lighting on the runways including embedded lights that run down the centerline of the runway and lights that help indicating the approach. Along taxiways, blue lights indicate the taxiway's edge, and some airports have embedded green lights that indicate the centerline. See Figure 9.12 as illustration.

Other light devices also help pilots approaching. This is the case of the Precision Approach Path Indicator (PAPI). The PAPI is a visual aid that provides guidance information to help a pilot acquire and maintain the correct approach (in the vertical plane) to an an airport. It consists of 4 lights display in a row located on the right side of the runway, approximately 300 meters beyond the landing threshold. These lights emit in the red spectrum below the gliding path and in the white spectrum above it. To use the guidance information provided by the aid to follow the correct glide slope, a pilot would maneuver the aircraft to obtain an equal number of red and white lights, i.e, 2 red lights on the left part and 2 white lights on the right part.

Instrumental Aids

Besides visual aids, that all airport have, a majority of big airports have also a number of radio-electric aids to assist aircraft and pilots:

A typical instrumental aid located in airport is the so-called VHF omnidirectional range (VOR), which help pilots finding a desired flying course. VORs are often installed together with a Distance Measuring Equipment (DME), which provides the distance between the aircraft and air (located in the airport). In this way, a pilot can use course and distance to proceed safely towards the runway head. These two equipments are also for en-route navigation and will be described with more detail in Chapter 10. On the contrary, there is one instrumental aid that is only used in airport operations: the Instrument Landing System (ILS).

Instrument Landing System (ILS): An ILS is a ground-based instrumental approach system that provides precision guidance to an aircraft approaching and landing on a runway. In poor visibility conditions (rain, fog, dark, etc.), pilots will be aided by an ILS to instrumentally find the runway and fly the correct approach and land safely, even if they cannot see the ground at some point of the approach procedure (or even during the whole approach procedure).

An ILS consists of two independent subsystems, one providing lateral guidance (localizer), and the other providing vertical guidance (glide slope or glide path). Aircraft guidance is provided by the ILS receivers in the aircraft by performing a modulation depth comparison. The localizer receiver on the aircraft measures the Difference in the Depth of Modulation[3] (DDM) of two signals, one of 90 Hz and the other of 150 Hz. The difference between the two signals varies depending on the position of the approaching aircraft from the centerline.

If there is a predominance of either 90 Hz or 150 Hz modulation, the aircraft is off the centerline. In the cockpit, the needle on the Horizontal Situation Indicator (HSI, the instrument part of the ILS), or Course Deviation Indicator (CDI), will show that the aircraft needs to fly left or right to correct the error to fly down the center of the runway. If the DDM is zero, the aircraft is on the centerline of the localizer coinciding with the physical runway centerline.

---

[3]It is based on the concept of space modulation, a radio amplitude modulation technique specifically used in ILS that incorporates the use of multiple antennas fed with various radio frequency powers and phases to create different depths of modulation within various volumes of three-dimensional airspace.

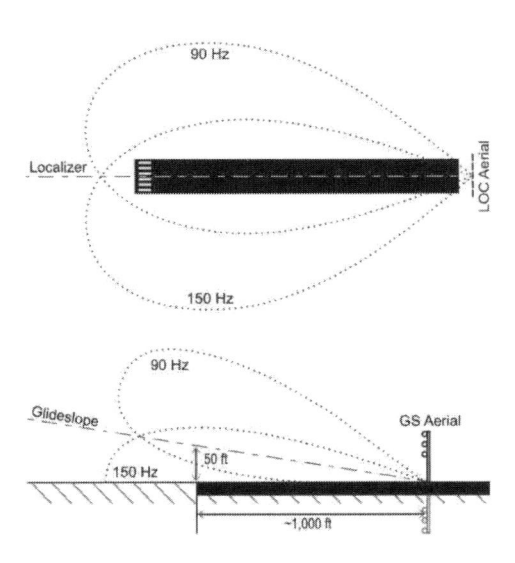

Figure 9.14: ILS: The emission patterns of the localizer and glide slope signals. Note that the glide slope beams are partly formed by the reflection of the glide slope aerial in the ground plane. © User:treesmill / Wikimedia Commons / CC-BY-SA-3.0.

Figure 9.15: ILS: Localizer array and approach lighting. Wikimedia Commons / Public Domain.

A glide slope (GS) or glide path (GP) antenna array is sited to one side of the runway touchdown zone. The GP signal is transmitted on a carrier frequency between 328.6 and 335.4 MHz. The centerline of the glide slope signal is arranged to define a glide slope of approximately 3° above horizontal (ground level). The pilot controls the aircraft so that the indications on the instrument (i.e., the course deviation indicator (see Chapter 5)) remain centered on the display. This ensures the aircraft is following the ILS centerline (i.e., it provides lateral guidance). The vertical guidance is shown on the instrument panel by the glide slope indicator, and aids the pilot in reaching the runway at the proper touchdown point. Many modern aircraft are able to embed these signals into the autopilot, allowing the approach to be flown automatically.

According to ICAO, there are different ILS categories attending at the visual range and the decision altitude: the visual range is the longitudinal distance at which the pilot

ILS category	Visual range	Decision altitude
Cat. I	$R \geq 760\ m\ (2500\ ft)$	$h_d \geq 61\ m\ (200\ ft)$
Cat. II	$760\ m > R \geq 365\ m\ (1200\ ft)$	$61\ m > h_d \geq 30\ m\ (100\ ft)$
Cat. III.a	$365\ m > R \geq 213\ m\ (700\ ft)$	$30\ m > h_d \geq 0\ m$
Cat. III.b	$213\ m > R \geq 46\ m\ (150\ ft)$	$15\ m > h_d \geq 0\ m$
Cat. III.c	$46\ m > R$	$h_d = 0\ m\ (0\ ft)$

Table 9.6: ILS categories. Data retrieved from ICAO [4].

is able to clearly distinguish the signs painted over the pavement or the lights if flying at night; the decision altitude is the minimum vertical altitude at which the pilot must abort the approach in case of not seeing any of the visual aids. See Table 9.6.

9.3.3 Safety management

Safety is the most important concern in airport operations. Thus, every airfield includes equipment and procedures for handling emergency situations. Commercial airfields include at least one emergency vehicle (with the corresponding crew) and a fire extinction unit specially equipped for dealing with airfield incidents and accidents.

Potential airfield hazards to aircraft include scattered fragments of any kind, nesting birds, and environmental conditions such as ice or snow. The field must be kept clear of any scattered fragment using cleaning equipment so that doesn't become a projectile and enter an engine duct. Similar concerns apply to birds nesting near an airfield which might endanger aircraft operations due to impact. To threaten birds, falconry is practiced within the airport boundaries. In adverse weather conditions, ice and snow clearing equipment can be used to improve traction on the landing strip. For waiting aircraft, special equipment and fluids are used to melt the ice on the wings.

9.3.4 Environmental concerns

The construction of new airports (or the enlargement of an existing one) has also a tremendous environmental impact, affecting on the countryside, historical sites, local flora and fauna. Also, vehicles operating in airports (aircraft but also surface vehicles) represent a major source of noise and air pollution which can be very disturbing and damaging for nearby residents and users. Moreover, operating aircraft have a dramatic impact on inhabiting birds colonies.

REFERENCES

[1] DE NEUFVILLE, R. and ODONI, A. (2003). *Airport systems: planning design, and management*, volume 1. McGraw-Hill New York.

[2] FRANCHINI, S., LÓPEZ, O., ANTOÍN, J., BEZDENEJNYKH, N., and CUERVA, A. (2011). *Apuntes de Tecnología Aeroespacial*. Escuela de Ingeniería Aeronáutica y del Espacio, universidad politécnica de madrid edition.

[3] GARCÍA CRUZADO, M. (2000). *ETSI Aeronáuticos*. Madrid .

[4] ICAO (1999). *Aerodrome Design and Operations*, volume Annex 14, Volume 1. International Civil Aviation Organization, third edition edition.

10
AIR NAVIGATION

Contents

10.1	Introduction	300
	10.1.1 Definition	300
	10.1.2 History	300
10.2	Technical and operative framework	306
	10.2.1 Communications, Navigation & Surveillance (CNS)	306
	10.2.2 Air Traffic Management (ATM)	308
10.3	Airspace Management (ASM)	311
	10.3.1 ATS routes	311
	10.3.2 Airspace organization in regions and control centers	314
	10.3.3 Restrictions in the airspace	316
	10.3.4 Classification of the airspace according to ICAO	317
	10.3.5 Navigation charts	317
	10.3.6 Flight plan	321
10.4	Technical support: CNS system	321
	10.4.1 Communication systems	322
	10.4.2 Navigation systems	325
	10.4.3 Surveillance systems	343
10.5	SESAR concept	348
	10.5.1 Single European Sky	348
	10.5.2 SESAR	348
	References	349

In this chapter we analyze air navigation as a whole, including an introduction and historical perspective in Section 10.1, the technical and operational framework, the so-called CNS-ATM concept, in Section 10.2, Section 10.3, and Section 10.4. Finally, in Section 10.5, we analyze the project SESAR, giving an overview of future trends in the air navigation system. A good introduction is given in SÁEZ and PORTILLO [4]. In depth studies that can be consulted include SÁEZ et al. [3], SAÉZ NIETO [5], PÉREZ et al. [2], and NOLAN [1].

10.1 Introduction

10.1.1 Definition

The air navigation is the process of steering an aircraft in flight from an initial position to a final position, following a determined route, and fulfilling certain requirements of safety and efficiency. The navigation is performed by each aircraft independently, using diverse external sources of information and proper on-board equipment.

Besides the primary goal above mentioned (safe and efficient controlled flight towards destination), three important goals can be mentioned:

- Avoid getting lost.
- Avoid collisions with other aircraft or obstacles.
- Minimize the influence of adverse meteorological conditions.

10.1.2 History (Sáez and Portillo [4])

The first navigation techniques at the beginning of the 20th century were rudimentary. The navigation was performed via terrain observation and pilots were provided with maps and compasses to locate the aircraft.

Dead reckoning

Some years later the *dead reckoning* was introduced. The dead reckoning consists in estimating the future position of the aircraft based on the current position, velocity, and course. Pilots were already provided with anemometers to calculate the airspeed of the aircraft and clock to measure time. The flights were design based on points (typically references in the terrain), and pilots had to follow the established track.

Obviously, when trying to fly a track from one point to another using *dead reckoning*, the errors were tremendous. This was due to three main reasons:

- Errors in the used instruments (anemometer, compass, and clock).
- Piloting errors.
- Wind effects.

The fist two are inherent to all types of navigation and they will always be to some extent. Nevertheless, errors in instruments are being reduced. Also piloting errors are being minimized due to automatic control systems. One can think that these two errors will converge to some bearable values. On the contrary, wind effects are more relevant, and still play a key role in the uncertainty of aircraft trajectories. Based on these errors, but in particular on wind effects, one can define the triangle of velocities and the track and course angles. See Figure 10.1.

(a) Triangle of velocities.

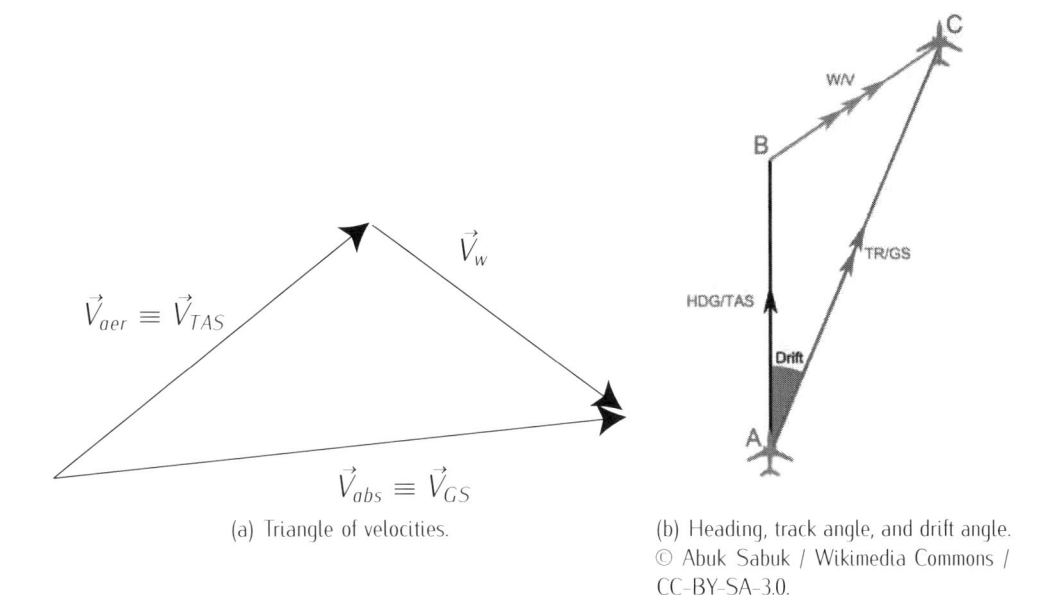

(b) Heading, track angle, and drift angle.
© Abuk Sabuk / Wikimedia Commons / CC-BY-SA-3.0.

Figure 10.1: Triangle of velocities. The absolute velocity (also referred to as ground speed) is equal to the vectorial sum of the aerodynamic velocity (also referred to as true air speed) and the wind speed.

Track angle (TR): Is the angle between the north (typically magnetic, but we can also use geographic) and the absolute velocity of the aircraft. It corresponds with the real track the aircraft is flying. The absolute velocity of the aircraft is the sum of the aerodynamic velocity and wind speed: $\vec{V}_{abs} = \vec{V}_{aer} + \vec{V}_w$. This is called triangle of velocities.

The course angle (HDG): The course corresponds to the heading and is the angle between the north (typically magnetic, but we can also use geographic) and the aerodynamic velocity vector. Notice that if we assume symmetric flight, it also coincides with the longitudinal axis of the aircraft. Typically it does not coincide with the track angle since the aircraft might have to compensate cross wind.

For instance, looking at Figure 10.1.b, the aircraft heading (aircraft course) is the vector joining A and B, but the real track is represented by the vector joining A and C. The corresponding angles would be calculated establishing a reference (typically the magnetic north). Notice that the difference between course and track is referred to as drift angle.

Some other elements that are important in defining the flight orientation are: the desired track, the cross-track error, and the bearing:

Desired Track (DTR) angle: Is the angle between the north (typically magnetic, but we can also use geographic) and the straight line joining two consecutive waypoints in the flight. Is the track we want to fly, which in ideal conditions would coincide with the track we are actually flying. Unfortunately, this rarely happens.

Cross-Track Error (XTE): Is the distance between the position of the aircraft and the line that represents the desired track. Notice that the distance between a point and a line is the perpendicular to the line passing through the point. Thus $XTE = d(DTR, TR)$, where d can be defined as the norm 2 (the euclidean distance).

Bearing: It is defined as the angle between the north (typically magnetic, but we can also use geographic) and the straight line (in a sphere or ellipsoid would not be exactly straight) connecting the aircraft with a reference point. Note that the bearing depends on the selected reference point. Nowadays, these points typically coincide either with navigational aids located on earth or waypoints calculated based on the information of at least two navigational aids located on earth.

Astronomic navigation

Besides the errors caused by wind effects, dead reckoning navigation had one fundamental drawback: It was required that the selected points acting as a reference were visible by the pilots in any circumstance. As the reader can intuitively imagine, these points were sometimes difficult to identify in case of adverse meteorological conditions (rain, fog, etc.) or at dark during night flights. Moreover, it was really difficult to obtain references over monotone landscapes as it is the case for oceans.

Therefore, pioneer aviators started to use astronomic devices. Devices such as the *astrolabe*[1] and the *sextant*[2] had been used since centuries for maritime navigation. Using these devices, pilots (helped by a man on board that was term *navigator*) were able to periodically determine the position and minimize errors.

Thanks to this combined type of navigation: the astronomic navigation used together with the dead reckoning navigation, the most important feats among the pioneers were given birth. Thus, in light of history, one can claim that the first oceanic flights in 1919 (Alcock and Brown) and 1929 (Linderbergh) were, in part, thanks to the implementation of the astronomic navigation, which allowed pilots to reach destination without getting lost.

---

[1] An astrolabe is an elaborate inclinometer, historically used by astronomers, navigators, and astrologers. Its many uses include locating and predicting the positions of the Sun, Moon, planets, and stars, determining local time given local latitude and vice-versa .

[2] The sextant is an instrument that permits measuring the angles between two objects, such for instance a star or planet and the horizon. Knowing the elevation of the sun the hour of the day, the latitude at which the observer is located can be determined.

10.1 INTRODUCTION

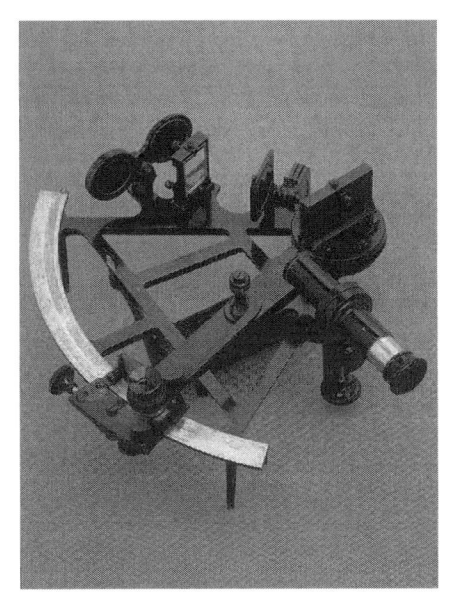

(a) A sextant. Author: User:Dodo / Wikimedia Commons / Public Domain.

(b) A modern persian astrolabe. © Masoud Safarniya / Wikimedia Commons / CC-BY-SA-3.0.

Figure 10.2: Astronomic navigation: sextant and astrolab.

Navigation aids

All the previously described navigation techniques did not require any infrastructure support on the ground, and therefore they can be considered as *autonomous navigation* techniques. However, such navigation was complicated and required lot of calculations on board. The *navigator* had to be continuously doing very complicated calculations and this was not operative at all.

As a consequence, there was a necessity of some type of earth-based navigation aids. The first to appear, in 1918, were the so-called aerial beacon (light beacon). This allowed night flight over networked areas, such the USA. However, these aids were limited. In 1919, the radio communications were started to be used. First, installing transmitters in the cockpits to communicate. Afterwards, using the radio-goniometry[3]. Radio-goniometers were installed on board and the navigation was performed determining determining the orientation of the aircraft with respect to two transmitter ground-stations which position was known.

Later on, in 1932, the Low Frequency Radio-Range (LFR) appeared, which was the main navigation system used by aircraft for instrument flying in the 1930s and 1940s until

---
[3]A goniometer is an instrument that either measures an angle or allows an object to be rotated to a precise angular position.

303

the advent of the VHF omnidirectional range (VOR) in the late 1940s. It was used for en route navigation as well as instrumental approaches. Based on a network of radio towers which transmitted directional radio signals, the LFR defined specific airways in the sky. Pilots navigated the LFR by listening to a stream of automated Morse codes. It was some sort of binary codification: hearing a specified tone meant to turn left (in analogy, 1 to turn left) and hearing a different specified tone meant to turn right (in analogy, 0 to turn right).

Since the 40s towards our days, the navigational aids have evolved significantly. To cite a few evolutions, the appearance of VOR and DME in the late 40s-early 50s, the concept of Area Navigation (RNAV) in the late 60s-early 70s, the fully automated ILS approach system in the late 60s, or even the satellite navigation (still to be fully operative) contributed to improve navigation performances. We will not describe them now, since all these types of navigation will be studied later on. As a consequence of the appearance of all these navigation aids throughout the years, nowadays the navigation is mainly performed using *instrumental navigation* techniques.

Navigation in the presence of other aircraft

Being able to fly, not getting lost, and avoiding terrain obstacles was at the beginning already a big challenge. At the beginning, due to the limited number of aircraft, the navigation did not consider the possibility of encountering other aircraft that might cause a collision. With the appearance of airports, attracting many aircraft to the same physical volume, the concept of navigation shifted immediately to that one of circulation.

Circulation can be defined as the *movement to and from or around something.* In air navigation, it appeared the necessity of making the aircraft circulate throughout certain structures defined in the air space or following certain rules. To avoid collisions, there were defined some rules based on the capacity of being able to see and to be seen. In the cases of approaches and departures in airports, it appeared the necessity of existence of someone with capability to assign aircraft a sequence to take off. These were the precursors of what we know today as air traffic controllers. In 1935, the first control center for air routes was created in the USA.

The air navigation as a system

As a result, juridic, operative, and technical support frameworks to regulate the air navigation were necessary.

The technical and operative frameworks should supply:

- **Information system prior departure**: related to meteorology, operative limitations, and limitations in the navigation aids.

- **Tactical support to pilots**: related to possible modifications in the conditions of the flight, specially to avoid potential conflicts with other aircraft or within regions under bad weather conditions.

- **Radio-electric infrastructures**: to provide aircraft navigation aids.

These three items have constituted the basic pillars in what is referred to as **system of air navigation** throughout its whole development. The technical and operative framework that conform the system (based on the so-called CNS-ATM[4] concept) will be studied in the forthcoming sections.

The juridic framework should take into account the following aspects: formation and licenses of the aeronautical personal; communication systems and procedures; rules about systems and performances on air, and air traffic control; air navigation requirements for aircraft certification, registration and identification for aircraft that carry out international flights; and aeronautical meteorology, maps, and navigation charts; among others. We give know a brief overview of the juridic framework. More in depth analysis will be undertaken in posterior courses regarding air law.

The International regulation (juridic) framework

The very first concern, at the beginning of the 20th century, was that of being able to fly. Within the following 30 years the concern changed to that of being able to fly anywhere (within the possibilities of the aircraft), even though in the presence of adverse navigation conditions, and avoiding collisions with other aircraft and the terrain.

The necessity of flying anywhere (including international flights) encouraged the development of an international regulation framework which could establish the rights and obligations when going beyond the domestic borders. Navigation systems and equipments should be also uniform, so that the crew could maintain the same *modus operandi* when trespassing borders.

In 1919, the International Commission for Air Navigation (ICAN) was created to provide international regulations. In practice, it was the principal organ of an international arrangement requiring administrative, legislative and judicial agents. At the end World War II, in november 1944, 55 states were invited to Chicago to celebrate a conference the Chicago Convention was signed. The Chicago Convention, as studied in Chapter 8, promoted the safe and orderly development of international civil aviation throughout the world. It set standards and regulations necessary for aviation safety, security, efficiency and regularity, as well as for aviation environmental protection. This obviously included air navigation regulations. The Chicago convention gave birth to ICAO, successor of ICAN.

---

[4]As it will be introduced later on, CNS stands for Communication, Navigation and Surveillance; ATM stands for Air Traffic Management.

The development of ICAO's regulation (see contents of Chicago convention in Chapter 8) has led to the creation of international juridic framework.

In Spain we count with AENA (Aeropuertos Españoles y Navegación Aérea), divided into two main directions: Airports and Air Navigation. Regarding the Air Navigation direction, its main function is to provide aircraft flying in what is termed as civil traffic (commercial flights and general aviation flights) all means so that aircraft are able to navigate and circulate with safety, fluidity, and efficiency over the air space under spanish responsibility. AENA is therefore the Air Navigation Service Provider (ANSP) in Spain. Also in Spain, the regulator organ is AESA (Agencia Estatal de Seguridad Aérea), which depends of the *Dir. General de Aviación Civil, Ministerio de Fomento*. AESA is the state body that ensures that civil aviation standards are observed in all aeronautical activity in Spain.

If we draw a parallelism, the European Civil Aviation Conference (ECAC) is the European regulator organ and the European Organization for the Safety of Air Navigation (EUROCONTROL) is the ANSP in Europe[5]. In the USA, both the function of regulator and ANSP is held by the FAA.

10.2 Technical and operative framework

The main goal of air navigation is to make possible air transportation day after day by means of providing the required services to perform operations safely and efficiently. These services are provided based on an organization, human resources, technical means, and a defined *modus operandi*.

The so constituted system is referred to as CNS-ATM (Communications, Navigation & Surveillance-Air Traffic Management). Therefore, CNS corresponds to the required technical means to fulfill the above mentioned air navigation's main goal, while ATM refers to the organizational scope and the definition of operational procedures.

10.2.1 Communications, Navigation & Surveillance (CNS)

Communications, navigation, and surveillance are essential technological systems for pilots in the air and air traffic controllers on the ground. They facilitate the process of establishing where the aircraft is and when and how it plans to arrive at its destination. It also facilitates the process of identifying and avoiding potential threats, e.g., potential conflicts with other aircraft or incoming storms.

---

[5] To be more precise, it is the ANSP in Belgium, Netherlands, Luxembourg, and north-west Germany

Communication

The communications are utilized to issue aeronautical information and provide flying aircraft with air transit services. The air transit services are provided form the different control centers (in which air traffic controllers operate), which communicate with aircraft to give instructions, or simply to inform about potential danger. On the other hand, aircraft must use the proper communication equipment (radios, datalink) to receive this service (by receiving this service it is meant to maintain bidirectional communication with control centers). Besides the communication aircraft-control center (the so-called mobile communications), there must be a communication network between ground stations, i.e., control centers, flight plan dispatchers, meteorological centers, etc. More details about the communication service will be given in Section 10.4.1.

Navigation

The navigation services refer to ground or orbital (satellites) infrastructures aimed at providing aircraft in flight with information to determine their positions and be able to navigate to the desired destination in the airspace. As already described in Chapter 5, the aircraft will have the required on-board equipment (navigation instruments and displays) to receive this service. More details about the navigation systems will be given in Section 10.4.2.

Surveillance

The objective of the surveillance infrastructure is to enable a safe, efficient, and cost-effective air navigation service. In airspaces with medium/high traffic density, the function of surveillance requires the use of specific systems that allow controllers to know the position of all aircraft that are flying under their responsibility[6] airspace. This service has been typically provided by radar stations in the ground. In this way the evolution of aircraft is monitored and potential threats can be identified and avoided. An instance of this would be two aircraft evolving in such a way that a potential conflict[7] is expected in the mid-term. The controller would advise instructions (using the above mentioned communication system) to the involved aircraft to avoid this threat. In the future, Automatic Dependent Surveillance-Broadcast (ADS-B) will be replacing radar as the primary surveillance method for controlling aircraft worldwide. There are also airborne systems that fulfill a surveillance function. That is the case of the Traffic Collision Avoidance System or Traffic alert and Collision Avoidance System (both abbreviated as TCAS). More details about the surveillance systems will be given in Section 10.4.3.

---

[6]Notice that each control center has the responsibility over a volume of airspace.

[7]In air navigation, a conflict is defined by a loss of separation minima. This separation minima is typically defined by a circle of 5 NM in the horizontal plane and a vertical distance of 1000 ft.

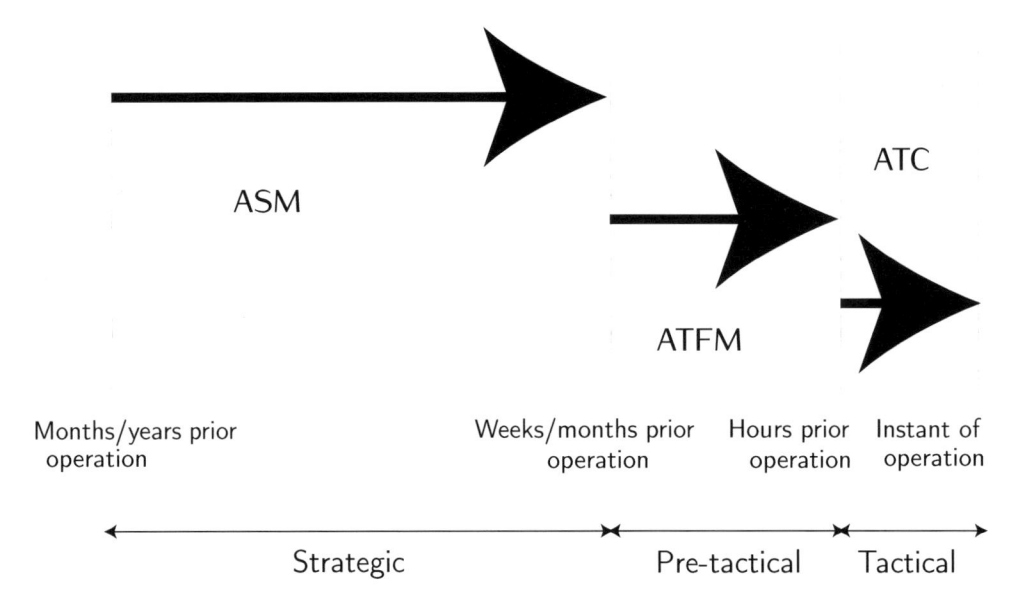

Figure 10.3: ATM levels.

10.2.2 Air Traffic Management (ATM)

The Air Navigation Service Providers (ANSPs), i.e., AENA in Spain, FAA in the USA, Eurocontrol in central Europe, must have technical capabilities to develop and support a technical CNS infrastructure, but on the other hand, it is also needed a highly structured organization with high skilled people, forming the operational support needed to provide transit, communication, and surveillance services. This operational infrastructure is referred to as Air traffic Management (ATM).

ATM is about the process, procedures, and resources which come into play to make sure that aircraft are safely guided in the skies and on the ground. If we consider the time-horizon between the management activity and the aircraft operation, we can identify three levels of systems:

- AirSpace management (ASM). (Strategic level).

- Air Traffic Flow and capacity Management (ATFM). (Pre tactical level).

- Air Traffic Control (ATC)[8]. (Tactical level)

---

[8]To be more precise, ATC belongs, together with the Flying Information Service (FIS) and the alert service (ALS), to the so-called Air Transit Services (ATS). Due to its importance, we restring ourselves to ATC.

Airspace management (ASM)

The first layer of the ATM system, the so-called Airspace Management (ASM), is performed at strategic level before aircraft departure, within months/years look-ahead time. The ASM is an activity which includes airspace modeling and design. As aircraft fly in the sky, they follow pre-planned routes conformed by waypoints, airways, departure and arrival procedures, etc. The route followed by an aircraft is selected by the company before departure based on the airspace design previously made by the ASM. The ASM activity includes, among others, the definition of the network or routes (referred to as ATS routes), the organization of the airspace in regions and control sectors, the classification (determined airspace is only flyable by aircraft fulfilling determined conditions) and the delimitation (some regions of the airspace might be limited/restringed/prohibited for civil traffic) of the airspace, and the publication of different navigation charts where all the previous information is reflected in a standard manner. Section 10.3 will be devoted to airspace management and organization, and all these issues will be tackled in detail.

Air traffic flow and capacity management (ATFM)

The second layer of the ATM system is the so called Air Traffic Flow and Capacity Management (ATFM[9]). It is performed at pre-tactical level before aircraft departure, within weeks up to one/two hours look ahead time. The idea is the following: Once the flight plan has been determined by the company according to its individual preferences and fulfilling the ASM airspace design and organization, the next step is to match the flight plan with all flights to be operating at the same time windows in the same areas in order to check whether the available capacity is exceeded. This is an important step as only a certain number of flights can be safely handled at the same time by each air traffic controller in the designated volumes of airspace under his/her responsibility. All flight plans for flights into, out of, and around a region, e.g., Europe, must be submitted to an air traffic flow and capacity management unit (the Eurocontrol's Central Flow Management Unit (CFMU) in Europe), where they are analyzed and processed. Matching the requested fights against available capacity is first done far in advance for planning purposes, then on the day before the flight, and finally, in real-time, on the day of the flight itself. If the available capacity is exceeded, the flight plans are modified, resulting in reroutings, ground delays, airborne delays, etc.

---

[9]The acronym ATFCM might be used as well. ATFM and ATFCM can be used interchangeably.

(a) Air route traffic controllers at work at the Washington Air Route Traffic Control Center. Author: User:Mudares / Wikimedia Commons / Public Domain.

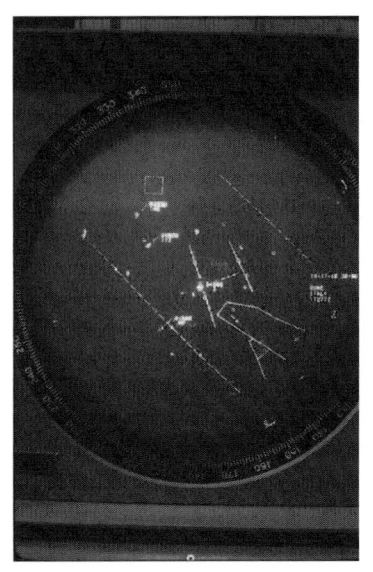

(b) Display for Air traffic Control Demonstration in Rome in 1972. © User:Dms489 / Wikimedia Commons / CC-BY-SA-3.0.

Figure 10.4: Air Traffic Control.

Air traffic control (ATC)

The third layer of the the ATM system is the so-called Air Traffic Control (ATC)[10]. It is performed at tactical level, typically during the operation of the aircraft or instants before departure. The idea is the following: once the flight plan has been approved by ATFM, it has to be flown. Unfortunately, there are many elements that introduce uncertainty in the system (atmospheric conditions, measurement errors, piloting errors, modeling errors, etc.) and the flight intentions, i.e., the flight plan, is rarely fully fulfilled. Thus, there must be a unit to ensure that all flight evolve safely, detecting and avoiding any potential hazard, e.g., a potential conflict, adverse meteorological conditions, by modifying the routes. This task is fulfilled by ATC. ATC is executed over different volumes of airspace (route, approximation, surface) in different dependences (Area Control Centers (ACC) and Control Tower) by different types of controllers. Controllers use communication services to advise pilots. Also, pilots are aided by the AIS and ALS. All these systems together increase the situational awareness of the pilot to circumvent any potential danger.

---

[10]As already mentioned before, to be more precise, the third layer is referred to as Air Transit Services (ATS), which is composed of the Flying Information Services (FIS), the Alert Services (ALS), and, fundamentally, the Air Traffic Control (ATC). Since ATC is by far the most important one, it has been typically used as the third layer by itself. We adopt the same criterion, but it is convenient not to forget the FIS and ALS.

10.3 Airspace Management (ASM)

The surrounding air is the fundamental mean in which aircraft fly and navigate. In aeronautics, the airspace is considered as the volume of air above the earth surface in which aircraft carry out their activity. The development of aviation has encouraged the organization of the airspace:

First, in the sense of sovereignty and responsibility of the different states. As a state has perfectly defined its territory and its territorial waters, where its laws apply, there must also exist a sovereign airspace. Notice that the international organizations (ICAO) define also the responsibility over the ocean[11].

Second, with the airspace politically delimited, the airspace must also be organized to allow the efficient and safe development of aircraft operations. It has been established a network of routes, equipped with navigation aids, so that the aircraft can navigate following the routes. Communication and surveillance services are also provided. This network of routes is referred to as ATS (Air Traffic Services) routes. These routes go through regions that determine the volumes of responsibility over which the functions of surveillance and control are executed. These regions are referred to as FIR/UIR (Flying Information Region/Upper Information Region). Inside each region, there are defined different areas of control depending on the phase of the flight.

10.3.1 ATS routes

A route is a description of the path followed by an aircraft when flying between airports. A complete route between airports often uses several airways connected by waypoints. However, airways can not be directly connected to airports. The transition from/to airports and airways is defined in a different way as we will see later on. Thus, the network of ATS routes refer only to the en-route part of the flight (excluding operations near airports).

Airways

An airway has no physical existence, but can be thought of as a motorway in the sky. In Europe, airways are corridors 10 nautical miles (19 km) wide. On an airway, aircraft fly at different flight levels to avoid collisions. The different flight levels are vertically separated 1000 feet. On a bi-directional airway, each direction has its own set of flight levels according to the course of the aircraft:

- Course to the route between 0° and 179°: east direction → Odd flight levels.
- Course to the route between 180° and 359°: west direction → Even flight levels.

---

[11] In aviation there is international air, in analogy with the international waters. The whole airspace is under responsibility of one or more states according to ICAO.

Airways should be identified by a route designator which consists of one letter of the alphabet followed by a number from 1 to 999. Selection of the letter should be made from those listed below:

- A, B, G, R, for routes which form part of the regional networks of ATS routes and are not area navigation routes (RNAV routes);
- L, M, N, P, for RNAV routes which form part of the regional networks of ATS routes;
- H, J, V, W, for routes which do not form part of the regional networks of ATS routes and are not RNAV routes;
- Q, T, Y, Z, for RNAV routes which do not form part of the regional networks of ATS routes.

Where applicable, one supplementary letter should be added as a prefix to the basic designator in accordance with the following:

- K to indicate a low level route established for use primarily by helicopters;
- U to indicate that the route, or a portion of it, is established in the upper airspace;
- S to indicate a route established exclusively for use by supersonic aircraft during acceleration, deceleration, and while in supersonic flight.

A supplementary letter may be added after the basic designator of the ATS route in question in order to indicate the type of service provided on the route:

- the letter F to indicate that on the route, or a portion of it, advisory service only is provided; or
- the letter G to indicate that on the route, or a portion of it, flight information service only is provided.

Each airway starts and finishes at a waypoint, and may contain some intermediate waypoints as well. Airways may cross or join at a waypoint, so an aircraft can change from one airway to another at such points. A waypoint is thus most often used to indicate a change in direction, speed, or altitude along the desired path. Where there is no suitable airway between two waypoints, ATC may allow a direct waypoint to waypoint routing which does not use an airway. Additionally, there exist special tracks known as ocean tracks, which are used across some oceans. Free routing is also permitted in some areas over the oceans.

Waypoints

A waypoint is a predetermined geographical position that is defined in terms of latitude/longitude coordinates (altitude is ignored). Waypoints may be a simple named

10.3 Airspace Management (ASM)

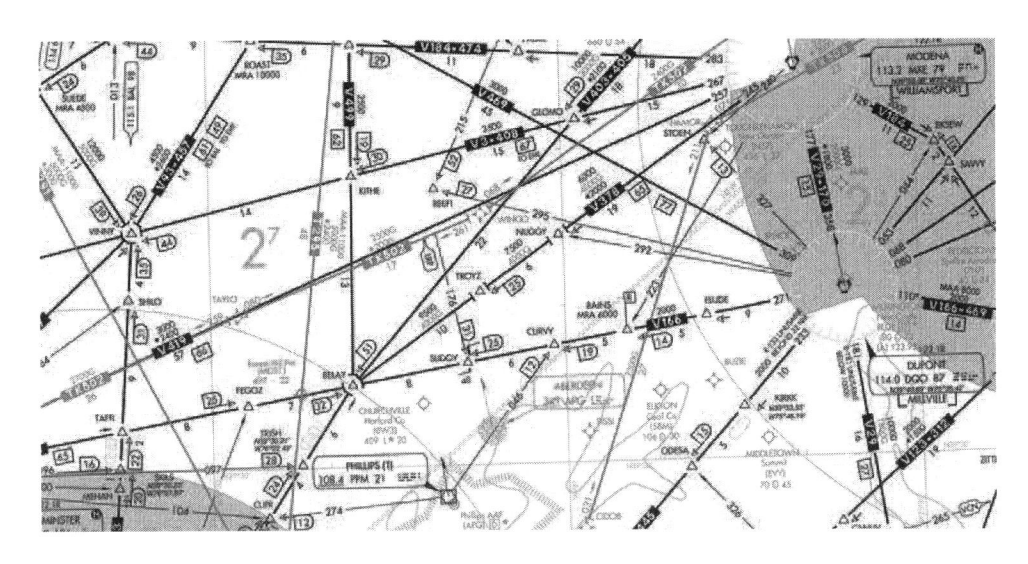

Figure 10.5: Air navigation chart: VOR stations as black hexagons (PPM, DQO, MXE), NDB station as brown spot (APG), VORNAV intersection fixes as black triangles (FEGOZ, BELAY, SAVVY), RNAV fixes as blue stars (SISSI, BUZIE, WINGO), VORNAV airways in black (V166, V499), RNAV airways in blue (TK502, T295), and other data. Author: User:Orion 8 / Wikimedia Commons / Public Domain.

point in space or may be associated with existing navigational aids, intersections, or fixes. Recently, it was typical that airways were laid out according to navigational aids such as VORs, NDBs, and therefore the position of the VORs or NDBs gave the coordinates of the waypoint (in this case, simply referred to as navaids). Nowadays, the concept of area navigation (RNAV) allows also to calculate a waypoint within the coverage of station-referenced navigation aids (VORs, NDBs) or within the limits of the capability of self contained aids, or a combination of these. Waypoints used in aviation are given five-letter names. These names are meant to be pronounceable or have a mnemonic value, so that they may easily be conveyed by voice.

ATS routes network

Summing up, the complete network of routes formed by airways and waypoints is referred to as ATS routes. The ATS routes are published in the basic manual for aeronautical information referred to as Aeronautical Information Publication (AIP). AIP publishes information for en-route and aerodromes in different charts (the so-called navigation charts), which are usually updated once a month coinciding with the Aeronautical Information Regulation and Control (AIRAC) cycle. Ocean tracks might change twice a day to take advantage of any favorable wind.

AIR NAVIGATION

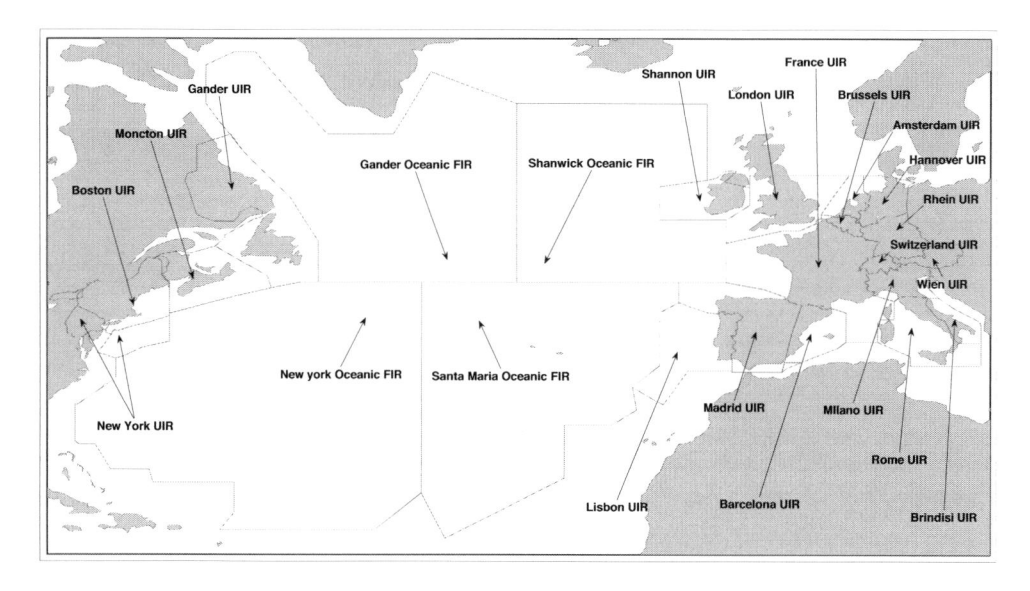

Figure 10.6: UIRs in the North Atlantic region and western Europe.

10.3.2 Airspace organization in regions and control centers

As mentioned in the introduction of the section, ATS routes go through regions that determine the volumes of responsibility over which the functions of surveillance and control are executed. These regions are referred to as FIR/UIR (Flying Information Region/Upper Information Region).

FIR regions cover an area of responsibility of a state up to 24500 feet (FL245). Over a FIR, it is defined an UIR, which covers flight levels above FL245. The UIR had to be defined when jets appeared. Notice that jets, opposite to propellers, flight more efficiently in upper flying levels. Typically the geographical area (surface on earth) for both FIRs and UIRs coincide. Figure 10.6 shows the FIR/UIR structure for the North Atlantic region and western Europe in the upper level. In the case of Spain, there are three FIR/UIR regions of responsibility:

- Barcelona FIR/UIR.
- Madrid FIR/UIR.
- Canarias FIR/UIR.

The need for surveillance and control inside FIR/UIR regions differs depending on the phase of the flight. Specifically, there are different needs when the aircraft is cruising (typically with stationary behavior) rather that when the aircraft is in the vicinity of an airport (either departing or arriving). Therefore, we can define different areas inside a FIR/UIR:

10.3 Airspace Management (ASM)

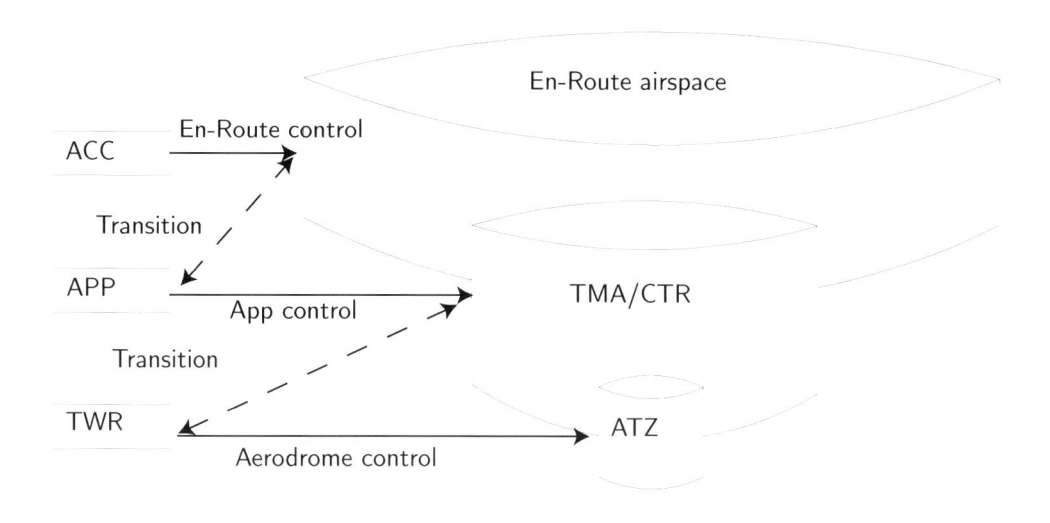

Figure 10.7: Control dependences, type of control and volumes of airspace under control.

- En-Route airspace: Volumes of airspace containing airways connecting airports' TMA (Terminal Maneuvering Areas), i.e., volumes containing the network of ATS routes.
- TMA: Volumes of airspace situated above one or more airports. In the TMA are located the SIDs[12] and STARs procedures that allow aircraft to connect airways with the runway when departing or arriving.
- CTR (Control zone, also referred to as transit zone): The CTR is inside a TMA, it is a volume of controlled airspace around an airport where air traffic is operating to and from that airport. Aircraft can only fly in it after receiving a specific clearance from air traffic control. CTRs contains ATZs.
- ATZ (Aerodrome Traffic Zone): ATZ are zones around an airport with a radius of 2 nm or 2.5 nm, extending from the surface to 2,000 ft (600 m) above aerodrome level. Aircraft within an ATZ must obey the instructions of the tower controller.

The labour of surveillance and control can be also divided in three levels as follows:

- En-Route control (also referred to as area control): executed over the En-Route airspace.
- Approximation control: executed over the CTR and TMA.
- Aerodrome control: executed over the ATZ.

---
[12]SID and STAR will be studied in Section 10.3.5.

315

This labour is carried out in different dependences:

- Control towers (TWR), where the aerodrome control is executed.
- Approximation offices (APP), where the approximation control is executed.
- Area Control Centers (ACC), where the en-route (area) control is executed.

Every airport has a control tower to host controllers and equipment so that surveillance, communications, and control is provided to aircraft when departing and arriving. Typically, the aerodrome control includes taxiing operations, take-off and landing, and its dependences are on the control tower (with the controller on the top with complete vision of the aerodrome). The initial climb and the approach phases are controlled within the approximation control level, which dependences are typically located in the base of the control tower. The area control level is provided from the so called ACC, which are not typically located in airports. For instance, in Spain there are five ACCs: Barcelona and Palma de Mallorca in FIR/UIR Barcelona; Madrid and Sevilla in FIR/UIR Madrid; Canarias in FIR/UIR Canarias.

Therefore, the airspace is divided and assigned to control dependences so that controllers can provide surveillance, communications, and control services (ATC services), and thus aircraft operate safely. When the traffic within an airspace portion assigned to a control dependence exceeds the capacities of the controller to carry out his/her duty with safety, such dependencies are divided in what is known as ATC sectors. In this way, the separation between all aircraft within an ATC sector is responsibility of a controller. He/She gives the proper instructions to avoid any potential conflict, and also coordinates the aircraft transfer between his/her sector and adjoining sectors. Obviously, the dimensions of the sector are determined by both the volume of traffic and the characteristics of it.

10.3.3 Restrictions in the airspace

The use of the airspace by ATS routes is limited to some areas of special use due to military operations, environmental policies, or simply security reasons. Therefore, different *special use* airspaces are defined and designated as:

- Prohibited (P).
- Restricted (R).
- Dangerous (D).

The prohibited zones contain a defined volume of airspace over a sovereign state in which the flight of any aircraft in prohibited, with the exception of those aircraft that are authorized by the ministry of defense. These areas are established due to national security reasons and are published in the navigation charts. The restricted zones contain a defined volume of airspace over a sovereign state in which the flight of any aircraft in restricted. In order to enter such zones, the aircraft must be authorized by ATC. The dangerous zones contain a define volume of airspace over a sovereign state in which there might be operations or activities considered of risk to other aircraft.

10.3.4 Classification of the airspace according to ICAO

There are two different rules under which aircraft can operate, both based on the instruments on board and the qualification of the crew:

- Visual Flight Rules (VFR).
- Instrumental Flight Rules (IFR).

In order operations to be carried out under VFR, the meteorological conditions must be good enough to allow pilots identify the visual references in the terrain and other aircraft. Such meteorological conditions are referred to as Visual Meteorological Conditions (VMC). VFR require a pilot to be able to see outside the cockpit, to control the aircraft's altitude, navigate, to maintain distance to surrounding clouds, and avoid obstacles and other aircraft. Essentially, pilots in VFR are required to *see and avoid*.

If the meteorological conditions are below the VMC threshold, the flight must be performed under IFR. Such meteorological condition are referred to as Instrumental Meteorological Conditions (IMC). Notice that IFR flights are under control by ATC services.

Since sometimes VFR and IFR flights must share the same airspace, it was necessary to regulate the operations. With that aim, ICAO has defined seven different classes of airspaces: A, B, C, D, E, F and G. The most restrictive one is Class A, where only IFR flights are permitted. The least restrictive is Class G, where both IFR and VFR flights are permitted. In any of the other airspace classes, sovereign authorities derive additional rules (based on the ICAO definitions) for VFR cloud clearance, visibility, and equipment requirements.

10.3.5 Navigation charts

The navigation charts are essential for air navigation. Their characteristics are internationally standardized in different ICAO documents, such for instance, the Annex 4: Aeronautical charts. According to the flight rules applied (IFR or VFR), the phase and level of flight, the characteristics of such charts differ. A complete route of an aircraft flying between two airports can be divided in three main parts: origin, en-route, and destination. Also, origin can be divided into take-off and initial climb, and destination can be divided into approach and landing. Figure 10.8 shows an schematic representation of the phases of a typical flight. Table 10.1 show the existing navigation charts according to ICAO's Annex 4.

Air navigation

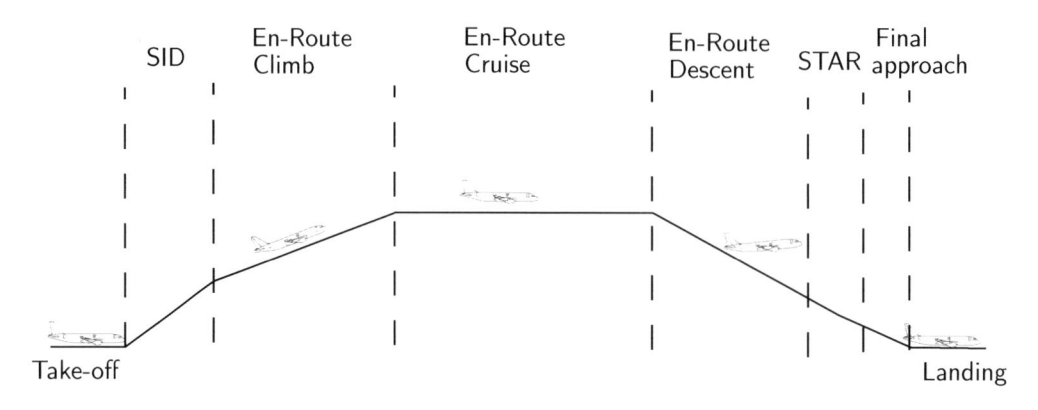

Figure 10.8: Phases in a flight.

Phase		Navigation chart
Aerodrome		Aerodrome chart
		Aerodrome ground movement chart
		Aircraft parking/Docking chart
		Aerodrome obstacles chart
Departures		Standard Instrumental Departure (SID) chart
En-route	Instrumental	En-route instrumental chart FIR
		En-route instrumental chart UIR
	visual	En-route visual chart UIR
Arrivals		Standard instrumental Arrival (STAR) chart
Approach	Instrumental	Precision Approach terrain chart
		Instrument approach chart
	Visual	Visual approach chart
Terminal Maneuvering Area (TMA)		Area chart

Table 10.1: Navigation charts. Data retrieved from Annex 4 ICAO.

The en-route phase is defined by a series of waypoints and airways (the already presented ATS routes). A section of an en-route lower navigation chart for the Washington D.C. area is given in Figure 10.9.

However, airports can not be directly connected by airways. Terminal Maneuvering Areas (TMA) are defined to describe a designated area of controlled airspace surrounding an airport due to high volume of traffic. Operational constraints and arrival and departure procedures are defined inside an airport TMA.

A flight departing from an airport must follow a Standard Instrument Departure (SID)

10.3 AIRSPACE MANAGEMENT (ASM)

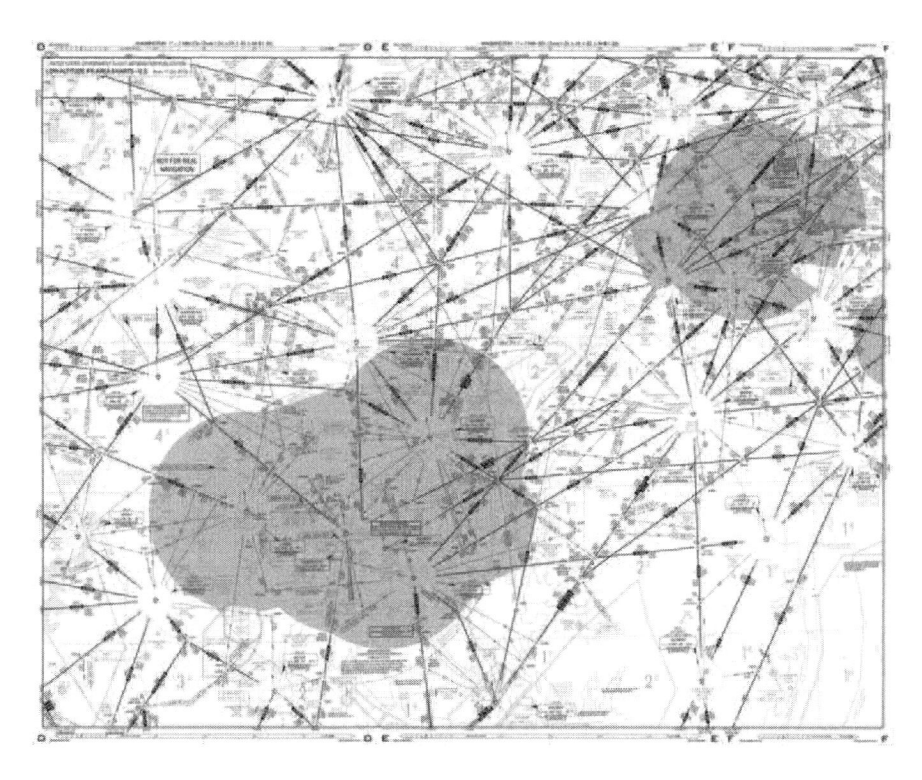

Figure 10.9: This low altitude section chart is issued for use in air navigation under IFR. The dimensions of this depicted area around Washington D.C. are nearly 280 and 220 nautical miles. Author: User:Orion 8 / Wikimedia Commons / Public Domain.

which defines a pathway from the runway to a waypoint or airway, so that the aircraft can join the en-route sector in a controlled manner. Before landing, an aircraft must follow two different procedures. It must follow first a Standard Terminal Arrival Route (STAR), which defines a pathway from a waypoint or airway to the Initial Approach Fix (IAF). Then, proceed from the IAF to runway following a final approach procedure. Figure 10.10 shows a final approach chart.

The reader is referred to the ANSP providers' AIP for more information on navigation charts[13]. For instance, AENA publishes its AIP in AIP AENA[14]. The en route charts can be consulted at AIP AENA EN-Route[15]. Airport charts can be consulted at AIP AENA Aerodromes[16].

---

[13]Notice that these documents are published in public access in the internet, but the ANSP holds the copyright on them, so they can not be published in this text book unless explicitly permitted.
[14]http://www.aena.es/csee/Satellite/navegacion-aerea/es/Page/1078418725020/
[15]http://www.aena.es/csee/Satellite/navegacion-aerea/es/Page/1078418725153//ENR-En-ruta.html
[16]http://www.aena.es/csee/Satellite/navegacion-aerea/es/Page/1078418725163//AD-Aerodromos.html

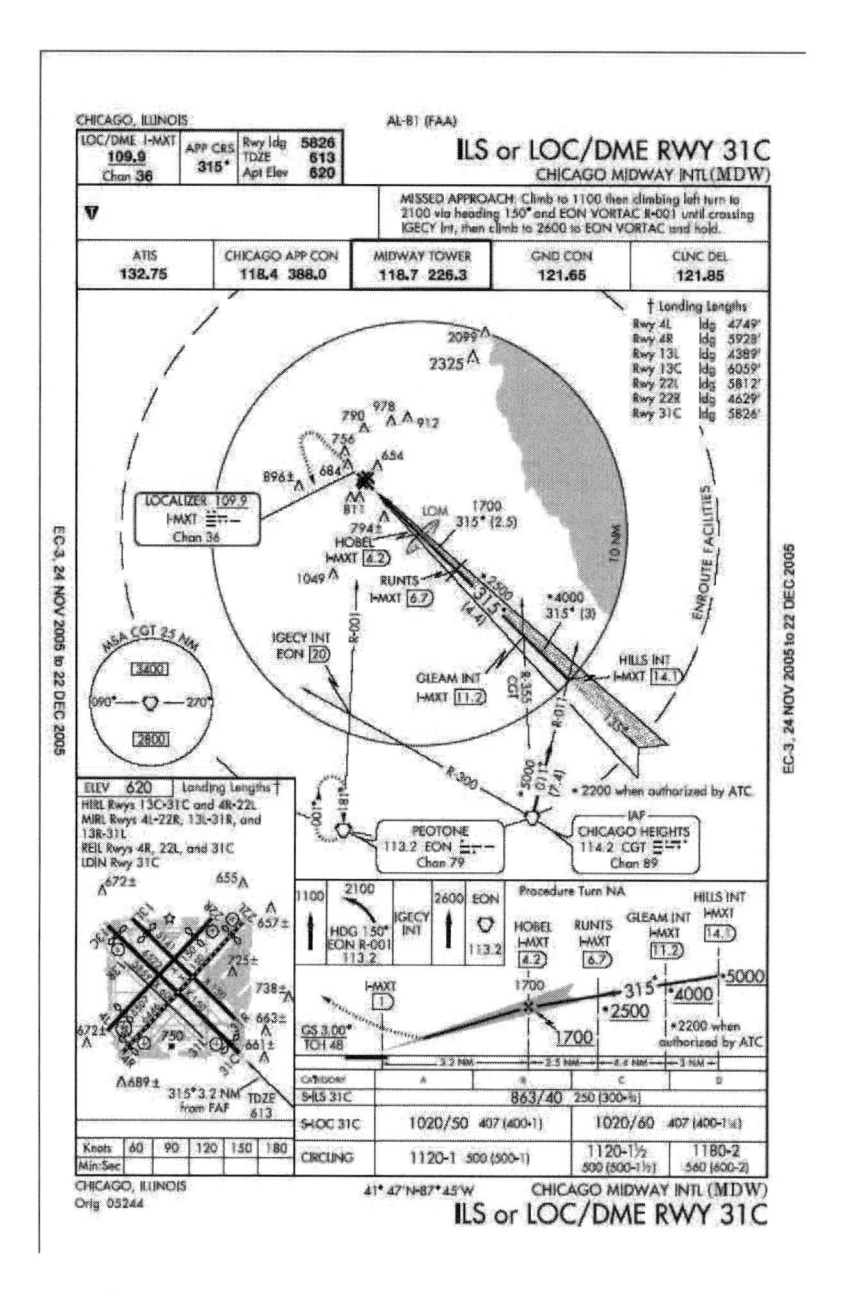

Figure 10.10: Instrumental approximation chart: ILS or LOC/DME RWY 31C approach chart for Chicago Midway. Minimums listed in lower right indicate with an operating Instrument Landing System a pilot can safely descend to 863' above sea level or 250' above ground level with a runway visual range of 4,000'. No longer valid for navigation. Author: User:Skywayman / Wikimedia Commons / Public Domain.

10.3.6 Flight plan

A flight plan is an aviation term defined by the International Civil Aviation Organization (ICAO) as:

> *Specified information provided to air traffic services units, relative to an intended flight or portion of a flight of an aircraft*"[17].

A flight plan is prepared on the ground and specified in three different manners: as a document carried by the flight crew, as a digital document to be uploaded into the Flight Management System (FMS), and as a summary plan provided to the Air Transit Services (ATS). It gives information on route, flight levels, speeds, times, and fuel for various flight segments, alternative airports, and other relevant data for the flight, so that the aircraft properly receives support from ATS in order to execute safe operations. Two safety critical aspects must be fulfilled: fuel calculation, to ensure that the aircraft can safely reach the destination, and compliance with Air Traffic Control (ATC) requirements, to minimize the risk of collision.

Flight planning is the process of producing a flight plan to describe a proposed aircraft flight. Flight planning requires accurate weather forecasts so that fuel consumption calculations can account for the fuel consumption effects of head or tail winds and air temperature. Furthermore, due to ATC supervision requirements, aircraft flying in controlled airspace must follow predetermined routes.

An effective flight plan can reduce fuel costs, time-based costs, overflight costs, and lost revenue from payload that can not be carried, simply by efficiently modifying the route and altitudes, speeds, or the amount of departure fuel.

10.4 Technical support: CNS system

As already mentioned, the term CNS (Communications, Navigation, and Surveillance) identifies the technical means through which air navigation is supported. Such technical means fulfill a key roll to ensure that air transportation is possible every day.

In order an aircraft to fly from one point to another in a safe way, it must keep continuos contact with the control services on earth by means of **communication systems**, it must use the **navigation systems** to continuously determine its position and address to the desired destination. In this whole process, the control services must use the **surveillance systems** to monitor aircraft and avoid any potential danger.

---

[17] ICAO Document 4444.

10.4.1 Communication systems

The technical means included under the term aeronautical communication fulfill a mission of spreading any information of interest to aircraft operations. This information is real-time information[18], and therefore is not included in the official documents published by the authorities. According to the provided service, ICAO has classified the communications in two main groups:

- Aeronautical Fixed Service (AFS): between terrestrial stations, i.e., fixed stations.
- Aeronautical Mobile Service (RR S1.32)[19]: between terrestrial stations and aircraft (mobile stations).

Aeronautical Fixed Service (AFS)

As defined by ICAO Standards documents in Annex 10 Vol II:

> The AFS is a telecommunication service between specified fixed points provided primarily for the safety of air navigation and for the regular, efficient, and economical operation of air services.

This service is typically on charge of spreading all the information prior departure (in NOTAMs), related with flight plans, meteorological information, operative state of the air space, etc. Such information must be transmitted to all fixed point or stations, e.g., control centers, that might need it to provide support to the aircraft in flight. This information is typically generated in one point and it is distributed using specific terrestrial networks. It is provided by voice and data networks and circuits, including:

- the Aeronautical Fixed Telecommunication Network (AFTN);
- the Common ICAO Data Interchange Network (CIDIN);
- the Air Traffic Services (ATS) Message Handling System (AMHS);
- the meteorological operational circuits, networks, and broadcast systems;
- the ATS direct speech networks and circuits;
- the inter-centre communications (ICC).

The major part of data message interchange in the AFS is performed by the Aeronautical Fixed Telecommunications Network, AFTN. This is a message handling network running according to ICAO Standards documented in Annex 10 to the ICAO Convention, in which it is defined as:

---
[18]information of two types: communications with pilots on real time, or information prior departure delivered in what is termed as NOTAM (NOtice To AirMen), which deals with information about the flight plan, meteorological conditions, operative conditions of navaids and/or ATS routes, etc.
[19]notation according to ICAO Annex 10-Vol II.

10.4 TECHNICAL SUPPORT: CNS SYSTEM

> *A worldwide system of aeronautical fixed circuits provided, as part of the aeronautical fixed service, for the exchange of messages and/or digital data between aeronautical fixed stations having the same or compatible communications characteristics.*

ATFN exchanges vital information for aircraft operations such as distress messages, urgency messages, flight safety messages, meteorological messages, flight regularity messages, and aeronautical administrative messages. One example could be the the spreading of the different flight plans that airlines must submit to authorities prior departure, which must be necessarily transmitted to the different control centers. The technology on which the AFTN is based is referred to as messages commutation. It transmits messages at low speed and therefore the network has low capacity. As a consequence, AFTN is completely outdated (however still widely used).

In order to create a technological upgrade to cope with the increasing volume of information, the CIDIN was conceived in the 1980's to replace the core of the AFTN. The technology on which the CIDIN is based is referred to as packages commutation and it is considered as a high speed, high capacity transmission network. Typically, most nodes which are part of the AFTN have also CIDIN capability, and thus the CIDIN can be considered as a data transport network which supports the AFTN.

Nevertheless, the volume of information needed is increasing more and more and CIDIN is about to be obsolete (if is not not already). The equipment and protocols upon which CIDIN (supporting also ATFN) is based need to be replaced by more modern technology with new messaging requirements. To meet these requirements, the ICAO has specified the ATS Message Handling System (AMHS), a standard for ground-ground communications not fully deployed yet. The AMHS is an integral part of the CNS/ATM concept, and it is associated to the Aeronautical Telecommunication Network (ATN) environment.

The goal of ATN is to be the *aeronautical internet*, a worldwide telecommunications network that allow any aeronautical actor (ATS services, airlines, private aircraft, meteorological services, airport services, etc.), exchange information in a safe way (control instructions, meteorological messages, flight parameters, position information, etc.), under standard message formats and standard communication protocols[20].

The European AMHS makes use of a TCP/IP network infrastructure, in line with the recent evolution of the ATN concept for ground communications. In addition to being the replacement for AFTN/CIDIN technology, the AMHS also provides increased functionality, in support of more message exchanges than those traditionally conveyed by the AFTN and/or CIDIN. This includes, for example, the capability to exchange binary data messages or to secure message exchanges by authentication mechanisms [21].

---

[20] The standards of the ATN can be consulted in the ICAO DOC 9705-AN/956: *Manual of Technical Provisions for the ATN.*

[21] The standards of the AMHS can be consulted in the ICAO Doc 9880-AN/466: *Manual on Detailed Technical Specifications for the Aeronautical Telecommunication Network (ATN).*

Aeronautical mobile service

On the other hand, the aeronautical mobile service includes all technical means required to support the communications between the aircraft and the ATS services (information, surveillance, and control) based on earth. These communications are typically pilot-controller.

As defined by ICAO Standards documents in Annex 10 Vol II:

> The aeronautical mobile service is a mobile service between aeronautical stations and aircraft stations, or between aircraft stations, in which survival craft stations may participate; emergency position-indicating radio-beacon stations may also participate in this service on designated distress and emergency frequencies.

These services are provided wireless, using radio channels. In the case of aeronautical communications, it is used the VHF (Very High Frequencies) band and HF (High Frequency) band. The channels in HF are only used for long-distance communications, when it is impossible to establish communication using VHF. VHF radio communications (for civil aviation) operate in the frequency range extending from 118MHz to 137MHz[22]. HF radio communications utilize practically the whole HF spectrum (3 MHz to 30MHz), depending on times of the day, seasonal variations, solar activity, etc.

The ultimate goal of this service is to allow communications between pilot and controller. In particular, in one control sector, the controller must be able to communicate with all aircraft inside the sector using only one of these radio channels (each sector has a unique frequency assigned). Therefore, the number and dimension of the sectors condition the location of the communication centers. The frequency assigned to each sector establish a double direction channel: pilot-controller; controller-pilot. That is the fundamental instrument in the functions of information, surveillance, and control of aircraft in flight.

The categories of messages handled by the aeronautical mobile service and the order of priority in the establishment of communications and the transmission of messages shall be as follows:

1. Distress calls, distress messages, and distress traffic (emergency messages).
2. Urgency messages.
3. Communications relating to direction finding (to modify the course).
4. Flight safety messages (movement and control).
5. Meteorological messages (meteorological information).
6. Flight regularity messages.

---

[22] Notice that the VHF range is 30MHz to 300MHz.

Aut. systems	Doppler Radar			
	Inertial navigation systems			
Non-aut. systems	Terrestrial	Puntual radio aids	NDB VOR DME ILS
		Zonal radio aids (hyperbolic)	Omega Loran Decca
	Spacial	GNSS	GPS GLONASS Galileo

Table 10.2: Navigational systems.

10.4.2 NAVIGATION SYSTEMS

The technical means included under the term navigation systems allow aircraft to know their positions at any time. It is important to distinguish between the systems that assist pilots (navigational aids) to steer their aircraft from origin to destination, and the techniques that pilots use to navigate.

The navigational aids constitute infrastructures capable to provide pilots all needed information in terms of position and guidance. On the other hand, the navigation techniques refer to the way in which pilots use these data about the position of the aircraft to navigate. In what follows, we are going to focus on the navigational aids systems.

The navigational aids systems can be classified in two main groups:

- Autonomous systems: Those systems that make only use of the means available in the aircraft to obtain information about its position.
- Non autonomous systems: Those external systems that provide the aircraft with the information about its position.

Table 10.2 provides a (non necessarily exhaustive) taxonomy.

Autonomous systems

Using only autonomous navigation systems, the most advanced navigation technique to be used is dead reckoning. As shown in Section 10.1, the dead reckoning consist in predicting the future position of the aircraft based on the current position, velocity, and course. Obviously, a reference (or initial) position of the aircraft must be known. In order

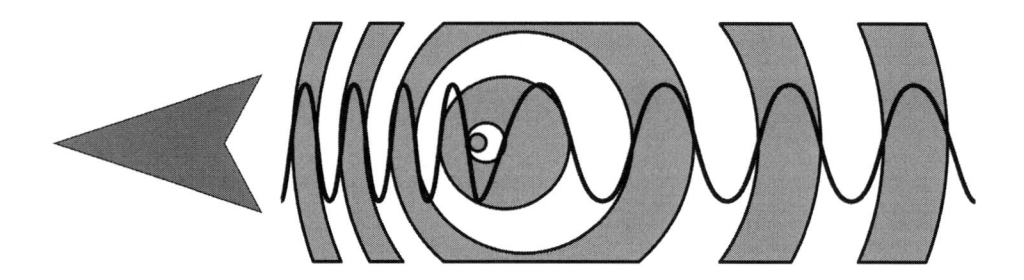

Figure 10.11: Doppler effect: change of wavelength caused by motion of the source. ©
User:Tkarcher / Wikimedia Commons / CC-BY-SA-3.0.

to determine this reference position, different means can be utilized, e.g., observing a point near the aircraft which position is known (very rudimentary), the observation of celestial bodies (also rudimentary), or the use of the so-called autonomous systems, which are also able to determine the velocity and course of the aircraft.

The two principal autonomous systems are:

- The Doppler radar.
- The Inertial Navigation System (INS).

Doppler radar: A Doppler radar is a specific radar that makes use of the Doppler effect to calculate the velocity of a moving object at some distance. It does so by beaming a microwave signal towards the target, e.g., a flying aircraft, and listening its reflection. Once the reflection has been listened, it is treated analyzing how the frequency of the signal has been modified by the object's motion. This variation gives direct and highly accurate measurements of the radial component of a target's velocity relative to the radar.

Inertial navigation system (INS): An inertial navigation system (INS) includes at least a computer and a platform or module containing accelerometers, gyroscopes, or other motion-sensing devices. The later is referred to as Inertial Measurement Unit (IMU). The computer performs the navigation calculations. The INS is initially provided with its position and velocity from another source (a human operator, a GPS satellite receiver, etc.), and thereafter computes its own updated position and velocity by integrating information received from the motion sensors. Figure 10.12 illustrates it schematically. The advantage of an INS is that it requires no external references in order to determine its position, orientation, or velocity once it has been initialized. On the contrary, the precision is limited, specially for long distances.

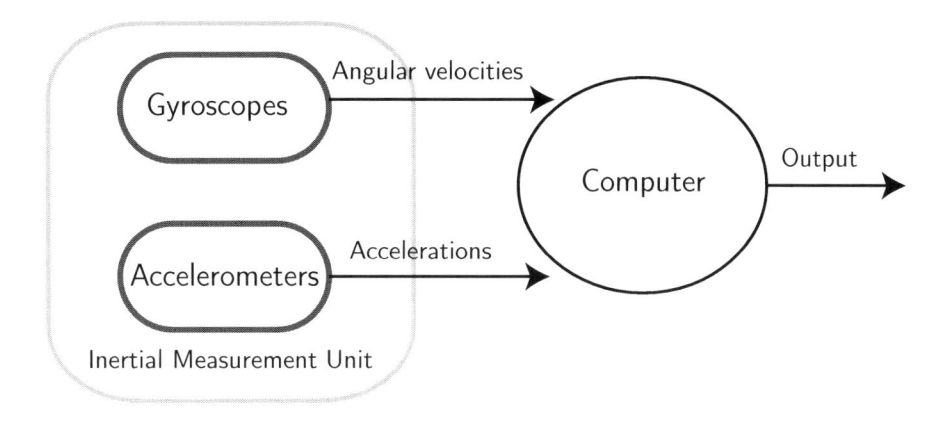

Figure 10.12: Scheme of an Inertial Navigation System (INS). The output refers to position, attitude, and velocity.

Gyroscopes measure the angular velocity of the aircraft in the inertial reference frame (for instance, the earth-based reference frame). By using the original orientation of the aircraft in the inertial reference frame as the initial condition and integrating the angular velocity, the aircraft's orientation (attitude) is known at all times.

Accelerometers measure the linear acceleration of the aircraft, but in directions that can only be measured relative to the moving system (since the accelerometers are fixed to the system and rotate with the system, but are not aware of their own orientation). Based on this information alone, it is known how the aircraft is accelerating relative to itself, i.e., in a non-inertial reference frame such as the wind reference frame, that is, whether it is accelerating forward, backward, left, right, upwards, or downwards measured relative to the aircraft, but not the direction (attitude) relative to the Earth. The attitude will be an input provided by the gyroscopes.

By tracking both the angular velocity of the aircraft and the linear acceleration of the aircraft measured relative to itself, it is possible to determine the linear acceleration of the aircraft in the inertial reference frame. Performing integration on the inertial accelerations (using the original velocity as the initial conditions) using the correct kinematic equations yields the inertial velocities of the system, and integration again (using the original position as the initial condition) yields the inertial position. These calculations are out of the scope of this course since one needs to take into account relative movement, which is to be studied in advance courses of mechanics. However, some insight is given in Chapter 7 and Appendix A.

Errors in the inertial navigation system: All inertial navigation systems suffer from integration drift: small errors in the measurement of acceleration and angular velocity are integrated into progressively larger errors in velocity, which are compounded into still

ACCURACY OF NAVIGATION SYSTEMS
(2-dimensional)

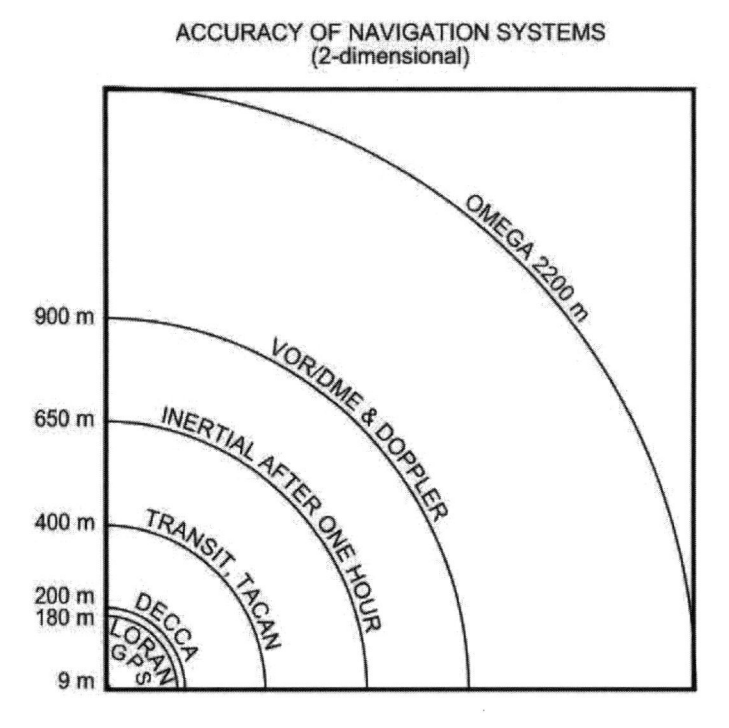

- OMEGA 2200 m
- 900 m
- VOR/DME & DOPPLER
- 650 m — INERTIAL AFTER ONE HOUR
- 400 m — TRANSIT, TACAN
- 200 m
- 180 m — DECCA
- 9 m — LORAN, GPS

Figure 10.13: Accuracy of navigation systems in 2d. © Johannes Rössel / Wikimedia Commons / CC-BY-SA-3.0

greater errors in position. Since the new position is calculated from the previous calculated position and the measured acceleration and angular velocity, these errors are cumulative and increase at a rate roughly proportional to the time since the initial position was input. Therefore the position must be periodically corrected by input from some other type of navigation system. The inaccuracy of a good-quality navigational system is normally less than 0.6 nautical miles per hour in position and on the order of tenths of a degree per hour in orientation. Figure 10.13 illustrates it in relation with other on-autonomous navigation systems (to be studied in what follows).

Accordingly, inertial navigation is usually supplemented with other navigation systems (typically non-autonomous systems), providing a higher degree of accuracy. The idea is that the position (in general, the state of the aircraft) is measured with some sensor, e.g., the GPS, and then, using filtering techniques (Kalman filtering, for instance), estimate the position based on a weighted sum of both measured position and predicted position (the one resulting from inertial navigation). The weighting factors are related to the magnitude of the errors in both measured and predicted position. By properly combining both sources the errors in position and velocity are nearly stable over time. The equation of the Kalman filter are not covered in this course and will be studied in more advanced courses.

Non autonomous systems

Non autonomous systems require the information generated by terrestrial stations or satellites to determine the position, course, and/or velocity of the aircraft. In this manner, the transmission station (transmitter) produces electromagnetic waves that are received in the reception point (receptor).

The supplied information is referred to as observables (it can be a distance, a course, etc.). Such information locates the aircraft inside a so-called situation surface, i.e., if the observable is the distance, one knows that the aircraft is located at some point on the surface of a sphere with center the transmission center.

More precisely, a situation surface is the geometric locus of the space which is compatible with the observables. The types of situation surfaces are:

- **Plane** perpendicular to the surface of earth where the aircraft is located. The observable is the course.

- **Spherical surface** with the center in the transmission station. The observable is the distance.

- **Hyperboloid of revolution**, being the focuses two external transmission centers exchanging information. The observable is the distance.

In general, using information coming only from one transmitter, it is impossible to locate an aircraft; more sources of information are needed. If two surfaces are intersected, one obtains a curve. If three (or more) surfaces are intersected, one obtains a point (if more than three, ideally also a point). Therefore, in order to locate an aircraft one needs either two transmitter (generating two surfaces) plus the altitude given by the altimeter, or three transmitters (generating three surfaces).

We turn now the discussion to analyze how such observables are obtained, i.e., how we are able to measure a distance or determine a course using terrestrial stations or satellites. These observables are obtained using different techniques based on electromagnetic fields, namely:

- Radiotelemetry.

- Radiogoniometry.

- Scanning beam.

- Spatial modulation.

- Doppler effect.

Radiotelemetry: It is based on the consideration that the electromagnetic waves travel at constant velocity, the velocity of light ($c = 300000$ km/s), and in straight line[23]. Under these assumptions, if we measure the time that the waves take from the instant in which the transmitter emits the wave and it receives it back after being rebooted by the receptor (the aircraft), by simple kinematic analysis one can obtain the distance.

Radiogoniometry: It is based on the consideration that the electric and magnetic field that constitute an electromagnetic wave are both perpendicular to the direction of propagation of the wave. Using this technique, by measuring the phases of the electric and magnetic fields of the wave, one can determine the angle that forms the longitudinal axis of the aircraft with the direction of the transmitted wave.

Scanning beam: It is based on the fact that the electromagnetic wave emitted by the transmitter has a dynamic radiation diagram[24], with a narrow principal lobe and very small secondary lobes. The receptor (aircraft) is only illuminated (radiated) if the principal lobe points to the aircraft. In this way, knowing the movement law of the radiation diagram, when the aircraft is illuminated one can obtain the direction between the transmitter and the aircraft.

Spatial modulation: This technique is original from air navigation. Two different electromagnetic waves are used. The first one is the reference signal, generating an omnidirectional magnetic field so that all points of the region receive the same information. These kind of antennas are referred to as isotropic antennas, and their radiation is referred to as isotropic or omnidirectional radiation. The second signal generates generates a (either static or dynamic) directional magnetic field. The comparison between the phases of the reference signal and the directional signal determines the direction of the aircraft.

Doppler effect: It is based on the change of frequency of a wave produced by the relative movement of the generating source (transmitter) with respect to the receiver (aircraft). In this way, one can obtain the distance between transmitter and aircraft.

Table 10.3 and Table 10.4 show a classification of the different navigation aids as a function of the different techniques and the different situation surfaces, and a classification of the different navigation aids as a function of the different flight phases, respectively.

---

[23]Notice that the fact that the light travels in a straight line was proven false in Einstein's theory of general relativity. Inside the atmosphere (using terrestrial transmitter with aircraft as receptors) one can assume as hypothesis a straight line. When using satellites, the trajectory is a curve and the straight line must be corrected.

[24]A diagram of radiation is a graphic representation of the intensity of a radiated signal in each direction. In some cases, there exist a principle lob and secondary lobs of less intensity.

10.4 TECHNICAL SUPPORT: CNS SYSTEM

(a) Directional radiation. © Timothy Truckle / Wikimedia Commons / CC-BY-SA-3.0.

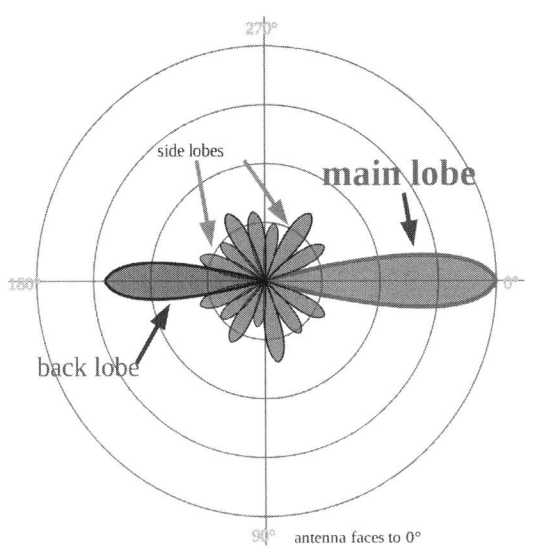

(b) Radar antenna. Author: User:Soerfm / Wikimedia Commons / Public Domain.

Figure 10.14: Scanning beam radiation diagram and a radar antenna that produces a directional radiation. Notice that the radar is a surveillance system.

Technique	Situation surface		
	Vertical plane	Spherical	Hyperbolic
Radiotelemetry	–	DME/GNSS/Radar	Loran-C
Radiogoniometry	NDB	–	–
Scanning beam	MLS/Radar	–	–
Spatial modulation	VOR/ILS/TACAN	–	–
Doppler effect	DVOR	–	–

Table 10.3: Navigation aids based on situation surface and the technique.

Aid	Flight phase			
	Climb Descent	Cruise	Non precision approach	Precision approach
NDB	✓	✓	✓	–
VOR	✓	✓	✓	–
DME	✓	✓	✓	–
INS	–	✓	–	–
Loran-C	–	✓	–	–
GNSS	✓	✓	✓	✓
ILS	–	–	–	✓

Table 10.4: Classification of the navigation aids based on the flight phase.

We continue now analyzing the different navigation aids that provide the required information to locate aircraft in flight. These have been already mentioned in Table 10.2 and Table 10.3.

The most important ones using the technique of **radiotelemetry** are of two kinds: those that locate the aircraft in spheres; those that locate the aircraft in revolution hyperboloids. The most important ones among the first ones are:

- DME: Distance Measurement Equipment.

- TACAN: TACtical Air Navigation equipment[25]. It is typically used in military aviation.

- GNSS: Global Navigation Satellites Systems.

- Radar[26]: Radio detection and ranging. This system is specific of the surveillance and it will be analyzed later on.

Due to its importance, we will just analyze more in depth the DME and the GNSS systems.

Distance Measurement Equipment (DME): Is a transponder-based radio navigation system that measures slant range distance by timing the propagation delay of radio signals. The DME system is composed of a transmitter/receiver (interrogator) in the aircraft and a receiver/transmitter (transponder) on the ground. Aircraft interrogate and the DME ground station responds.

Aircraft use DMEs to determine their distance from a land-based transponder (terrestrial station) by sending and receiving pulse pulses of fixed duration and separation. The ground stations are typically co-located with VORs. A low-power DME can also be co-located with an ILS glide slope antenna installation where it provides an accurate distance to touchdown. The simultaneous syntonization of two terrestrial DME stations by the aircraft allows us to locate the aircraft in two dimensions: latitude and longitude. By means of an altimeter, the aircraft is located in the 3D space. When co-located with a VOR, it also provides the direction of flight (read at the on-board ADF equipment). Therefore, the duple VOR-DME provides both position and course. It's important to understand that DME provides the physical distance from the aircraft to the DME transponder. This distance is often referred to as *slant range* and depends trigonometrically upon both the altitude above the transponder and the ground distance from it. DME operation will continue and possibly expand as an alternate navigation source to space-based navigational systems such as GPS and Galileo.

---

[25]Equivalent to the use together of a VOR and a DME.
[26]This system uses both radiotelemetry and scanning beam techniques.

10.4 TECHNICAL SUPPORT: CNS SYSTEM

(a) D-VOR/DME ground station. © User:Yaoleilei / Wikimedia Commons / CC-BY-SA-2.0.

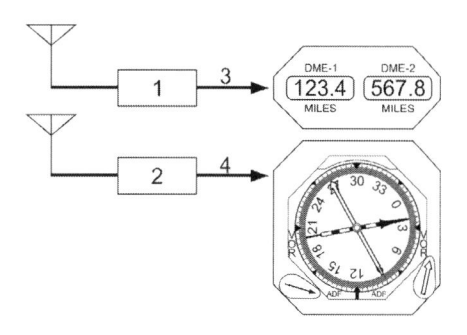

(b) DME on-board receiver, together with ADF unit (reading the VOR information). © User:Tosaka / Wikimedia Commons / CC-BY-SA-3.0.

Figure 10.15: VOR-DME.

Global Navigation Satellite Systems (GNSS): GNSS are global systems that use a medium high constellation of satellites describing quasi-circular orbits inclined with respect to the terrestrial equator. Currently, just two systems are in practice active: The american GPS and the Russian GLONASS. Some other constellations are in development, such as the European Union Galileo positioning system, the Chinese Compass navigation system, and Indian Regional Navigational Satellite System.

GNSS systems are based on the transmission of an electromagnetic wave by the satellite that is captured and de-codified by the receiver (the aircraft). The basic information that we can obtain is the time that the signal takes while traveling. This time provides a so-called pseudo-distance. This is due to the fact that there exists a synchrony error between the time of the aircraft and the time of the satellites, and therefore, we can not know with certainty the real distance. These systems provide location and time information anywhere on or near the Earth where there is a line of sight to four or more GPS satellites[27].

This system is intended to offer higher precision (with an error of about 10m in determining the position), global coverage, and continuos navigation. However, one of the fundamental drawbacks that have made so far these systems impractical for air navigation is their strategic character in terms of national security. GLONASS was only open to limited civilian use in 2007. The GPS is maintained by the United States government and is freely accessible by anyone with a GPS receiver. However, its reliability is not complete in terms of precision and continuity in the coverage, i.e., the signal has no integrity due to its military character.

---

[27]Notice that in order to determine the position of the receptor (the aircraft) we need in this case 4 satellites. This is because an extra satellite is needed to determine the synchrony error between the time of the aircraft and the time of the satellites.

Air navigation

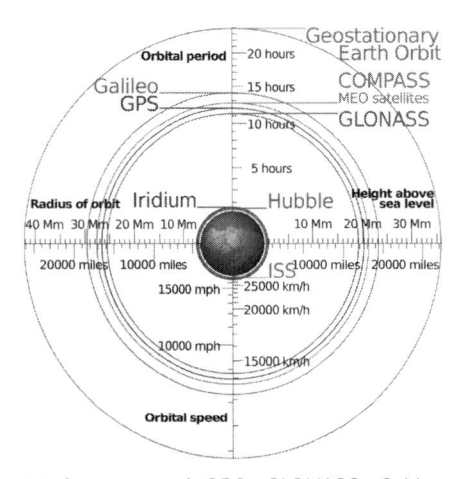

(a) Comparison of GPS, GLONASS, Galileo and COMPASS (medium earth orbit satellites) orbits with International Space Station, Hubble Space Telescope and geostationary orbits, and the nominal size of the earth. © Geo Swan / Wikimedia Commons / CC-BY-SA-3.0.

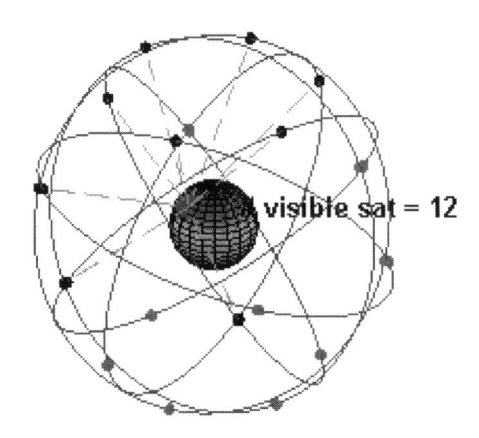

(b) Simulation of the original design of the GPS space segment, with 24 GPS satellites (4 satellites in each of 6 orbits), showing the evolution of the number of visible satellites from a fixed point (45N) on earth (considering "visibility" as having direct line of sight). Author: User:El pak / Wikimedia Commons / Public Domain.

Figure 10.16: GNSS systems.

Therefore, in order GNSS to be used (still in a limited way and always with back-up systems) in some phases of the flight, a first generation of GNSS (the so-called GNSS-1) was conceived as a combination between the existing satellite navigation systems, i.e., GPS and GLONASS, and some type of augmentation system. Augmentation of a global navigation satellite system (GNSS) is a method of improving the navigation system's attributes, such as accuracy, reliability, and availability, through the integration of external information into the calculation process. These additional information can be for example about sources of satellite error (such as clock drift, ephemeris, or ionospheric delay), or about additional aircraft information to be integrated in the calculation process. There are three types of augmentation systems, namely:

- The **Satellite-Based Augmentation System (SBAS)**: a system that supports wide-area or regional augmentation through the use of additional satellite-broadcast messages. SBAS systems are composed of multiple, strategically located ground stations. The ground stations take measurements of GNSS satellite signals, which are used to generate information messages that are sent back to the satellite constellation, which finally broadcasts the messages to the end users (aircraft). Regional SBAS include WAAS (US), EGNOS (EU), SDCM (Russia), MSAS (Japan), and GAGAN (India).

- The **Ground-Based Augmentation System (GBAS)**: a system that supports augmentation through the use of terrestrial radio messages. As for SBAS, terrestrial

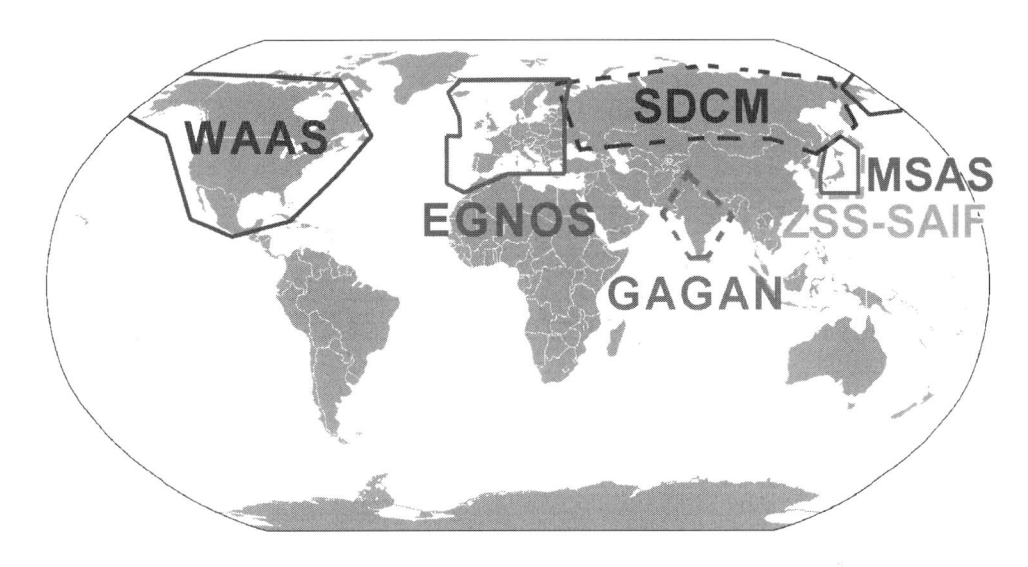

Figure 10.17: Service Areas of Satellite Based Augmentation Systems (SBAS). © User:Persimplex / Wikimedia Commons / CC-BY-SA-3.0.

stations take GNSS signal measurements and generate information messages, but in this case these messages are directly transmitted to the end user (the aircraft). GBAS include, for instance, the LAAS (US).

- The **Airborne Based Augmentation System (ABAS)**: in this augmentation system the ground stations analyze only information coming from the aircraft. This information is transmitted back to the aircraft.

GNSS-2 is the second generation of satellite systems that will provide a full civilian satellite navigation system. These systems will provide the accuracy and integrity necessary for air navigation. These fully civil satellite systems include the European Galileo, which is expected to be fully operative in 2020. Also a civil GPS version is under development.

Radiotelemetry is also used in another way by the so called **hyperbolic systems** (those systems that locate the aircraft in revolution hyperboloids): LORAN-C, Omega, DECCA. Omega and DECCA are now not being used. We focus now on LORAN-C.

LORAN-C: The LOng RAnge Navigation system is constituted by a chain of stations that allow a wide coverage range using low frequency radio signals. It is an evolution of its precursor: Loran-A, developed during World War II. LORAN is based on measuring the time difference between the receipt of signals from a pair of radio transmitters. A given constant time difference between the signals from the two stations can be represented by a hyperbolic line of position. If the positions of the two synchronized stations are known,

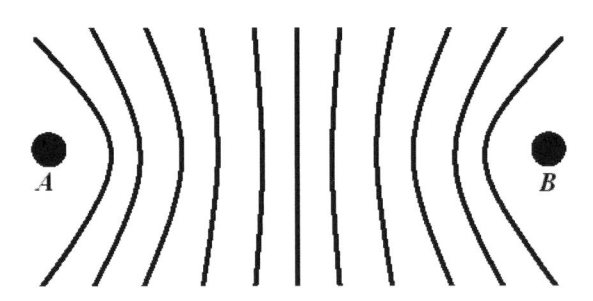

(a) LORAN diagram: the difference between the time of reception of synchronized signals from radio stations A and B is constant along each hyperbolic curve. Author: User: Massimiliano Lincetto / Wikimedia Commons / Public Domain.

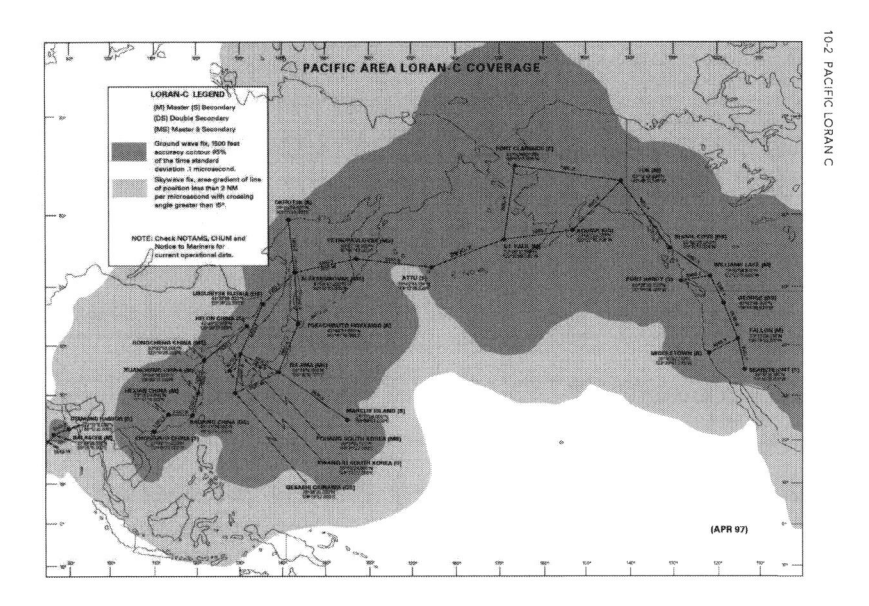

(b) LORAN coverage over the Pacific. Author: User:Sv1xv / Wikimedia Commons / Public Domain.

Figure 10.18: LORAN.

then the position of the raircraft can be determined as being somewhere on a particular hyperbolic curve where the time difference between the received signals is constant. In ideal conditions, this is proportionally equivalent to the difference of the distances from the aircraft to each of the two stations.

An aircraft which only receives signals form a pair of LORAN stations cannot fully fix its position. The aircraft must receive and calculate the time difference between a second pair of stations. This allows to be calculated a second hyperbolic line on which

10.4 TECHNICAL SUPPORT: CNS SYSTEM

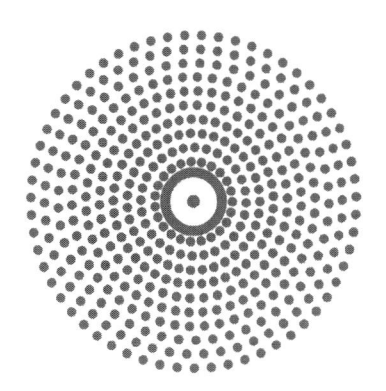

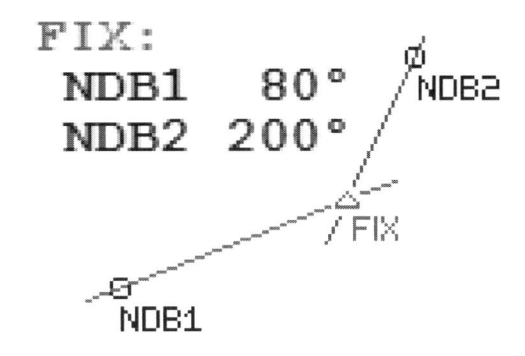

(a) NDB symbol on a navigation chart. Author: User: Denelson83 / Wikimedia Commons / Public Domain.

(b) NDB fix calculation. Author: Jed Smith / Wikimedia Commons / Public Domain.

Figure 10.19: Non Directional Beacon.

the aircraft is located. In practice, one of the stations in the second pair also may be (and frequently is) in the first pair. This means signals must be received from at least three LORAN transmitters to locate exactly the aircraft. By determining the intersection of the two hyperbolic curves, the location of the aircraft can be determined.

LORAN has been widely used to navigate when overflying oceans, where DME and VOR coverage ranges are insufficient. In recent decades LORAN use has been in steep decline, with the GNSS systems as primary replacement. However, there have been attempts to enhance LORAN, mainly to serve as a backup and land-based alternative to GNSS systems.

We turn now the discussion to analyze the navigational aids that use the technique of **radiogoniometry**. The most important one is:

- NDB: Non-Directional (radio) Beacon[28].

Non-Directional Beacon (NDB): A NDB is a radio transmitter at a terrestrial location that is used to obtain the course or position of an aircraft. Due to the fact that NDB uses radiogoniometry, its signals are affected (more than other aids) by atmospheric conditions, mountainous terrain, coastal refraction, and electrical storms, particularly at long range. The navigation based on NDB aids consists of two fundamental parts: the automatic direction finder (ADF), which is the equipment on-board the aircraft that detects the

---

[28]The on-board equipment that captures the information is called Automatic Direction Finder (ADF). Thus, sometimes, these equipments are named as a whole as NDB-ADF.

NDB's signal, and the NDB transmitter. ADF equipment determines the direction to the NDB station relative to the aircraft, which is presented to the pilot on a Radio Magnetic Indicator (RMI). In this way, in a simple, intuitive manner pilots know if the aircraft is addressing towards an NDB; if not, they now de deviation and can correct the course.

NDBs are also used to determine airways of fixes. NDB bearings[29] provide a method for defining a network of routes aircraft can fly. In this way, the network of terrestrial NDB stations (also VORs) can uniquely define a network of fixes (connected by airways, i.e., the bearings) in the sky. Indeed, 20-30 years ago, the routes aircraft followed to complete a flight plan were only based on NDB/VOR stations. In a navigation chart a NDB is designated by a symbol as in Figure 10.19.a. More recently, another way of navigation has arisen: the so-called RNAV. It is based on calculating fixes based on the information provided by two aids. For instance, using the information coming from two NDBs, fixes are computed by extending lines through known navigational reference points until they intersect. In this manner, many fictitious (in the sense that are not related to an existing terrestrial station) fixes or waypoints have been defined, increasing the network of routes and thus the capacity and efficiency of the system. See Figure 10.19.b.

We turn now the discussion to analyze the navigational aids that use the technique of **Spatial modulation**. The most important ones are:

- VOR: VHF Omnidirectional Radio range.
- ILS: Instrument Landing System.

The ILS has been already studied in Chapter 9. We study now the VOR[30].

VOR: It is a type of short-range radio navigation system for aircraft, enabling aircraft to determine their position and/or course by receiving VHF radio signals transmitted by a network of fixed ground radio stations. The VOR was developed in the US during World War II and finally deployed by 1946. VORs can be considered all fashioned, but they have played a key role in the development of the modern air navigation. As we pointed out in the case of NDBs, VORs have been traditionally used as intersections along airways, and thus, to configure airways. Many people have been claiming throughout years that the GNSS system will sooner rather than later substitute them (as well as NDBs, DMEs, etc.), but however VORs still play a fundamental role in air navigation. Indeed, VOR is the standard air navigational system in the world, used by both commercial and general aviation.

The way a fix or a direction can be obtained based on VOR information is identical as what have been exposed for NDBs. However, VOR's signals provide considerably greater accuracy (90 meters approx.) and reliability than NDBs due to a combination of factors.

---

[29]A bearing is a line passing through the station that points in a specific direction.
[30]Source: Wikipedia: VOR

10.4 TECHNICAL SUPPORT: CNS SYSTEM

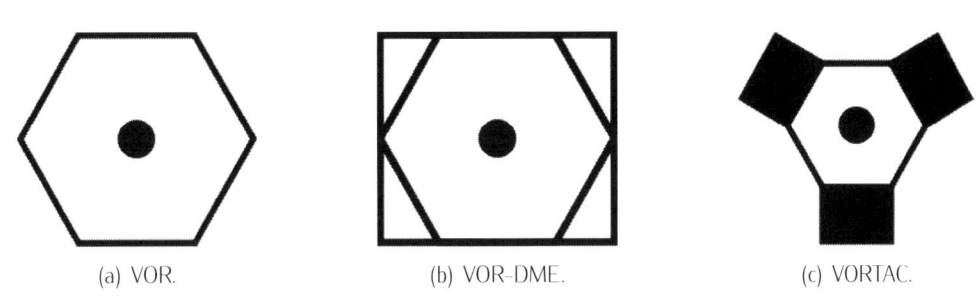

(a) VOR.   (b) VOR-DME.   (c) VORTAC.

Figure 10.20: VOR (Author: Denelson83 / Wikimedia Commons / Public Domain), VOR-DME (Author: User:mamayer / Wikimedia Commons / / CC0 1.0), and VORTAC (Author: User:Denelson83 / Wikimedia Commons / Public Domain) symbols on a navigation chart.

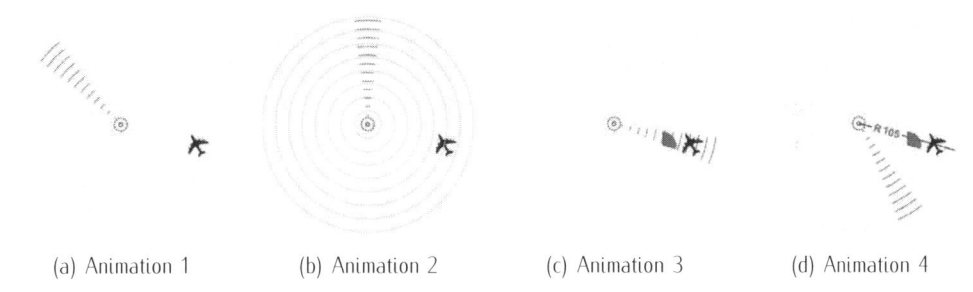

(a) Animation 1   (b) Animation 2   (c) Animation 3   (d) Animation 4

Figure 10.21: Animation that demonstrates the spatial modulation principle of VORs. A radio beam sweeps around (actually it's done 30 times per second). When the beam is at the local magnetic north direction, the station transmits a second, omni-directional signal. The time between the omni-directional signal and instant in which the aircraft is receives the directional beam gives the angle from the VOR station (105 deg in this case). © User:Orion 8 / Wikimedia Commons / CC-BY-SA-3.0.

VHF radio is less vulnerable to diffraction (course bending) around terrain features and coastlines. Phase encoding suffers less interference from thunderstorms.

Typically, VOR stations have co-located DME or military TACAN. A co-located VOR and TACAN is called a VORTAC. A VOR co-located only with DME is called a VOR-DME. A VOR radial with a DME distance allows a one-station position fix. VOR-DMEs and TACANs share the same DME system. The different symbols that identify this co-inhabiting systems are illustrated in Figure 10.20.

A VOR ground station emits an omnidirectional signal, and a highly directional second signal that varies in phase 30 times a second compared to the omnidirectional one. By comparing the phase of the directional signal to the omnidirectional one, the angle (bearing) formed by the aircraft and the station can be determined. Figure 10.21 illustrates it. This line of position is called the "radial" from the VOR. This bearing is then displayed in the cockpit of the aircraft in one of the following four common types of indicators:

(a) Onmi-Bearing Indicator: the VOR intercepts the aircraft in radial 250 (approx.); the CDI (white vertical bar) indicates that the aircraft is flying is the direction of the radial; The To–From indicator (indicating From) says that the aircraft has already overflown the VOR station and keeps going. Author: User:Quistnix / Wikimedia Commons / Public Domain.

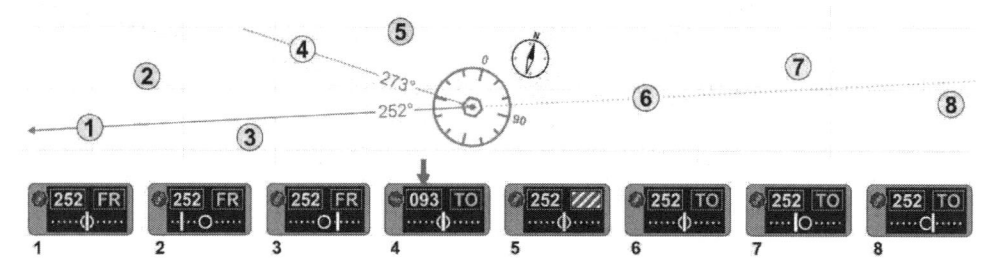

(b) VOR CDI Explanation: On the position (1) the aircraft is on the radial 252, and the direction flag marks FR (course 252 away FRom the VOR station). In situation (2) and (3) if you fly away FRom the station then the CDI's needle shows the direction (left or right) to the 252 radial and the distance in degrees (the needle scale is of 2 degrees). (4) is exactly like (1) but approaching the VOR. If the aircraft is wide away from the selected radial then the striped flag (5) warns the pilot that the aircraft is out of the segment where meaningful indication can be done. On the position (6) the aircraft is on the back course of radial 252. You see that if you fly with heading 252 degrees then you fly exactly TO the station. At (7) and (8) can be explained as (2) and (3). © User:Orion 8 / Wikimedia Commons / CC-BY-SA-3.0.

Figure 10.22: VOR displays interpretation.

1. Omni-Bearing Indicator (OBI): is the typical light-airplane VOR indicator. It consists of a knob to rotate an "Omni Bearing Selector" (OBS), and the OBS scale around the outside of the instrument, used to set the desired course. A "course deviation indicator" (CDI) is centered when the aircraft is on the selected course, or gives left/right steering commands to return to the course. An *ambiguity* (TO-FROM) indicator shows whether following the selected course would take the aircraft to, or away from the VOR station. A thorough explanation on how this instrument works is given in Figure 10.22.

2. Radio Magnetic Indicator (RMI): features a course arrow superimposed on a rotating card which shows the aircraft's current heading at the top of the dial. The "tail" of the course arrow points at the current radial from the station, and the "head" of the arrow points at the inverse (180 deg different) course to the station.

3. Horizontal Situation Indicator (HSI): is considerably more expensive and complex than a standard VOR indicator, but combines heading information with the navigation display in a much more user-friendly format, approximating a simplified moving map.

4. An Area Navigation (RNAV) system is an onboard computer with display and up-to-date navigation database. At least two VOR stations (or one VOR/DME station) is required for the computer to plot aircraft position on a moving map, displaying the course deviation relative to a VOR station or waypoint.

Finally, we analyze the navigational aids that use the technique of **Scanning beam**. This technique is based on concentrating the radiation of electromagnetic waves in a particular direction. Big antennas and high frequencies must be used. The most important systems are:

- MLS: Microwave landing system.

- Radar: radio detection and ranging.

As mentioned before, the radar is the fundamental system among the surveillance equipments and it will be studied later on. We analyze now the MLS.

MLS: A microwave landing system (MLS) is a precision landing system originally intended to replace or supplement instrument landing systems (ILS). MLS has a number of operational advantages when compared to ILS, for instance, including a wide selection of channels to avoid interference with other nearby airports, excellent performance in all weather conditions, less influence of the orography in the quality of the signal, and more flexible range of vertical and horizontal descent angles, which in principle would allow for efficient descents. The system may be divided into five functions: approach azimuth, back azimuth, approach elevation, range and data communications.

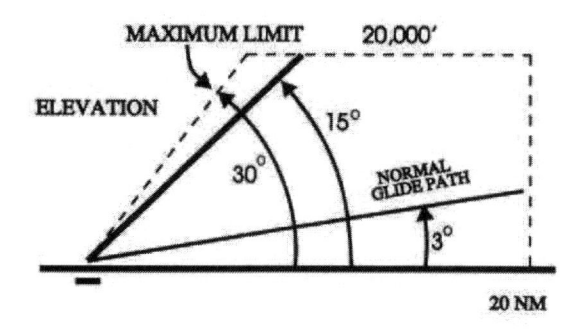

(a) Coverage Volumes of the Elevation station. Author: User:BetacommandBot / Wikimedia Commons / Public Domain.

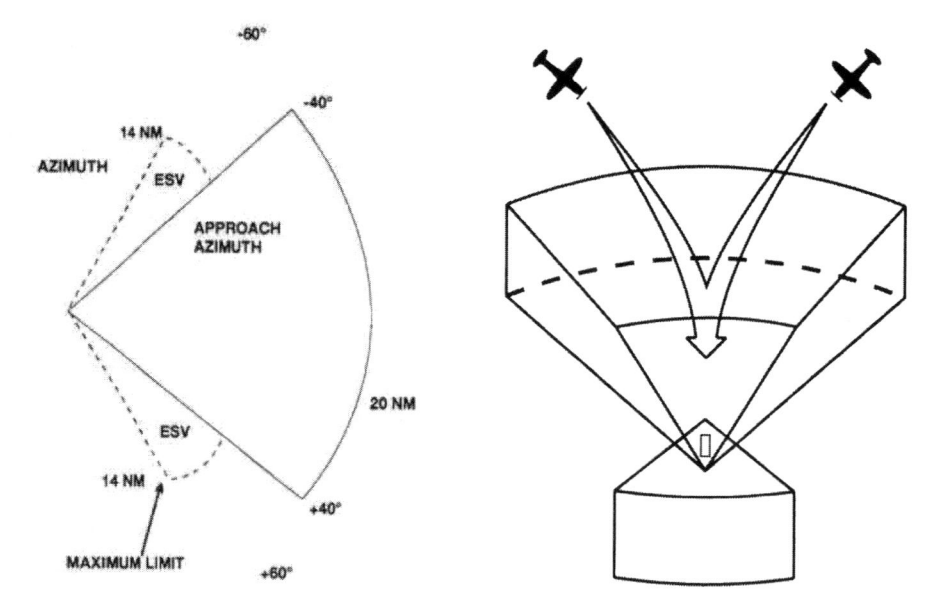

(b) Coverage Volume of the Azimuth station. Author: User:BetacommandBot / Wikimedia Commons / Public Domain.

(c) Coverage Volumes 3-D Representation. © User:Epolk / Wikimedia Commons / Public Domain.

Figure 10.23: MLS coverage.

MLS systems became operational in the 1990s. However, it has not been used much. This is due to two main reasons: first, the ILS has evolved and it is now more robust; second, and more important, GNSS systems allowed the expectation of the same level of positioning detail with no equipment needed at the airport. However, the GNSS navigation is still not a reality and therefore, and MLS continues to be of some interest in Europe.

10.4.3 Surveillance systems

The technical means included under the term aeronautical surveillance fulfill a mission of providing real-time information over the position of the aircraft to ATC function, i.e., controllers, with the aim of ensuring safety by properly separating them and avoiding thus any potential conflict. The surveillance has been traditionally (and still is nowadays) carried out in the different control dependences as shown in Figure 10.7, i.e., Area Control Centers (ACC) centers, Approximation (APP) dependences, and Control Tower (TWR), using the radar. However, within the use of satellite communications, most likely Automatic Dependent Surveillance Broadcast (ADSB) will be replacing (sooner rather than later) radar as the primary surveillance method for controlling aircraft worldwide. ADSB will increase the situational awareness of pilots with cockpit displays, enabling them more autonomy in self-separation. There are also airborne systems that fulfill a surveillance function. That is the case of the Traffic Collision Avoidance System or Traffic alert and Collision Avoidance System (both abbreviated as TCAS), which acts as an automatic advisory back-up system in case of imminent threat, i.e., when the human-based ATC layer has failed.

Radar

Radar was developed before and during World War II, where it played a key role in all aerial battles. The term RADAR was coined in 1941 by the United States Navy as an acronym for radio detection and ranging.

A radar system has a transmitter that emits electromagnetic radio signals in predetermined directions. When these come into contact with an object they are usually reflected back towards the receiver. A radar receiver is usually in the same location as the transmitter. By using using radiotelemetry techniques, the position of the radiated object can be determined and displayed. If the object is moving, there is a slight change in the frequency of the radio waves due to the Doppler effect.

In aviation, two radar techniques are applied:

- The original technique described above, that detects the objects due to its finite magnitude. This kind of radars are referred to as simply primary radars (PSR). In this case the aircraft is a passive object.
- The secondary radar (SSR): in this case, the radar requires the aircraft to carry an on board equipment called transponder. The transponder is interrogated from earth, responding with coded values such as flight level, flight code, direction, or velocity. This version was standardized by ICAO in the 80s with the aim at supporting air traffic control and surveillance.

The presentation of data in the screen that use controllers is very different in both cases. In the primary radar, only points (called targets) are presented with no identification,

(a) Radar antenna. The parabolic antenna is the PSR and the rectangular one above it is the SSR. © Magnus Manske / Wikimedia Commons / CC-BY-SA-3.0.

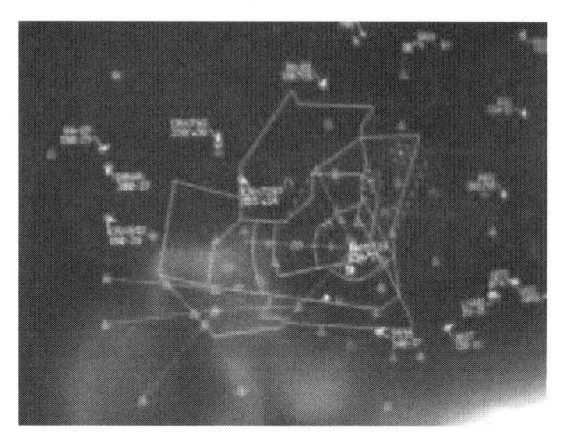

(b) Approach radar screen (90's generation). Image is based on primary and secondary datas. © Magnus Manske / Wikimedia Commons / CC-BY-SA-2.5.

Figure 10.24: Radar antenna and ATC display.

nor any information. Fixed targets can be mountains or any other orographic accident, while mobile targets can be identified with aircraft. Thus, the PSR is more interpretative. In the case of the secondary radar, the targets that are presented in the screen have a identification code, and provide also data such as flight level or velocity. Obviously, this information is much more useful for a controller to fulfill the surveillance function since each aircraft has a unique transponder.

The use of radar to supply the information required to fulfill surveillance is performed in three different scenarios with three different types of equipment: Long range secondary radar for En-route control; primary radar and short range secondary radar for approaches; surface radar (primary) at the airport.

TCAS

Traffic Collision Avoidance System (TCAS)[31] is an airborne aircraft collision avoidance system designed to reduce the incidence of Mid-Air Collision (MAC) between aircraft. It acts as the last safety back-up layer. Based on SSR transponder signals, it monitors the airspace around an aircraft for other aircraft equipped with a corresponding active transponder (independently of air traffic control) and advises instructions to pilots in case

---

[31] also refereed to as Aircraft Collision Avoidance System (ACAS).

10.4 TECHNICAL SUPPORT: CNS SYSTEM

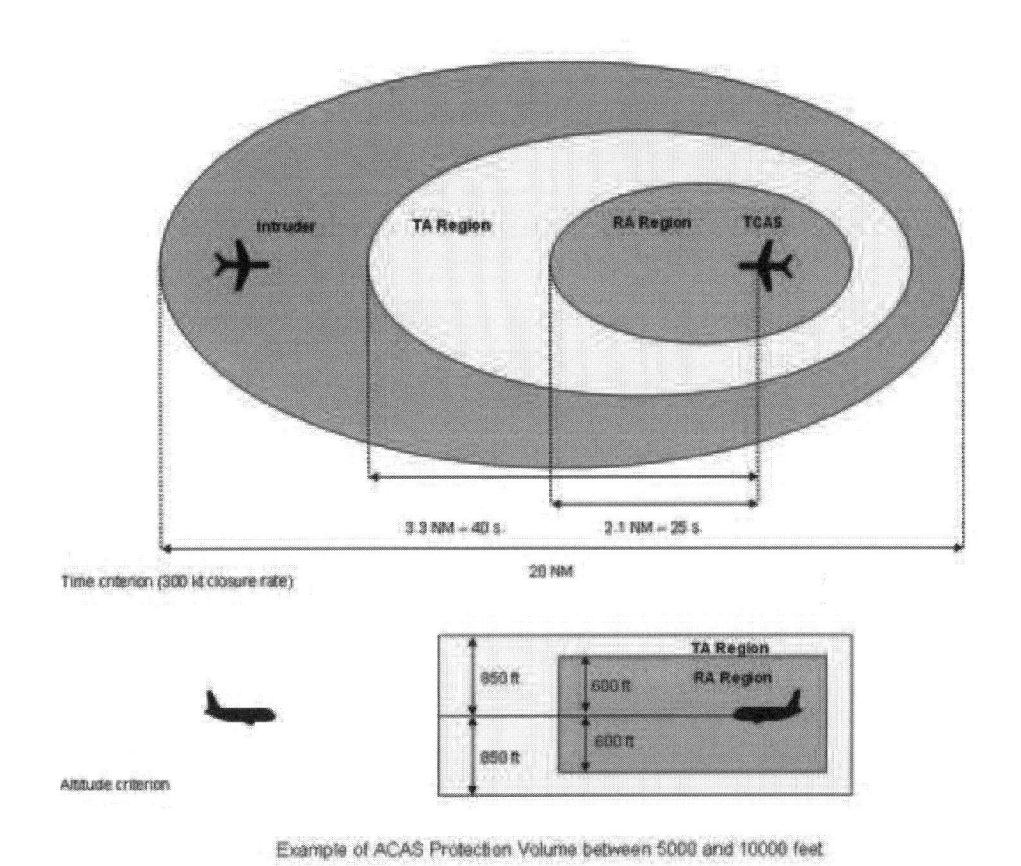

Figure 10.25: TCAS protection volume and traffic advisories. Author: User:Jemr69 / Wikimedia Commons / Public Domain.

of the presence of other aircraft which may present a threat.

A TCAS installation consists of the following components: Telecommunication systems (antennas, transponder, etc.), TCAS computer unit, and cockpit presentation. In modern aircraft, the TCAS cockpit display may be integrated in the Navigation Display (ND).

TCAS issues the following types of advisories: Traffic advisory (TA) and Resolution advisory (RA). Traffic advisory is a situational awareness advisory, i.e., pilots must be aware of conflicting aircraft to either maintain separation in visual rules or coordinate with ATC to avoid the thread in instrumental rules. The RA is the last safety layer. It is advised when a mid-air collision is to occur within less than 35 seconds. In this case, pilots are expected to respond immediately and the controller is no longer responsible for separation of the aircraft involved in the RA until the conflict has been resolved. Typically the RA will involve coordinated instructions to the two aircraft involved, e.g., flight level up and flight level down.

ADSB

Automatic Dependent Surveillance-Broadcast (ADS-B) is a GNSS based surveillance technology for tracking aircraft. It is still under development and most likely will replace radar as main surveillance system.

ADS-B technology consists of two different services: *ADS-B Out* and *ADS-B In*. *ADS-B Out* periodically broadcasts information about aircraft, such as identification code, position, course, and velocity, through an onboard transmitter. *ADS-B In* is the reception by aircraft of traffic information, flight information, and weather information, as well as other ADS-B data such as direct communication from nearby aircraft. The system relies on two fundamental components: a satellite navigation system (GPS nowadays; in the future a GNSS system with more integrity is desirable) and a datalink (ADS-B unit). With all this information, two fundamental issues will be acquired: first, controllers will be able to position and separate aircraft with improved precision and timing (since the information is more accurate); second, pilots will increase their situational awareness.

The potential benefits of ADS-B are:

- Improve situational awareness: Pilots in an ADS-B equipped cockpit will have the ability to see, on their in-cockpit flight display, other traffic operating in the airspace as well as access to clear and detailed weather information. They will also be able to receive pertinent updates ranging from temporary flight restrictions (TFR's) to runway closings.

- Improve visibility: aircraft will be benefited by air traffic controllers ability to more accurately and reliably monitor their position. Fully equipped aircraft using the airspace around them will be able to more easily identify and avoid conflict with ADS-B out equipped aircraft. ADS-B provides better surveillance in fringe areas of radar coverage.

- Others such as: Reduce environmental impact (more efficient trajectories), increase safety (by increasing situational awareness and visibility as mentioned above), increase capacity and efficiency of the system (enhance visual approaches, closely spaced parallel approaches, reduced spacing on final approach, reduce aircraft separations, improve ATC services in non-radar airspace (such oceans, enabling free routes), etc.).

Nowadays, most airliners are equipped with ADS-B. However, since the equipment is very expensive, most regional aircraft do not have it. Therefore, still ADS-B can not be used as primary surveillance system due to its low degree of implantation. Nevertheless, there is a road map both in Europe and the US to increasingly equip all aircraft with ADSB by 2020. Another issue is the low integrity of GPS as main satellite system. The implementation of the GNSS-2 will circumvent this problem.

10.4 TECHNICAL SUPPORT: CNS SYSTEM

Figure 10.26: ADS-B sketch. Author: User:AuburnADS-B / Wikimedia Commons / Public Domain.

10.5 SESAR concept

10.5.1 Single European Sky

Contrary to the United States, Europe does not have a single sky, one in which air navigation is managed at the European level. Furthermore, European airspace is among the busiest in the world with over 33,000 flights on busy days and high airport density. This makes air traffic control even more complex. The EU Single European Sky is an ambitious initiative launched by the European Commission in 2004 to reform the architecture of European air traffic management. It proposes a legislative approach to meet future capacity and safety needs at a European rather than a local level. The Single European Sky is the only way to provide a uniform and high level of safety and efficiency over Europe's skies. The key objectives are to :

- Restructure European airspace as a function of air traffic flows;
- Create additional capacity; and
- Increase the overall efficiency of the air traffic management system.

The major elements of this new institutional and organizational framework for ATM in Europe consist of: separating regulatory activities from service provision, and the possibility of cross-border ATM services; reorganizing European airspace that is no longer constrained by national borders; setting common rules and standards, covering a wide range of issues, such as flight data exchanges and telecommunications.

10.5.2 SESAR

As part of the Single European Sky initiative, SESAR (Single European Sky ATM Research) represents its technological dimension. It will help create a paradigm shift, supported by state-of-the-art and innovative technology. The SESAR program will give Europe a high-performance air traffic control infrastructure which will enable the safe and environmentally friendly development of air transport. The goals of SESAR are:

- triple capacity of the system;
- increase safety by a factor of 10;
- reduce environmental impact by 50%; and
- reduce the overall cost of the system by 50%.

We will not cover more details of SESAR in this course. The reader is referred to SESAR[32], and the master plan in SESAR Consortium [6] for more information on SESAR.

---
[32] http://www.sesarju.eu/

REFERENCES

[1] Nolan, M. S. (2010). *Fundamentals of air traffic control*. Cengage Learning.

[2] Pérez, L., Arnaldo, R., Saéz, F., Blanco, J., and Gómez, F. (2013). *Introducción al sistema de navegación aérea*. Garceta.

[3] Sáez, F., Pérez, L., and Gómez, V. (2002). *La navegación aérea y el aeropuerto*. Fundación AENA.

[4] Sáez, F. and Portillo, Y. (2003). *Descubrir la Navegación Aérea*. Aeropuertos Españoles y Navegación Aérea (AENA).

[5] Saéz Nieto, F. J. (2012). *Navegación Aérea: Posicionamiento, Guiado y Gestión del Tráfico Aéreo*. Garceta.

[6] SESAR Consortium (April 2008). SESAR Master Plan, SESAR Definition Phase Milestone Deliverable 5.

Part IV

Appendixes

A

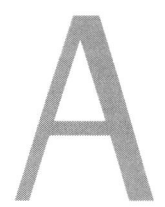

6-DOF Equations of Motion

Contents

A.1	Reference frames	354
A.2	Orientation between reference frames	355
	A.2.1 Wind axes–Local horizon orientation	357
	A.2.2 Body axed–Wind axes orientation	358
A.3	General equations of motion	358
	A.3.1 Dynamic relations	358
	A.3.2 Forces acting on an aircraft	360
A.4	Point mass model	361
	A.4.1 Dynamic relations	361
	A.4.2 Mass relations	362
	A.4.3 Kinematic relations	362
	A.4.4 Angular kinematic relations	364
	A.4.5 General differential equations system	364
References		367

This appendix is devoted to the deduction of the 6DOF general equations of motion of the aircraft. The reader is referred to GÓMEZ-TIERNO et al. [1] for a thorough and comprehensive overview. Other references on mechanics of flight are, for instance, HULL [2] and YECHOUT et al. [3].

6-DOF Equations of Motion

A.1 Reference frames

Definition A.1 (*Inertial Reference Frame*). According to classical mechanics, a inertial reference frame $F_I(O_I, x_I, y_I, z_I)$ is either a non accelerated frame with respect to a quasi-fixed reference star, or either a system which for a punctual mass is possible to apply the second Newton's law:

$$\sum \vec{F}_I = \frac{d(m \cdot \vec{V}_I)}{dt}$$

Definition A.2 (*Earth Reference Frame*). An earth reference frame $F_e(O_e, x_e, y_e, z_e)$ is a rotating topocentric (measured from the surface of the earth) system. The origin O_e is any point on the surface of earth defined by its latitude θ_e and longitude λ_e. Axis z_e points to the center of earth; x_e lays in the horizontal plane and points to a fixed direction (typically north); y_e forms a right-handed thrihedral (typically east).

Such system it is sometimes referred to as *navigational system* since it is very useful to represent the trajectory of an aircraft from the departure airport.

Hypothesis A.1. Flat earth: *The earth can be considered flat, non rotating and approximate inertial reference frame.* Consider F_I and F_e. Consider the center of mass of the aircraft denoted by CG. The acceleration of CG with respect to F_I can be written using the well-known formula of acceleration composition from the classical mechanics:

$$\vec{a}_I^{CG} = \vec{a}_e^{CG} + \vec{\Omega} \wedge (\vec{\Omega} \wedge \vec{r}_{O_I CG}) + 2\vec{\Omega} \wedge \vec{V}_e^{CG}, \tag{A.1}$$

where the centripetal acceleration and the Coriolis acceleration are neglectable if we consider typical values: $\vec{\Omega}$, the earth angular velocity is one revolution per day; $\vec{r}$ is the radius of earth plus the altitude (around 6380 [km]); $\vec{V}_e^{CG}$ is the velocity of the aircraft in flight (200-300 [m/s]). This means $\vec{a}_I^{CG} = \vec{a}_e^{CG}$ and therefore F_e can be considered inertial reference frame.

Definition A.3 (*Local Horizon Frame*). A local horizon frame $F_h(O_h, x_h, y_h, z_h)$ is a system of axes centered in any point of the symmetry plane (assuming there is one) of the aircraft, typically the center of gravity. Axes (x_h, y_h, z_h) are defined parallel to axes (x_h, y_h, z_h).

In atmospheric flight, this system can be considered as quasi-inertial.

Definition A.4 (*Body Axes Frame*). A body axes frame $F_b(O_b, x_b, y_b, z_b)$ represents the aircraft as a rigid solid model. It is a system of axes centered in any point of the symmetry plane (assuming there is one) of the aircraft, typically the center of gravity. Axis x_b lays in to the plane of symmetry and it is parallel to a reference line in the aircraft (for instance, the zero-lift line), pointing forwards according to the movement of the aircraft. Axis z_b also lays in to the plane of symmetry, perpendicular to x_b and pointing down according to *regular aircraft performance. Axis y_b is perpendicular to the plane of symmetry forming a right-handed thrihedral (y_b points then the right wing side of the aircraft).*

Definition A.5 (***Wind Axes Frame***). *A wind axes frame $F_w(O_w, x_w, y_w, z_w)$ is linked to the instantaneous aerodynamic velocity of the aircraft. It is a system of axes centered in any point of the symmetry plane (assuming there is one) of the aircraft, typically the center of gravity. Axis x_w points at each instant to the direction of the aerodynamic velocity of the aircraft $\vec{V}$. Axis z_w lays in to the plane of symmetry, perpendicular to x_w and pointing down according to regular aircraft performance. Axis y_b forms a right-handed thrihedral.*

Notice that if the aerodynamic velocity lays in to the plane of symmetry, $y_w \equiv y_b$.

A.2 ORIENTATION BETWEEN REFERENCE FRAMES

According to the classical mechanics, to orientate without loss of generality a reference frame system F_I with respect to another F_F: if both have common origin, it is necessary to perform a generic rotation until axis coincide; if the origin differs, it is necessary, together with the mentioned rotation, a translation to make origins coincide.

There are different methods to orientate two systems with common origin, such for instance, *directors cosinos, quaternions* or the *Euler angles*, which indeed will be used in this dissertation.

Definition A.6 (***Euler angles***). *Euler angles represent three composed and finite rotations given in a pre-establish order that move a reference frame to a given referred frame. This is equivalent to saying that any orientation can be achieved by composing three elemental and finite rotations (rotations around a single axis of a basis), and also equivalent to saying that any rotation matrix can be decomposed as a product of three elemental rotation matrices.*

Remark A.1. The pre-establish order of elemental rotations is usually referred to as Convention. In aeronautics and space vehicles it is universally utilized the **Tailt-Bryan Convention**. Such convention is also referred to as **Convention 321**.

Definition A.7 (***Transformation or rotation matrix***). *If the three components of a vector $\vec{A}$ in F_I are known, the transformation or rotation matrix L_{FI} expresses a vector $\vec{A}$ in the reference system F_F as follows:*

$$\vec{A}_F = L_{FI}\vec{A}_I \tag{A.2}$$

Remark A.2. L_{FI} can be obtained by simply obtaining the three individual rotation matrixes and properly multiplying them.

6-DOF Equations of Motion

Example A.1 (*Convention 321*). Given two reference systems, F_I and F_F, with common origin, we want to make F_I coincide with F_F: first we rotate F_I around axis z_I an angle δ_3, obtaining the first intermediate reference systems F_1. Second, we rotate system F_1 around axis y_1 an angle δ_2, obtaining the second intermediate reference system F_2. Third, we rotate the system F_2 around axis x_2 an angle δ_1, obtaining the final reference system F_F.

First, we express the unit vector of F_1 as a function of unit vector of F_I:

$$\begin{bmatrix} \vec{i}_1 \\ \vec{j}_1 \\ \vec{k}_1 \end{bmatrix} = \begin{bmatrix} \cos\delta_3 & \sin\delta_3 & 0 \\ -\sin\delta_3 & \cos\delta_3 & 0 \\ 0 & 0 & 1 \end{bmatrix} \begin{bmatrix} \vec{i}_I \\ \vec{j}_I \\ \vec{k}_I \end{bmatrix}. \tag{A.3}$$

The rotation matrix will be:

$$L_{1I} = R_3(\delta_3) = \begin{bmatrix} \cos\delta_3 & \sin\delta_3 & 0 \\ -\sin\delta_3 & \cos\delta_3 & 0 \\ 0 & 0 & 1 \end{bmatrix}, \tag{A.4}$$

where $R_3(\delta_3)$ is the notation of the individual matrix of rotation of an angle δ_3 around the third axis (axis z).

Therefore, the vector $\vec{A}$ expressed in the first intermediate reference frame F_1 (Notated $\vec{A}_1$) will be:

$$\vec{A}_1 = L_{1I}\vec{A}_I. \tag{A.5}$$

Operating analogously for the second individual rotation:

$$\begin{bmatrix} \vec{i}_2 \\ \vec{j}_2 \\ \vec{k}_2 \end{bmatrix} = \begin{bmatrix} \cos\delta_2 & 0 & -\sin\delta_2 \\ 0 & 1 & 0 \\ \sin\delta_2 & 0 & \cos\delta_2 \end{bmatrix} \begin{bmatrix} \vec{i}_1 \\ \vec{j}_1 \\ \vec{k}_1 \end{bmatrix}. \tag{A.6}$$

$$L_{21} = R_2(\delta_2) = \begin{bmatrix} \cos\delta_2 & 0 & -\sin\delta_2 \\ 0 & 1 & 0 \\ \sin\delta_2 & 0 & \cos\delta_2 \end{bmatrix}. \tag{A.7}$$

$$\vec{A}_2 = L_{21}\vec{A}_1. \tag{A.8}$$

Finally, for the third individual rotation:

$$\begin{bmatrix} \vec{i}_F \\ \vec{j}_F \\ \vec{k}_F \end{bmatrix} = \begin{bmatrix} 1 & 0 & 0 \\ 0 & \cos\delta_1 & \sin\delta_1 \\ 0 & -\sin\delta_1 & \cos\delta_1 \end{bmatrix} \begin{bmatrix} \vec{i}_2 \\ \vec{j}_2 \\ \vec{k}_2 \end{bmatrix}. \tag{A.9}$$

A.2 Orientation between reference frames

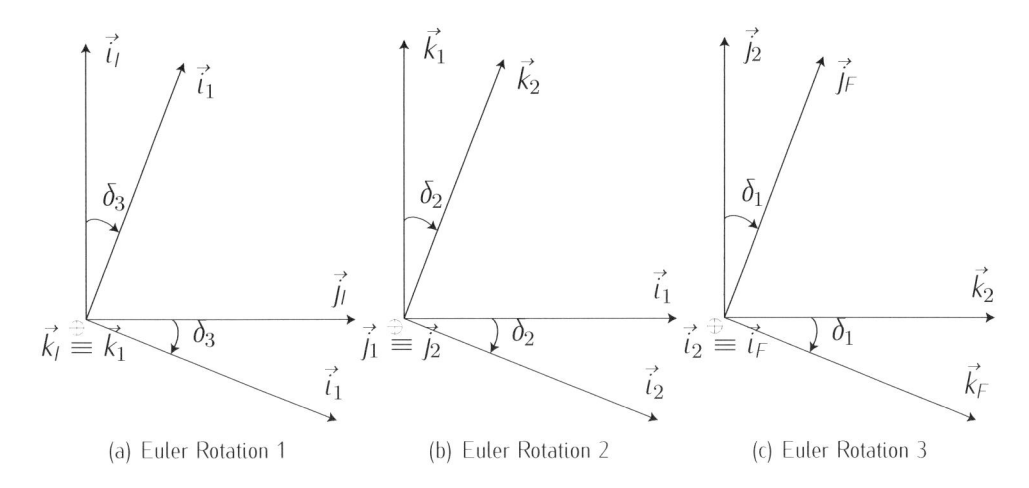

(a) Euler Rotation 1   (b) Euler Rotation 2   (c) Euler Rotation 3

Figure A.1: Euler angles

$$L_{F2} = R_1(\delta_1) = \begin{bmatrix} 1 & 0 & 0 \\ 0 & \cos\delta_1 & \sin\delta_1 \\ 0 & -\sin\delta_1 & \cos\delta_1 \end{bmatrix}. \tag{A.10}$$

$$\vec{A}_F = L_{F2}\vec{A}_2. \tag{A.11}$$

Composing:

$$\vec{A}_F = L_{F2}L_{21}L_{1I}\vec{A}_I, \tag{A.12}$$

and the global rotation matrix will be:

$$L_{FI} = \begin{bmatrix} \cos\delta_2\cos\delta_3 & \cos\delta_2\sin\delta_3 & -\sin\delta_2 \\ \sin\delta_1\sin\delta_2\cos\delta_3 - \cos\delta_1\sin\delta_3 & \sin\delta_1\sin\delta_2\sin\delta_3 + \cos\delta_1\cos\delta_3 & \sin\delta_1\cos\delta_2 \\ \cos\delta_1\sin\delta_2\cos\delta_3 + \sin\delta_1\sin\delta_3 & \cos\delta_1\sin\delta_2\sin\delta_3 - \sin\delta_1\cos\delta_3 & \cos\delta_1\cos\delta_2 \end{bmatrix}. \tag{A.13}$$

A.2.1 Wind axes–Local horizon orientation

To situate the wind axis reference frame with respect to the local horizon reference frame, the general form given in Example (A.1) is particularized for:

- $F_I \equiv F_h$; $F_F \equiv F_w$,
- $\delta_3 \equiv \chi \to$ Yaw angle,
- $\delta_2 \equiv \gamma \to$ Flight path angle,

6-DOF Equations of Motion

- $\delta_1 \equiv \mu \to$ Bank angle.

The transformation matrix will be:

$$L_{wh} = \begin{bmatrix} \cos\gamma\cos\chi & \cos\gamma\sin\chi & -\sin\gamma \\ \sin\mu\sin\gamma\cos\chi - \cos\mu\sin\chi & \sin\mu\sin\gamma\sin\chi + \cos\mu\cos\chi & \sin\mu\cos\gamma \\ \cos\mu\sin\gamma\cos\chi + \sin\mu\sin\chi & \cos\mu\sin\gamma\sin\chi - \sin\mu\cos\chi & \cos\mu\cos\gamma \end{bmatrix}. \quad (A.14)$$

A.2.2 Body axed–Wind axes orientation

To situate the wind axis reference frame with respect to the local horizon reference frame, the general form given in Example (A.1) is particularized for:

- $F_I \equiv F_w$; $F_F \equiv F_b$,
- $\delta_3 \equiv -\beta \to$ Sideslip angle,
- $\delta_2 \equiv \alpha \to$ Angle of attack,
- $\delta_1 = 0$.

The transformation matrix will be:

$$L_{wh} = \begin{bmatrix} \cos\alpha\cos\beta & -\cos\alpha\sin\beta & -\sin\alpha \\ \sin\beta & \cos\beta & 0 \\ \sin\alpha\cos\beta & -\sin\alpha\sin\beta & \cos\alpha \end{bmatrix}. \quad (A.15)$$

A.3 General equations of motion

The physic–mathematical model governing the movement of the aircraft in the atmosphere are the so-called *general equations of motion*: three equations of translation and three equations of rotation. The fundamental simplifying hypothesis is:

Hypothesis A.2. 6-DOF model: *The aircraft is considered as a rigid solid with six degrees of freedom, i.e., all dynamic effects associated to elastic deformations, to degrees of freedom of articulated subsystems (flaps, ailerons, etc.), or to the kinetic momentum of rotating subsystems (fans, compressors, etc), are neglected.*

A.3.1 Dynamic relations

The dynamic model governing the movement of the aircraft is based on two fundamental theorems of the classical mechanics: the theorem of the quantity of movement and the theorem of the kinetic momentum:

Theorem A.1. Quantity of movement: *The theorem of quantity of movement establishes that:*

$$\vec{F} = \frac{d(m\vec{V})}{dt}, \quad (A.16)$$

where $\vec{F}$ is the resulting of the external forces, $\vec{V}$ is the absolute velocity of the aircraft (respect to a inertial reference frame), m is the mass of the aircraft, and t is the time.

Remark A.3. For a conventional aircraft holds that the variation of its mass with respect to time is sufficiently slow so that the term $\dot{m}\vec{V}$ in Equation (A.16) could be neglected.

Theorem A.2. Kinematic momentum: *The theorem of the kinematic momentum establishes that:*

$$\vec{G} = \frac{d\vec{h}}{dt}, \quad (A.17)$$

$$\vec{h} = I\vec{\omega}, \quad (A.18)$$

where $\vec{G}$ is the resulting of the external momentum around the center of gravity of the aircraft, $\vec{h}$ is the absolute kinematic momentum of the aircraft, I is the tensor of inertia, and $\vec{\omega}$ is the absolute angular velocity of the aircraft.

Definition A.8 (*Tensor of Inertia*). *The tensor of inertia is defined as:*

$$I = \begin{bmatrix} I_x & -J_{xy} & -J_{xz} \\ -J_{xy} & I_y & -J_{yz} \\ -J_{xz} & -J_{yz} & I_z \end{bmatrix}, \quad (A.19)$$

where I_x, I_y, I_z are the inertial momentums around the three axes of the reference system, and J_{xy}, J_{xz}, J_{yz} are the corresponding inertia products.

The resulting equations from both theorems can be projected in any reference system. In particular, projecting them into a body-axes reference frame (also to a wind-axes reference frame) have important advantages.

Theorem A.3. Field of velocities *Given a inertial reference frame denoted by F_0 and a non-inertial reference frame F_1 whose related angular velocity is given by $\vec{\omega}_{01}$, and given also a generic vector $\vec{A}$, it holds:*

$$\left\{\frac{\partial \vec{A}}{\partial t}\right\}_0 = \left\{\frac{\partial \vec{A}}{\partial t}\right\}_1 + \vec{\omega}_{01} \wedge \vec{A}_1 \quad (A.20)$$

The three components expressed in a body-axes reference frame of the total force, the

6-DOF Equations of Motion

total momentum, the absolute velocity, and the absolute angular velocity are denoted by:

$$\vec{F} = (F_x, F_y, F_z)^T, \tag{A.21}$$
$$\vec{G} = (L, M, N)^T, \tag{A.22}$$
$$\vec{V} = (u, v, w)^T, \tag{A.23}$$
$$\vec{\omega} = (p, q, r)^T. \tag{A.24}$$

Therefore, the equations governing the motion of the aircraft are:

$$F_x = m(\dot{u} - rv + qw), \tag{A.25a}$$
$$F_y = m(\dot{v} + ru - pw), \tag{A.25b}$$
$$F_z = m(\dot{w} - qu + pv), \tag{A.25c}$$
$$L = I_x\dot{p} - J_{xz}\dot{r} + (I_z - I_y)qr - J_{xz}pq, \tag{A.25d}$$
$$M = I_y\dot{q} - (I_z - I_x)pr - J_{xz}(p^2 - r^2), \tag{A.25e}$$
$$N = I_z\dot{r} - J_{xz}\dot{p} + (I_x - I_y)pq - J_{xz}qr, \tag{A.25f}$$

System (A.25) is referred to as Euler equations of the movement of an aircraft.

A.3.2 Forces acting on an aircraft

Hypothesis A.3. Forces acting on an aircraft: *The external actions acting on an aircraft can be decomposed, without loss of generality, into propulsive, aerodynamic and gravitational, notated respectively with subindexes* $((\cdot)_T, (\cdot)_A, (\cdot)_G)$:

$$\vec{F} = \vec{F}_T + \vec{F}_A + \vec{F}_G, \tag{A.26}$$
$$\vec{G} = \vec{G}_T + \vec{G}_A, \tag{A.27}$$

The gravitational force can be easily expressed in local horizon axes as:

$$(\vec{F}_G)_h = \begin{bmatrix} 0 \\ 0 \\ mg \end{bmatrix}, \tag{A.28}$$

where g is the acceleration due to gravity.

Hypothesis A.4. Constant gravity: *The acceleration due to gravity in atmospheric flight of an aircraft can be considered constant* ($g = 9.81[m/s^2]$), *due to a small altitude of flight when compared to the radius of earth. Therefore, the little variations of g as a function of h are neglectable.*

To project the force due to gravity into wind-axes reference frame:

$$(\vec{F}_G)_w = L_{wh}(\vec{F}_G)_h = \begin{bmatrix} -mg\sin\gamma \\ mg\cos\gamma\sin\mu \\ mg\cos\gamma\cos\mu \end{bmatrix}, \qquad (A.29)$$

Introducing the propulsive, aerodynamic and gravitational actions in System (A.25):

$$-mg\sin\gamma + F_{T_x} + F_{A_x} = m(\dot{u} - rv + qw), \qquad (A.30a)$$
$$mg\cos\gamma\sin\mu + F_{T_y} + F_{A_y} = m(\dot{v} + ru - pw), \qquad (A.30b)$$
$$mg\cos\gamma\cos\mu + F_{T_z} + F_{A_z} = m(\dot{w} - qu + pv), \qquad (A.30c)$$
$$L_T + L_A = I_x \dot{p} - J_{xz}\dot{r} + (I_z - I_y)qr - J_{xz}pq, \qquad (A.30d)$$
$$M_T + M_A = I_y \dot{q} - (I_z - I_x)pr - J_{xz}(p^2 - r^2), \qquad (A.30e)$$
$$N_T + N_A = I_z \dot{r} - J_{xz}\dot{p} + (I_x - I_y)pq - J_{xz}qr. \qquad (A.30f)$$

The three aerodynamic momentum of roll, pitch and yaw (L_A, M_A, N_A) can be controlled by the pilot through the three command surfaces, ailerons, elevator and rudder, whose deflections can be respectively notated by $\delta_a, \delta_e, \delta_r$. Notice that such deflection have also influence in the three components of aerodynamic force, and therefore the 6 equations are coupled and must be solved simultaneously.

A.4 POINT MASS MODEL

Hypothesis A.5. Point mass model: *The translational equations (A.25a)-(A.25c) are uncoupled from the rotational equations (A.25d)-(A.25f) by assuming that the airplane rotational rates are small and that control surface deflections do not affect forces. This leads to consider a 3 Degree Of Freedom (DOF) dynamic model that describes the point variable-mass motion of the aircraft.*

Under this hypothesis, the translational problem (performances) can be studied separately from the rotational problem (control and stability).

A.4.1 DYNAMIC RELATIONS

Therefore, the dynamic equations governing the translational motion of the aircraft are uncoupled:

$$-mg\sin\gamma + F_{T_x} + F_{A_x} = m(\dot{u} - rv + qw), \qquad (A.31a)$$
$$mg\cos\gamma\sin\mu + F_{T_y} + F_{A_y} = m(\dot{v} + ru - pw), \qquad (A.31b)$$
$$mg\cos\gamma\cos\mu + F_{T_z} + F_{A_z} = m(\dot{w} - qu + pv), \qquad (A.31c)$$

The aerodynamic forces, expressed in wind axes, are as follows:

$$(\vec{F}_A)_w = \begin{bmatrix} -D \\ -Q \\ -L \end{bmatrix}, \qquad (A.32)$$

where D is the aerodynamic drag, Q is the aerodynamic lateral force, and L is the aerodynamic lift.

The propulsive forces, expressed in wind axes, are as follows:

$$(\vec{F}_T)_w = \begin{bmatrix} T\cos\epsilon\cos\nu \\ T\cos\epsilon\sin\nu \\ -T\sin\epsilon \end{bmatrix}, \qquad (A.33)$$

where T is the thrust, ϵ is the thrust angle of attack, and ν is the thrust sideslip.

Hypothesis A.6. Fixed engines: *We assume the aircraft is a conventional jet airplane with fixed engines. Almost all existing aircrafts worldwide have their engines rigidly attached to their structure.*

A.4.2 Mass relations

Hypothesis A.7. Variable mass: *The aircraft is modeled as variable mass particle.*

The variation of mass is given by the consumed fuel during the flight:

$$\dot{m} + \phi = 0. \qquad (A.34)$$

A.4.3 Kinematic relations

Hypothesis A.8. Moving Atmosphere: *The atmosphere is considered moving, i.e., wind is taken into consideration. Vertical component is neglected due its low influence. Only kinematic effects are considered, i.e., dynamic effects of wind are also neglected due its low influence.* The wind velocity $\vec{W}$ can be expressed in local horizon axes as:

$$\vec{W}_h = \begin{bmatrix} W_x \\ W_y \\ 0 \end{bmatrix} \qquad (A.35)$$

Considering the earth axes reference system as a inertial system, and assuming that earth axes are parallel to local horizon axes, the absolute velocity $\vec{V}^G = \vec{V}^A + \vec{W}$ can be expressed referred to a wind axes reference as follows:

$$\vec{V}_e^G = \begin{bmatrix} \dot{x}_e \\ \dot{y}_e \\ \dot{z}_e \end{bmatrix} = L_{hw}\vec{V}_w^A + \vec{W}_h = L_{hw}\begin{bmatrix} u \\ v \\ w \end{bmatrix} + \begin{bmatrix} W_x \\ W_y \\ 0 \end{bmatrix} = L_{wh}^T\begin{bmatrix} V \\ 0 \\ 0 \end{bmatrix} + \begin{bmatrix} W_x \\ W_y \\ 0 \end{bmatrix}. \quad (A.36)$$

Remark A.4. Notice that the absolute aerodynamic velocity $\vec{V}^A$ expressed in a wind axes reference frame is $(V, 0, 0)$.

Therefore, the kinematic relations are as follows:

$$\dot{x}_e = V \cos\gamma \cos\chi + W_x, \quad (A.37a)$$
$$\dot{y}_e = V \cos\gamma \sin\chi + W_y, \quad (A.37b)$$
$$\dot{z}_e = -V \sin\gamma, \quad (A.37c)$$

Equations (A.37a)–(A.37c) provide the movement law and the trajectory of the aircraft can be determined.

Notice that Equation (A.37c) is usually rewritten as

$$\dot{h}_e = V \sin\gamma, \quad (A.38a)$$

Remark A.5. If one wants to model a flight over a spherical earth, since the radius of earth is sufficiently big and the angular velocity of earth is sufficiently small, it holds that the rotation of earth has very low influence in the centripetal acceleration and it is thus neglectable. Therefore, the hypothesis of flat earth holds in the dynamics of an aircraft moving over an spherical earthwith the following kinematic relations:

$$\dot{\lambda}_e = \frac{V \cos\gamma \cos\chi + W_x}{(R+h)\cos\theta}, \quad (A.39a)$$
$$\dot{\theta}_e = \frac{V \cos\gamma \sin\chi + W_y}{R+h}, \quad (A.39b)$$
$$\dot{h}_e = V \sin\gamma, \quad (A.39c)$$

where λ and θ are respectively the longitude and latitude and R is the radius of earth.

6-DOF Equations of Motion

A.4.4 Angular kinematic relations

In what follows the three components of absolute angular velocity of the aircraft are related with the orientation angles of the aircraft with respect to a local horizon reference system:

$$\vec{\omega}_I \approx \vec{\omega}_h = \begin{bmatrix} p \\ q \\ r \end{bmatrix} = \dot{\mu}\vec{i}_w + \dot{\gamma}\vec{j}_1 + \dot{\chi}\vec{k}_h. \tag{A.40}$$

Projecting the unit vectors in wind axes using the appropriate transformation matrices:

$$p = \dot{\mu} - \dot{\chi}\sin\gamma \tag{A.41a}$$
$$q = \dot{\gamma}\cos\mu + \dot{\chi}\cos\gamma\sin\mu \tag{A.41b}$$
$$r = -\dot{\gamma}\sin\mu + \dot{\chi}\cos\gamma\cos\mu \tag{A.41c}$$

A.4.5 General differential equations system

For a point mass model, the general differential equations system governing the motion of an aircraft is stated as follows:

$$-mg\sin\gamma + T\cos\epsilon\cos\nu - D = m(\dot{V}), \tag{A.42a}$$
$$mg\cos\gamma\sin\mu + T\cos\epsilon\sin\nu - Q =$$
$$\quad -mV(\dot{\gamma}\sin\mu + \dot{\chi}\cos\gamma\cos\mu), \tag{A.42b}$$
$$mg\cos\gamma\cos\mu - T\sin\epsilon - L =$$
$$\quad -mV(\dot{\gamma}\cos\mu - \dot{\chi}\cos\gamma\sin\mu), \tag{A.42c}$$
$$\dot{x}_e = V\cos\gamma\cos\chi + W_x, \tag{A.42d}$$
$$\dot{y}_e = V\cos\gamma\sin\chi + W_y, \tag{A.42e}$$
$$\dot{z}_e = -V\sin\gamma, \tag{A.42f}$$
$$\dot{m} + \phi = 0. \tag{A.42g}$$

If we assume the following hypothesis:

Hypothesis A.9. Symmetric flight: *We assume the aircraft has a plane of symmetry, and that the aircraft flies in symmetric flight, i.e., all forces act on the center of gravity and the thrust and the aerodynamic forces lay on the plane of symmetry. This leads to non sideslip, i.e., $\beta = \nu = 0$, and non lateral aerodynamic force, i.e., $Q = 0$, assumptions.*

Hypothesis A.10. Small thrust angle of attack: *We assume the thrust angle of attack is small $\epsilon \ll 1$, i.e., $\cos\epsilon \approx 1$ and $\sin\epsilon \approx 0$. For commercial aircrafts, typical performances do not exceed $\epsilon = \pm 2.5 [deg]$ ($\cos 2.5 = 0.999$); in taking off rarely can go up to $\epsilon = 5 - 10 [deg]$, but still $\cos 10 = 0.98$.*

ODE system (A.42) is as follows:

$$-mg \sin \gamma + T - D = m(\dot{V}), \tag{A.43a}$$
$$mg \cos \gamma \sin \mu = -mV(\dot{\gamma} \sin \mu - \dot{\chi} \cos \gamma \cos \mu), \tag{A.43b}$$
$$mg \cos \gamma \cos \mu - L = -mV(\dot{\gamma} \cos \mu + \dot{\chi} \cos \gamma \sin \mu), \tag{A.43c}$$
$$\dot{x}_e = V \cos \gamma \cos \chi + W_x, \tag{A.43d}$$
$$\dot{y}_e = V \cos \gamma \sin \chi + W_y, \tag{A.43e}$$
$$\dot{z}_e = -V \sin \gamma, \tag{A.43f}$$
$$\dot{m} + \phi = 0. \tag{A.43g}$$

Operating Equation (A.43b)· cos μ - Equation (A.43c)· sin μ it yields:

$$L \sin \mu = mV\dot{\chi} \cos \gamma. \tag{A.44}$$

Operating Equation (A.43b)· sin μ + Equation (A.43c)· cos μ it yields:

$$L \cos \mu - mg \cos \gamma = mV\dot{\gamma}. \tag{A.45}$$

REFERENCES

[1] GÓMEZ-TIERNO, M., PÉREZ-CORTÉS, M., and PUENTES-MÁRQUEZ, C. (2009). *Mecánica de vuelo*. Escuela Técnica Superior de Ingenieros Aeronáuticos, Universidad Politécnica de Madrid.

[2] HULL, D. G. (2007). *Fundamentals of Aiplane Flight Mechanics*. Springer.

[3] YECHOUT, T., MORRIS, S., and BOSSERT, D. (2003). *Introduction to Aircraft Flight Mechanics: Performance, Static Stability, Dynamic Stability, and Classical Feedback Control*. AIAA Education Series.

INDEX

A320neo, 245
AC–Alternating Current, 135
ACARE: Advisory Council for Aeronautics Research in Europe, 12, 258
accelerometer, 327
ADF–Automatic Direction Finder, 128
ADS-B–Automatic Dependent Surveillance-Broadcast, 307, 346
AESA: Agencia Estatal de Seguridad Aérea, 12, 270, 305
Airbus, 7, 241
airway, 311
ANSP: Air Navigation Service Providers, 10
APU–Auxiliary Power Unit, 135
ASM–Airspace management, 309
ASM: Air Space Management, 11
ATFM–Air traffic flow and capacity management, 309
ATFM: Air Traffic Flow Management, 11
ATIS–Automatic Terminal Information Service, 289
ATM–Air Traffic Management, 308
ATM: Air Traffic Management, 11
ATS routes, 311
ATS: Air Traffic Services, 11

B737 MAX, 246
Boeing, 241
Bonbardier, 242

CDI–Course Deviation Indicator, 128, 293
CDS–Cockpit Display Systems, 131
CFMU–Central Flow Management Unit, 309
CFRP–carbon-fibre-reinforced plastics, 112
Chicago Convention, 11
Clean Sky, 12, 258
CNS–Communications, Navigation & Surveillance, 306
CNS: Communication, Navigation, and Surveillance, 10
Comac C919, 247
contrails, 261
CPR–Combuster Pressure Ratio, 157
CPR–Compressor Pressure Ratio, 156
CWY–clear way, 277

DC–Direct Current, 135
Deregulation Act, 7
DLR: German Aerospace Center, 11
DME–Distance Measurement Equipent, 304, 332
DME–Distance Measurement Equipment, 293
DOC–Direct Operational Costs, 247
DOF–Degrees Of Freedom, 171

EASA: European Aviation Safety Agency, 12
ECAM–Electronic Centralized Aircraft Monitor, 134
EICAS–Engine Indication and Crew Alerting System, 134
Embraer, 242

Index

EPR–Engine Pressure Ratio, 161
ESA: European Space Agency, 9
ETR–Engine Temperature Ratio, 162
Eurocontrol, 12

FAA: Federal Aviation Administration, 12
FCU–Flight Control Unit, 132
FIR–Flight Information Region, 314
Flag companies, 7
flight plan, 321
FMS–Flight Management System, 131
FO–Factor of Occupancy, 249
FW–Fuel Weight, 184

GLONASS–Global Navigation Satellite System, 125
GPS–Global Position System, 125
GPU–Ground Power Unit, 135
GRP–glass-reinforced plastic, 112
gyroscope, 327

IATA, 239
IATA: International Air Transport Association, 11, 270
ICAO, 234
ICAO–International Civil Aviation Organization, 129
ICAO: International Civil Aviation Organization, 11, 270, 305
IFR–Instrumental Flight Rules, 289
ILS–Instrumental Landing System, 125, 128, 293, 304
inertial navigation system, 326
INTA: Instituto Nacional de Técnica Aeroespacial, 11
IOC–Indirect Operational Costs, 247
IPCC: Intergovernmental Panel on Climate Change, 13
IPR–Inlet Pressure Recovery, 153

Low cost companies, 8
LW–Landing Weight, 184

MAC–Mid-Air Collision, 307, 345
MCP–Mode Control Panel, 131
MFW–Maximum Fuel Weight, 185
MLS–Microwaves Landing System, 125, 304, 341
MLW–Maximum Landing Weight, 185
MPL–Maximum PayLoad, 184
MTOW–Maximum Take-Off Weight, 185
MZFW–Maximum Zero Fuel Weight, 185

NASA: National Aeronautics and Space Administration, 9, 11
ND–Navigation Display, 131
NDB–Non-Directional Beacon, 125, 128
NextGen: Next Generation of air transportation system, 12, 258
NPR–Nozzle Pressure Ratio, 161

OEW–Operating Empty Weight, 184
ONERA: The French Aerospace Lab, 11

PAPI–Precision Approach Path Indicator, 292
PFD–Primary Flight Display, 131
PKC–Passenger Kilometer Carried, 249
PL–Payload, 184

radar, 326
RAT–Ram Air Turbines, 135
RF–Reserve Fuel, 184
RMI–Radio Magnetic Indicator, 128
RNAV–Area Navigation, 304

SESAR, 348
SESAR: Single European Sky ATM Research, 12, 258
SKO–Seat Kilometer Offered, 249
SWY–stop way, 277

TCAS–Traffic Collision Avoidance System, 307, 345
Technical terms english-spanish
    Range-Rango, 182

Index

Acceleremoter-Acelerómetros, 123
Aerodynamic efficiency-Eficiencia aerodinámica., 65
Aerodyne-Aerodino, 20
Aerospace vehicles-Vehículos aeroespaciales, 20
Aerostat-Aeroestato, 20
Afterburning-Postcomubustores, 162, 166
Ailerons-Alerones, 31
Aircraft-Aeronave, 20
Airfoil-Perfil aerodinámico, 28
Airship-Dirigible, 20
Airspeed indicator-Indicador the velocidad, 126
Aluminum alloys-Aleaciones de aluminio, 110
Angle of attack-Ángulo de ataque, 60
Artificial horizon-Horizonte artificial, 122, 126
Attitude-Actitud del avión, 123
Autogyro-Autogiro, 24
Bending moment-Momento Flector, 103
Blades-palas, 24
Boundary layer-Capa límite, 53
Brittleness-Fragilidad, 108
Buckling-Pandeo, 104
Camber-Curvatura, 59
Chord-Cuerda aerodinámica, 55, 59
Coefficient of lift-Coeficiente de Sustentación., 71
Combustor-Cámara de combustión, 150, 156
Composite materials-Materiales compuestos, 111
Composite Monocoque-Monocasco, 115
Composite Semi-monocoque-Semi-monocasco, 115
Compressible flow-Flujo compresible, 50
Compressor-Compresor, 150, 154
Continuity equation-Ecuación de continuidad, 49
Corrosion-Corrosión, 109
cross-track error-desviación lateral, 302
dead reckoning-navegación a estima, 300
desired track angle-ángulo de rumbo, 302
Direccional gyro-Girodireccional, 122, 127
doppler effect-efecto doppler, 329
Drag-Resistencia, 58
Drag-Resitencia aerodinámica, 28
Ductility-Ductilidad, 108
Earth reference frame-Ejes tierra, 170
Eddy-Remolino, 55
Elevator-Timón de profundidad, 31
Empennage-Cola, 30
Endurance-Autonomía, 182
Enlargment-Alargamiento, 68
Fatigue-Fatiga, 109
Fixed-wing aircraft-Aeroplano, 20
Frame-Cuaderna, 115
Glider-Planeador, 21
Gliding-Planeo, 176
Gyrodino-Girodino, 24
Gyroscope-Giróscopo, 123
High subsonic-Susónico alto, 22
High-lift devices-Dispositivos hipersustentadores, 74
Horizontal stabilizer-Estabilizador horizontal, 27, 30
Hover-Vuelo en punto fijo, 24
Incompressible flow-Flujo incompresible, 50
Induced coefficient-Coeficiente

371

inducido, 72
Inlet-Difusor, 150, 153
Jet engine-Motor de reacción, 150
Landing gear-Tren de aterrizaje, 28
Landing-Aterrizaje, 181
Leading edge-Borde de ataque, 28, 59, 68
Lift-Sustentación, 58
Lift-SutentaciÙn, 28
Low subsonic-Subsónico bajo, 22
Magnetic Compass-Brújula, 127
Magnetic compass-Brújula, 122
Mass flow-Flujo másico, 149
Momentum-Momento, 48
Noozle-tobera, 33
Nozzle-Tobera, 150, 160
Parabolic polar-Polar parabólica, 72
Parasite coefficient-Coeficiente parásito, 72
Pitch-Cabeceo, 170
Plasticity-Plasticidad, 109
Probe-Sonda, 25
Propeller propulsion-Propulsión a hélice, 148
Quantity of movement equation-Ecuación de cantidad de movimiento, 50
radiogoniometry-radiogonometría, 329
radiotelemetry-radiotelemetría, 329
Reynolds number-Número de Reynolds, 54
Rib-Costilla, 117
Roll-Alabeo, 170
Root chord-Cuerda en la raíz, 68
Rotorcraft-Aeronave de alas giratorias, 20
Rudder-Timón de dirección, 31
scanning beam-haz explorador, 329
Shear stress-Esfuerzo de cortadura, 53
Shock wave-Onda de choque, 57
Sideslip-Resbalamiento, 170

Skin-Revistimiento, 115
Space Launcher-Lanzadera espacial, 25
Spar-Larguero, 117
spatial modulation-modluaciòn espacial, 329
Speed of sound-Velocidad del sonido, 56
Stagnation point-Punto de remanso, 52
Stiffening elements-Elementos rigidizadores, 115
Stiffness-Rigidez, 109
Strain-Elongación, 105
Stream line-Línea de corriente, 48
Stream tube-Tubo de corriente, 48
Strength-Resistencia, 108
Stress-Esfuerzo, 52, 104
Stringer-Larguerillo, 116
Supersonic-Supersónico, 22
Sweep-Flecha, 68
Take off-Despegue, 179
Tip chord-Cuerda en la punta, 68
Torsion-Torsión, 104
Toughness-Tenacidad, 108
track angle-ángulo de derrota, 301
Traction-Tracción, 104
Trailing edge-Borde de salida, 28, 59, 68
Turbine-Turbina, 150, 158
Turbofan-Turbofán, 162, 164
Turbojet-Turborreactor, 162
Turboprop-Turbopropulsor, 162, 165
Turn indicator-Bastón y bola, 128
Variometer-Variómetro, 128
Vertical stabilizer-Estabilizador vertical, 27, 30
Vortex-Vórtice, 55
Wind axis frame-Ejes viento, 170
Wingspan-Envergadura, 68
Yaw-Guiñada, 170
aerodrome-aeródromo, 270

INDEX

air side–lado aire, 269, 273
airfield–campo de vuelo, 270
airport–aeropuerto, 270
apron–plataforma, 273
asphalt–asfalto, 276
asphalt–zona de parada, 277
baggage carousels–hipódromos de recogida de equipaje, 283
block time–tiempo bloque, 248
check in–facturación, 273
clear way–zona libre de obstáculos, 277
concourse–sala de espera antes de embarque, 273
concrete–hormigón, 276
customs–aduana, 287
gate–puerta de embarque, 273
Gross Domestic Product (GDP)–Producto Interior Bruto (PIB), 271
land side–lado tierra, 269, 273
master plan–plan maestro o plan director, 269, 273
non operational costs–costes no operativos, 247
operational costs–costes operativos, 247
runway–pista de despegue/aterrizaje, 273
taxiway–calle de rodadura, 273
terminal building–edificio terminal, 273
windscck–manga de viento, 290
TF–Trip Fuel, 184
TOW–Take-Off Weight, 184
TPR–Turbine Pressure Ratio, 158

UIR–Upper Information Region, 314

VFR–Visual Flight Rules, 289
VOR–VHF Omnidirectional Radio range, 125, 128, 293, 304, 338

waypoints, 312

XTE–Cross-Track Error, 302

ZFW–Zero Fuel Weight, 184

373